T0201969

Introduction to Dynamical Wave Function Collapse

OXFORD SERIES ON REALISM IN QUANTUM PHYSICS

Introduction to Dynamical Wave Function Collapse

Philip Pearle
Hamilton College, Clinton NY, USA

OXFORD
UNIVERSITY PRESS

Great Clarendon Street, Oxford, OX2 6DP,
United Kingdom

Oxford University Press is a department of the University of Oxford.
It furthers the University's objective of excellence in research, scholarship,
and education by publishing worldwide. Oxford is a registered trade mark of
Oxford University Press in the UK and in certain other countries

Published in the United States of America by Oxford University Press
198 Madison Avenue, New York, NY 10016, United States of America

British Library Cataloguing in Publication Data

Data available

Library of Congress Control Number: 2023920625

ISBN 9780198901372

DOI: 10.1093/oso/9780198901372.001.0001

Printed and bound by
CPI Group (UK) Ltd, Croydon, CR0 4YY

MIX
Paper | Supporting
responsible forestry
FSC® C013604

For Betty. You are the wind beneath my wings.

Acknowledgements

This book began life, quite a few years ago, as a joint venture with Antony Valentini and Simon Saunders, to co-author a book entitled *Three Roads to Quantum Reality*. The "measurement problem" highlights the inability of standard quantum theory to provide a logically sound entity that can correspond to the state of reality in Nature.

There are three alternative theories that, arguably, are the leading contenders permitting a realist interpretation. Antony is an advocate for the Pilot-Wave theory, of deBroglie and Bohm, where reality consists of particles and fields, whose behavior is governed by the wave function, Simon works with the Everettian theory where reality consists of many worlds, described by the wave function. I support the CSL (continuous spontaneous localization) dynamical collapse theory I invented, where reality is a single world is described by the wave function.

There were to be four chapters in the book. The first was to be an introduction to the field of Foundations of Quantum Theory written by Antony. This was to be followed by a chapter written by each of us on the theory we espouse.

At the conclusion of our writing, we realized that we had each written too much to appear in one book and, while our subjects truly reflected the proposed book title, our chapters reflected quite different styles and approaches.

We thought it fitting to abandon the one-book idea, and embrace a three-book idea, to be published as a series entitled 'Realism and Quantum Physics.' Our editor, Sonke Adlung, was quite supportive of our conclusion, and this book (and two others) is the result. Antony's book is *Introduction to Quantum Foundations and Pilot-Wave Theory*. (I particularly recommend reading Antony's 'Part I: Quantum Foundations' for an excellent overview of the salient issues in the whole field.) The title of Simon's book is *Everett, Decoherence, and Quantum Mechanics: the Case for Many Worlds*.

So, my heartfelt thanks to Antony, Simon and Sonke, without whom this book would never have seen the light.

I also thank the physicists who have performed experiments to test the consequences of dynamical collapse, and the physicists who have involved themselves in dynamical collapse theories. I am, of course, specially gratified where my own work has found a place in the work of others. For me, this enterprise has been a great adventure!

I also thank Toru Ohira, Kelvin McQueen, and Daniel Sudarsky, who volunteered to read and comment on earlier versions of what turned out to be this book.

Preface

Bohr: Those who are not shocked when they first come across quantum theory cannot possibly have understood it.[1]

Feynman: I think I can safely say that nobody understands quantum mechanics.[2]

How, you might ask, did I come to care about the subject that led to my discovering the theory that is the topic of this book?

I entered MIT in 1953, majoring in electrical engineering. Every undergraduate took a sequence of physics courses and, in my first two years, I had enjoyed my courses on Newtonian physics and electricity and magnetism, taught by the physics department. So, in my junior year, I thought I would like a bit more physics. I therefore enrolled in the next course required for physics majors. It changed my life.

The course, taught by Professor Felix Villars, used a rather unusually structured and thoughtful textbook *Quantum Theory*, by David Bohm, published in 1951. After a few weeks of the assigned reading I became impatient. There was a lot of history of the early days of quantum theory, followed by subjects I hadn't expected to see there, waves and probability. I thought this was a theory about particles. When were they going to tell me how to predict the locations of particles?

A third of the way hrough the semester, it dawned on me that they weren't ever going to tell me how to predict the locations of particles! Instead, they were fixated upon this thing called a wave function, and the best they could do was use it to predict the probabilities of the positions of particles.

Well, OK, if that's all they could do, but something bothered me.

With Newton's second law, say applied to a ball, you can see where the ball is and see how fast it is moving initially, and so you can calculate its future position. With this new law, Schrödinger's equation, you can't see the initial wave function, whatever that might mean, so how do you calculate its future?

I had no idea how to answer this question. It bothered me so much I went to Professor Villar's office during office hours and asked him the question I thought might stump him as it had stumped me: "Where do you get the initial wave function?" His answer came back immediately, "From the previous wave function."

At first I was stunned. Such a simple answer. But, it didn't make sense. Today's wave function depends on yesterday's, and so forth: where does it ever end? But, that's not what Professor Villars meant, as he explained.

---

[1] Heisenberg (1971).

[2] 1964 Cornell Messenger lectures, *The Character of Physical Law*, snippet available on YouTube, https://www.youtube.com/watch?v=w3ZRLllWgHI, ModernLibrary/BBC, London 1965, reprinted in 2017 by MIT Press, Cambridge, p. 129.

He was talking about a particle in an experimental context. An experiment is designed so that the particle is forced to choose a definite eigenstate of an operator. One doesn't know beforehand what choice will be made but, afterwards, one finds the particle in a particular state. Then, one regards that as initial state for the next experiment, designed to force the particle to choose the eigenstate of a different operator. The initial state's wave function is used to predict the probabilities of the next experiment's possible outcomes, and so forth.

I was directed to the soon-to-be-read Chapter 6, Section 6.3, entitled "Effects of Process of Observation on the Wave Function," where Bohm, describing a measurement of the position of a particle writes in part: " ... when the position of the electron was observed, the wave function suffered a collapse from a broad front down to a narrow region. ... This type of collapse of the wave function does not occur in any classical theory.[3] Why does it occur here?" And, Bohm gives an answer in Section 6.3.

His answer is that the wave function describing the apparatus and the particle being measured evolves, via Schrödinger's equation, into a superposition of states, each state corresponding to one of the possible outcomes of the measurement. He then argues that a random phase factor multiplies each state in the superposition. He states that the collapse, the replacement of a superposition of states by just one of the states with appropriate probability, is only justified because, each time the measurement is repeated, the random phase factors of the associated superpositions ensure there will be no difference if one uses the ensemble of uncollapsed superpositions or the ensemble of associated collapsed states to make predictions.,

It was only by reading more on this that I learned to doubt the soundness of the argument (Section 2.4). Just because no experiment can tell the difference between a collection of superpositions with random phases and a collection of truly collapsed state vectors does not mean that they are the same.

Schrödinger himself regarded quantum theory's "collapse of the wave function"' postulate as anathema:[4]

Schrödinger: If all this damned quantum jumping were really here to stay, I should be sorry I ever got involved with quantum theory.[5]

"Quantum jump" is another name for the collapse, for which there is no explanation: somehow a superposition of state vectors miraculously chooses to become just one of them. The conditions under which this is to occur are vaguely specified: "after a measurement." So, the existence of this poorly defined feature of quantum theory is called the "measurement problem."

Because of taking this course, I became excited by this flaw in standard quantum theory. Surely, I thought, working on this problem had to be rewarding. I decided,

---

[3]As far as I am aware, this was the first time the word "collapse" was applied to describe what Schrödinger called "the abrupt change by measurement," what Dirac called a "jump" in chapter 10 of his textbook, which others have called "reduction" or "projection."

[4]See Chapter 1 for extensive quotes on this issue from his famous 1935 paper "The Present Status of Quantum Mechanics."

[5]Heisenberg (1971).

after graduating with BS and MS degrees in electrical engineering, to literally and figuratively change course and go to graduate school in physics, with the idea of tackling this problem.

There was encouragement from some of the greatest minds in physics.

Einstein: The latest and most successful creation of theoretical physics, namely Quantum Mechanics, is fundamentally different in its principles from the two pro- grammes which we will briefly call Newton's and Maxwell's. For the quantities that appear in its laws make no claim to describe Physical Reality itself, but only the prob- abilities for the appearances of a particular physical reality on which our attention is fixed ... [6]

Dirac: It seems clear that the present quantum mechanics is not in its final form. Some further changes will be needed, just about as drastic as the changes made in passing from Bohr's orbit theory to quantum mechanics. ...It might very well be that the new quantum mechanics will have determinism in the way that Einstein wanted.[7]

However, once in MIT graduate school, I soon learned that the subject had a stigma in the physics community at large and, if I wished to graduate, I had better work at something useful, say particle physics, and not even mention my interest. So I did.

But, after graduating, I started thinking again about collapse, which led to my first paper.[8] In it, I criticized the collapse postulate for being ill-defined. I concluded that the only logical course was to never collapse the wave function and to adopt Einstein's view:

Einstein: One arrives at very implausible theoretical conceptions, if one attempts to maintain the thesis that the statistical quantum theory is in principle capable of produc- ing a complete description of an individual physical system. On the other hand, those difficulties of theoretical interpretation disappear, if one views the quantum-mechanical description as the description of ensembles of systems. ...the ψ function is to be un- derstood as the description not of a single system but of an ensemble of systems. ...For, if the statistical quantum theory does not pretend to describe the individual system completely, it appears unavoidable to look elsewhere for a complete description of the individual system ... [9]

Where elsewhere to look? Could one add a term to Schrödinger's equation? Could the wave function evolution be changed so collapse occurs dynamically, as a smooth

---

[6] In *James Clerk Maxwell: A Commemoration Volume 1831–1931*, The Macmillan Company, New York, 1932.

[7] In *Albert Einstein: Historical and Cultural Perspectives: The Centennial Symposium in Jerusalem (1979)*, p.85, edited by G. Holton and Y. Elkana, Princeton University Press, Princeton NJ, 1982.

[8] Pearle (1967)

[9] In *Albert Einstein: Philosopher Scientist* Vol II, P. A. Schilpp (ed), Harper & Row, New York, 1959.

evolution of the wave function? Then, no jumps, goodbye collapse postulate. The wave function could be taken to represent the individual system.

It took a while to find a way to approach the problem. I envisioned the usual Hamiltonian evolving the wave function into a superposition of macroscopically different states, and then the added term taking over. It would evolve the coefficients of the wave function so that eventually one coefficient would have magnitude 1, and the rest magnitude 0.

The added term needs to have in it a variable that fluctuates randomly, so that each time the modified Schrödinger equation is solved one gets a different outcome. And, the different outcomes have to be obtained with the observed probabilities predicted so well by standard quantum theory.

It took about a decade before I figured how to do this, My next paper[10] gave necessary and sufficient conditions the evolving squared amplitudes must obey in order for this behavior to occur (these are given in Section 1.4).

I had to learn about variables that fluctuate randomly. I soon decided to use the best known, Brownian motion (whose time derivative is white noise), which made a lot of mathematical tools available. And so I was able to construct a modified Schrödinger equation that would do the job.[11] But, as the consequences of that theory were explored, problems emerged that I did not know how to resolve.

Einstein: A person who never made a mistake never tried anything new.

Eventually, after another decade I figured out how to resolve those problems (including help from ideas that appeared in 1986 from another dynamical collapse theory, created by Ghirardi, Rimini, and Weber), and arrived at the theory presented here,[12] which I named the Continuous Spontaneous Localization theory (CSL for short).

The purpose of this book is to make accessible the ideas behind and the consequences of CSL. I have tried to make it readable by a theoretically inclined undergraduate student who has taken the senior level course in quantum theory. I envision a varied audience: such undergraduate students, graduate students, interested physicists and philosophers of physics, maybe experimentalists who might think of testing the theory or theorists who might think of incorporating some of its ideas in work of their own.

The first 7 chapters of this book contain its main content, "the book proper."

Chapter 1 presents the rationale for the theory.

Chapter 2 gives the simplest example of a modified Schrödinger equation that satisfactorily collapses the wave function.

Chapters 3 and 4 consider richer modifications until the non–relativistic CSL theory is arrived at, whose consequences are to be tested by experiment.

Chapter 5 is a bit of a digression, explaining a different dynamical collapse theory, the SL (Spontaneous Localization) theory of Ghirardi, Rimini and Weber.

---

[10]Pearle (1976)

[11]Pearle (1979)

[12]Pearle (1989). The personal story described above is repeated in Pearle (2007) and continued forward to the discovery of CSL.

Chapter 6, the longest of these chapters, discusses a wide range of experimental tests of CSL.

Finally, Chapter 7 discusses interpretation: using the benefits of CSL, a way is suggested to go from the abstract wave function to the description of observed reality.

The following 7 chapters contain supplementary material to teach chapter of the book proper. They contain calculations I deemed too long or too complicated, whose results are cited in the book proper.

Chapter 15 provides an introduction for physicists to Stochastic Differential Equations (SDEs). CSL has an alternative form in terms of SDE's, completely equivalent to the one used in the book proper. The form used in the book proper is simpler and more intuitive than the SDE form. But, since I introduced CSL in SDE language, and the SDE formulation appears in the literature, I have included an introduction to the subject for those interested.

Chapter 16 shows how to express CSL in SDE form, and its use is illustrated in chapters 17 and 18 where two problems are solved using it.

I hope you enjoy reading this book. I have enjoyed writing it!

Contents

1
Introduction

It's something unpredictable, but in the end it's right.[1]

Standard quantum theory is arguably the best, most productive, set of ideas ever created. And yet, there is something terribly wrong with it. This book describes one way to fix it.

The notion of a real world is a primitive concept. We are comfortable with the notion because we have sensory, empirical evidence for it. We see its traces around us. And, wonderfully, we have the ability to think coherently of that reality. These thoughts are realized in theories containing mathematical equations. Some of the symbols in these equations are taken to correspond to the reality itself. They are associated to mental pictures of reality.

For classical Newtonian physics, one picture is a collection of point–like particles moving about in space. Others are a single point, representing all the particles, moving about in configuration space or in phase space. Classical electrodynamics provides a picture of electric and magnetic vectors tied to each point of space, changing length and direction with time. General relativity's picture is curved space–time. In these theories there are symbols corresponding to particle locations or fields or the metric. They take on numerical values capable of empirical verification in certain circumstances, which is what gives credibility to the pictures.

For standard quantum theory, the situation is different because, while there *are* pictures, they *cannot* correspond to reality.

One picture is of that most important construct, the state vector $|\psi, t\rangle$, rotating about in Hilbert space. Other pictures arise from projections of the state vector on particular bases. The ever-popular position basis gives the wave function picture, a time-varying complex number, that is, a complex field, in configuration space. What then goes wrong with these pictures?

1.1 "The abrupt change by measurement ... is the most interesting point of the entire theory."

Suppose the "state" of the state vector is taken to mean that it corresponds to the physical state of reality, to what we see. The problem is, if so initially, the Schrödinger equation readily evolves the state vector into something that does *not* represent reality, does not represent what we see.

---

[1] Green Day (1997).

In a letter dated August 8, 1935, Einstein made this point dramatically to Schrödinger[2], with an example of "... a substance, in chemically unstable equilibrium, perhaps a charge of gunpowder ... ," for which:

> In the beginning the ψ-function characterizes a reasonably well-defined macroscopic state. But, according to your equation, after the course of a year this is no longer the case at all. Rather, the ψ-function then describes a sort of blend of not-yet and of already exploded systems. Through no art of interpretation can this ψ-function be turned into an adequate description of a real state of affairs; in reality there is just no intermediary between exploded and not-exploded.

Schrödinger wrote back on August 19, 1935, "In a lengthy essay that I have just written I give an example that is very similar to your exploding powder keg ... " and he proceeded to give his own dramatic example, the now famous "cat paradox," where a state vector ends up as the sum of a "cat alive" state vector and a "cat dead" state vector with equal amplitudes.

Indeed, it is quite common, in measurement situations set up by experimenters, for the observed measurement outcome to be one of a number of possibilities, because nature is fundamentally quantum mechanical. Which of these possible outcomes is actually seen cannot be predicted. Indeed, the outcomes of repeated identical experiments appear randomly. For such situations, the corresponding Schrödinger's equation description is a state vector that does *not* correspond to one of the physical states we actually see. Instead, the state vector evolves into a *superposition* of such states:

$$|\psi\rangle = \sum_{n=1}^{N} c_n |\text{system and apparatus state denoting the } n\text{th outcome}\rangle.$$

The creators of quantum theory wanted the state vector to correspond to the reality we see around us. The Schrödinger equation doesn't give this. Accordingly, they postulated an additional, different kind of state vector evolution. This is called variously the "collapse" or "reduction"' or "projection"' or "jump" postulate. It is stated in textbooks as a necessary rule for how to apply standard quantum theory. For example, in Dirac's textbook:[3] " ... a measurement always causes the system to jump into an eigenstate of the dynamical variable that is being measured ... ."

The postulate entails more. Which eigenstate the system jumps into is governed by random statistics. And, the statistics obeys the "Born Rule." This says the nth result will be observed to occur with probability $|c_n|^2$. (This is why the amplitudes c_n are often given the sobriquet "probability amplitudes," to pay homage to this interpretive role.)

Does this removal of the problematic superposition by fiat then allow us to interpret the state vector as representing reality?

Schrödinger argued *no*. He strongly objected to adding, as a basic element of the theory, the collapse postulate's evolution, which requires occasionally abandoning the evolution in his equation. In his 1935 "lengthy essay,"[4A] he described the situation as follows:

---

[2]Fine (1986).
[3]Dirac (1958).
[4]Schrödinger (1935).

For each measurement one is required to ascribe to the psi-function (= the prediction-catalog) a characteristic, quite sudden change, which depends on the measurement result obtained, and so cannot be foreseen; from which alone it is already quite clear that this second kind of change of the psi-function has nothing whatever in common with its orderly development between two measurements. The abrupt change by measurement ... is the most interesting point of the entire theory. It is precisely the point that demands the break with naive realism. For this reason one can not put the psi-function directly in place of the model or of the physical thing. And indeed not because one might never dare impute abrupt unforeseen changes to a physical thing or to a model, but because in the realism point of view observation is a natural process like any other and cannot per se bring about an interruption of the orderly flow of natural events.

Schrödinger thought this was so important that he repeated it:

... any measurement suspends the law that otherwise governs continuous time-dependence of the psi-function and brings about in it a quite different change, not governed by any law but rather dictated by the result of the measurement. But laws of nature differing from the usual ones cannot apply during a measurement, for objectively viewed it is a natural process like any other, and it cannot interrupt the orderly course of natural events. Since it does interrupt that of the psi-function, the latter can not serve, like the classical mode, as an experimentally verifiable representation of an objective reality.

Thus, Schrödinger objected to abandoning his equation for the ad hoc collapse rule. He argues that, in physics, variables representing real things obey the same dynamical equations in all situations, including measurement situations. If, on account of measurement, the state vector changes dynamics in mid-stream, that is unacceptable behavior in a theory of nature/reality.

So, without the collapse postulate there are unreal superpositions. With the collapse postulate there is unacceptable dynamics. Thus, with or without the collapse postulate, the state vector cannot be taken to represent reality.[5][6]

The collapse postulate may be regarded as the Achilles heel of an otherwise spectacularly successful theory.[7]

Our best theory, quantum theory, spectacularly fails to describe what goes on all the time in nature. *Things happen, events occur, and quantum theory does not describe that.*

1.2   Guessing a Law of Nature

I think equation guessing might be the best method to proceed to obtain the laws for the part of physics which is presently unknown.[8]

---

[5] There is more to which one may object about the collapse postulate. It is ill-defined. It specifies that the state vector undergoes its abrupt change "when a measurement is completed." However, no one has ever been able to define a measurement. And, even given a specific example of a measurement, no one been able to specify when it is completed.

[6] Pearle (1967), Ballentine (1970, 1990).

[7] In the literature, this failure is called the "measurement problem." I have expressed the opinion that this is a misnomer. It should be called the "reality problem." Using standard quantum theory, we know how to describe measurements. What we don't know how to do is describe reality.

[8] Feynman (1965).

In this book we explore a way to maintain the point of view that the state vector represents the state of nature in our world. Schrödinger's objection to this point of view is that its application to a measurement situation is unacceptable because it requires the collapse postulate, which "suspends the law that otherwise governs continuous time-dependence of the psi-function." The law he objects to suspending is his own, Schrödinger's equation for the time evolution of the state vector.

But, this objection would disappear if there was a *different* law, call it a modified Schrödinger's equation, which provides a "continuous time-dependence of the psi-function" in all situations. And, in particular, when applied to a measurement situation, the dynamical evolution of the state vector is such that it ends up consistent with what we see, consistent with the outcome of the collapse postulate.

To come right down to it, we wish to guess a new law of nature. Its purpose is to describe collapse as a physical process.

This is what the first part of this book, Chapters 1–4 is about. I will take the reader through steps leading to what I called the continuous spontaneous localization (CSL) theory. (The second part of this book discusses consequences of CSL). It is my hope that it will be seen as plausible and even attractive, for at its core it is rather simple, and does its job remarkably well.

Two equations completely comprise CSL.

The first equation is a modified Schrödinger equation. To the usual Hermitian Hamiltonian operator $\hat{H}$ governing the state vector's time evolution is added a non-Hermitian operator $i\hat{H}'$ (where $\hat{H}'$ is Hermitian, so $i\hat{H}'$ is called "anti-Hermitian"). It is postulated that $\hat{H}'$ depends upon a field $w(\mathbf{x},t)$ that pervades all of three-dimensional space. It is assumed that *any* field $w(\mathbf{x},t)$ (that is, any space-time dependence of the field) at all is possible. Additionally, $\hat{H}'$ depends upon certain operators whose effect is to determine the end-states of the collapse process.

The state vector then evolves according to the modified Schrödinger equation

$$\frac{d}{dt}|\psi,t\rangle = \frac{1}{\hbar}[-i\hat{H} + \hat{H}']|\psi,t\rangle.$$

For each specific field $w(\mathbf{x},t)$ the modified Schrödinger equation has a solution. Since the modified Hamiltonian is different for each different field, the solutions are different. Thus, one obtains an *ensemble* of possible state vector solutions.

Because the modified Hamiltonian is not Hermitian, the squared state vector norm $\langle\psi,t|\psi,t\rangle$ changes with time. It does *not* remain at the value 1 as in standard quantum theory. This is crucial for the working of the second equation of the theory.

That second equation is called the "probability rule." The field $w(\mathbf{x},t)$ provided by nature does not obey an evolution equation. Instead, the probability that nature chooses any particular field needs to be specified.

Suppose you have an initial state vector $|\psi,0\rangle$ and you put into the modified Schrödinger equation *any* field $w(\mathbf{x},t')$, where $0 \leq t' \leq t$, and solve for $|\psi,t\rangle$. The probability rule says that the probability this $w(\mathbf{x},t')$ actually appears in nature is proportional to $\langle\psi,t|\psi,t\rangle$. (There is a proportionality factor that ensures the total probability is 1 when all possible fields are summed over.)

Qualitatively, the probability rule says, *only fields that give rise to large norm state vectors are likely to occur.*

That's it! To repeat, to describe what is happening in any circumstance, begin with the initial state vector, select an arbitrary $w(\mathbf{x}, t')$, solve the modified Schrödinger equation and find the state vector $|\psi, t\rangle$ that thereby evolves. Calculate $\langle \psi, t | \psi, t \rangle$, thus obtaining the probability this $w(\mathbf{x}, t')$ and $|\psi, t\rangle$ occur. When this is achieved for all possible $w(\mathbf{x}, t')$, the job is done.

How does this get rid of macroscopic superpositions? We will see that in great detail further in this book, but here is an illustration of how this works.

Consider an experiment with two possible outcomes, each of equal likelihood. To fix ideas, it might be a Stern–Gerlach experiment, where the initial state describes a spin-1/2 particle in an eigenstate of spin in the x-direction, and the apparatus measures the spin in the z-direction. Denote the possible z-spin eigenstates by $|1\rangle, |2\rangle$. The initial state vector expressed in this basis is

$$|\psi, 0\rangle = \frac{1}{\sqrt{2}} \Big[|1\rangle + |2\rangle \Big] |\text{initial state of apparatus}\rangle.$$

For a *particular* set of $w(\mathbf{x}, t)$, it is found that the result of the Schrödinger equation evolution, after a brief time interval T, is negligibly different from

$$|\psi, T\rangle \sim |1\rangle |\text{apparatus state has detected } 1\rangle,$$

where, using the probability rule, the sum of the probabilities associated to this set is very close to .5, say .4999999999 ... .

For a second, different, particular set of $w(\mathbf{x}, t)$, it is found that the result of the Schrödinger equation evolution at time T is negligibly different from

$$|\psi, T\rangle \sim |2\rangle |\text{apparatus state has detected } 2\rangle.$$

where, using the probability rule, the sum of the probabilities associated to this set is also .4999999999 ... .

Now, there remain lots of $w(\mathbf{x}, t)$ that are not in either particular set. What about them? Each leads to a state vector at time T that is the sum of the above two evolved state vectors, where coefficients may be comparably sized. So, such state vectors do *not* describe collapse. However, the size of their coefficients is extremely tiny, and the sum at time T of all their probabilities has, according to the probability rule, a net value .000000000 .... , experimentally indistinguishable from 0.

Thus, the theory produces collapse, describes the occurrence of one detection event or the other, and does so with the Born probabilities.[B]

Although all this entails a modification of Schrödinger's own equation, since it resolves the problem of the reality of the state vector that concerned Schrödinger, I like to imagine that he would approve!

1.3 Four Remarks

This chapter concludes by presenting a simple gambling game called the "Gambler's Ruin" game. Its behavior is strikingly analogous to that needed in a theory of dynamical state vector collapse, providing a hint as to how one may go about constructing such a theory.

Before doing so, I think this is as good a place as any to insert four comments I wish to make.

The first is to call the reader's attention to the Supplement. Chapters 1–7 comprise the tale told in this book. However, following this, for each chapter, there is a supplementary chapter.

The second is to mention a mathematical formulation of CSL that differs from the one presented here, but is completely equivalent.

The third is to make some introductory remarks concerning what might be called "interpretation," a subject discussed in detail in Chapter 7.

The last might be called ruminations on the idea of a guessed law of nature.

1.3.1   Supplement

The CSL theory is simple to write down. And, its consequences can often be fairly simple to arrive at, especially, as often happens, when one can simplify or neglect altogether the effect of the Hamiltonian $\hat{H}$ and just focus on the effects of $\hat{H}'$.

I have tried to keep the mathematical manipulations in chapters 1–7 as simple as possible. Indeed, the reader will find that much of the math involves just integrals of Gaussian functions, equations arrived at in Appendix A, Eqs. (A.2), (A.3), (A.5). I add that I have often put more steps than usual in derivations, to make them easier to follow.

However, sometimes the need for more complicated mathematics arises. For example, the simplest interesting problems in standard quantum theory, the free particle in one dimension and the one-dimensional harmonic oscillator, get more complicated in CSL, although here too they are completely solvable problems.

When the math entailed by a discussion becomes, in my judgment, too complicated or lengthy, so that a full exposition would impede the flow of the argument, I summarize the results of the needed calculation, but I believe readers should have the complete calculation available, if they want to see it. That is the purpose of the supplementary chapters 8–14, to provide the details of all such calculations for each chapter. Additional material sometimes is offered as well.

The purpose of chapters 15–18 is explained next.

1.3.2   A Single Schrödinger Equation

When I first began describing dynamical collapse by incorporating a random quantity in a modified Schrödinger equation, I introduced the idea of writing the modified Schrödinger equation as a stochastic differential equation (SDE).[9]

The simplest example of an SDE is

$$dx(t)/dt = a[x(t)] + b[x(t)]w(t).$$

Here, a and b are functions of $x(t)$. And, $w(t)$ is a special kind of random function of time, called "white noise" (a tutorial on white noise appears in Appendix D). Roughly speaking, $w(t)$ can be *any* function of time, with each function having associated a well-defined probability (of Gaussian form, given in Eq.(D.6)). Such a differential

---

[9]Pearle (1984a)

equation may be thought of as many differential equations, a different equation for each different $w(t)$, each with a different solution $x(t)$.

This idea,[10] that the modified Schrödinger equation be formulated as an SDE, has been extensively used and has been very fruitful.

SDEs are a well-developed area of mathematics. The above equation is typically considered solved by finding the probability distribution of $x(t)$.

In the years before I arrived at CSL, I considered other dynamical models. All were couched in SDE language. When I first introduced CSL, as a single modified Schrödinger equation, it was expressed as an SDE.[11]

An advantage of the SDE formulation of CSL is that it is expressed as a single equation. The probability rule for the white noise that appears in the SDE does not depend upon the state vector, it is just the standard probability rule for mathematical white noise.

Another advantage is that the modified Schrödinger equation can be written in terms of a state vector $|\psi, t\rangle_N$ whose norm is 1 (that is what the subscript N is there to indicate) as in standard quantum theory.

A disadvantage of the SDE formulation of CSL is that, unlike the usual Schrödinger equation, the modified Schrödinger equation is non-linear in the state vector. it depends on matrix elements of operators, containing terms of the form $\langle \psi, t | \hat{O} | \psi, t \rangle$.

The advantage of the formulation of CSL presented in this chapter is that, as discussed, the modified Schrödinger equation is linear in the state vector. We are more familiar with linear equations, their behavior is much easier to understand than the behavior of a non-linear equation.

Another issue is non-locality. Bell's inequality says that one cannot get a theory with "hidden variables" (extra variables beyond the state vector) that agrees with the predictions of standard quantum theory unless there are non-local effects: the part of the state vector describing "here" must depend upon the part of the state vector describing "there."

CSL is a prime example of the issue raised by Bell. It has a "hidden" (extra) variable, the white noise field $w(\mathbf{x}, t)$ mentioned earlier. It does agree with the predictions of standard quantum theory. And indeed, as Bell's theorem demands, it has non–locality built in.

In the form of CSL discussed here, the source of the non–locality is clear. The modified (linear) Schrödinger equation is *completely local*. The non–locality arises *totally* from the (non–linear) probability rule: a high probability field (giving rise to a large norm state vector) has correlations between the values of the field "here" and "there." In the SDE formulation of CSL, the non–locality arises from the non–linearity, but how that works isn't so obvious.

Another benefit of the formulation presented here is that one might consider there is an "action–reaction" relationship, lacking in the SDE formulation. The field determines the state vector (through the modified Schrödinger dynamics) while the state vector affects the field (the state vector norm determines the field's probability through the probability rule).

---

[10] Pearle (1979).
[11] Pearle (1989).

The two formulations of CSL are equivalent. The white noise field in the SDE formulation and the random field in the formulation presented here are related by a precise mathematical expression (to be given later). The state vectors evolving under these related fields are themselves related by

$$|\psi, t\rangle_N \equiv \frac{|\psi, t\rangle}{\sqrt{\langle \psi, t | \psi, t \rangle}}. \tag{1.1}$$

That is, they are equal up to a normalization factor, $|\psi, t\rangle_N$ being the normalized state vector solution of the SDE and $|\psi, t\rangle$ the un-normalized state vector we utilize here.

We will not use the SDE approach in this chapter, since a linear, local Schrödinger equation is simpler and more familiar than a non–linear non–local one.

However, readers desiring to work professionally with CSL might wish to learn the SDE formulation, since it commonly appears in the literature. For them, or for anybody who just wants to learn about SDEs, I have provided an SDE tutorial for physicists in Chapter 15. This is written without any mention of its application to CSL. I think it is a valuable pedagogical contribution!

Then, for those interested in the SDE formulation of CSL, this appears in Chapter 16, with two applications in Chapters 17 and 18..

1.3.3 Interpretational Remarks

Next, some remarks about the CSL description of reality I have come to espouse (a more complete discussion appears in Chapter 7).

The physical state of reality in Nature has the state vector in Hilbert space as its corresponding theoretical entity (as well as $w(\mathbf{x}, t)$, which must also be considered as having a physical counterpart). How do we obtain from that abstract thing, the state vector, a description of what we see, observable reality (chairs, trees, etc.)?

A clue may be taken from classical mechanics, where the physical state of Nature is also represented by an abstract mathematical entity, a vector in phase space, which might also be dubbed a state vector.

Here, observable reality can be obtained from the particle positions. These are projections of the vector on the position axes: the objects in observable reality are constructible from the clumping of such particles. We note that observable reality needs much less information than is contained in the full description of reality, much less than the phase space vector is capable of providing.

One could say that obtaining observable reality in classical mechanics consists of three entities and two steps. Entity one is the abstract vector. From it, one takes the the first step of constructing entity two, the particle positions. From those, one takes the second step of constructing entity three, what is observable.

For the collapse theory, let us proceed structurally similarly, with three entities and two steps.

Starting with the first entity, the state vector, we begin to take the first step by constructing the squared magnitudes of the state vector's projections on *any* set of orthonormal basis vectors. We will say these represent an *aspect of reality*. Thus, for a single particle, such things as $|\psi(\mathbf{x})|^2$, $|\psi(\mathbf{p})|^2$, $|\psi(E, \ell, \ell_z)|^2$, ... , all represent aspects of reality.

In homage to Bell, I like to borrow the word "stuff" he used to describe the way Schrödinger visualized his wave function:[12]

In the beginning, Schrödinger tried to interpret his wavefunction as giving somehow the density of stuff of which the world is made.

I give the name "stuff" or "stuff-distributions" to the second entity, the $|\langle a_n|, \psi \rangle|^2$. Thus, listed above are respectively position-stuff,[13] momentum-stuff, energy and angular momentum-stuff, ... Note that standard quantum theory does not allow speaking of a particle having position and momentum at the same time but, here, analogously to classical Newtonian physics, a particle has position-stuff and momentum-stuff at the same time.

One can consider the stuff corresponding to any basis. However, of special interest, are bases that correspond to what exists in space, since that is the purview of our senses. So, an electron's position-stuff provides us visual pictures. Whether constructed from a fairly localized wave function like that of an electron in the ground state of a hydrogen atom, or a spread out wave function, like that of an electron emerging from a two-slit screen to form the interference pattern that repeated position measurements bring to fruition, we regard the position-stuff as representing an aspect of reality that exists in space. So, from this point of view, the two-slit interference pattern of an electron comes about because something real actually goes through each slit: the electron's position-stuff.

Now, to describe our observable world, we must consider the description of many particles. It is therefore of particular interest to conclude the first step by specializing the basis we choose, to have stuff corresponding to the number of the most common particles, electrons, protons and neutrons, in *any* volume V. This particle number in V-stuff is the second entity we construct from the first entity, the state vector. If we can promote knowledge of this stuff to knowledge of the numbers of particles in V, we would be able to say where atoms are. By their clumping, we would have a description of observable reality, of what we see.

So, we now proceed to the second step, to construct the third entity. This entity is a *special subset* of the particle number in V-stuff. It provides what we will call *observability*. This subset enables us to *assert the actual numbers* of various kinds of particles in a volume V.

How is this to be done? To illustrate, suppose for simplicity there is only one kind of particle, the "nucleon." One can construct an operator representing nucleon number density and then, by integrating it over V, construct an operator $\hat{N}_V$ representing the nucleon number in V. Once we have the operator, we can construct its eigenbasis, which satisfies $\hat{N}_V |n\rangle_V = n|n\rangle_V$, where the integer n denotes a possible value for the number of nucleons in V. Given a many-particle (normalized) state vector $|\psi, t\rangle$ at time t, one may construct $|_V\langle n|\psi, t\rangle|^2$, the n nucleon-stuff in V.

---

[12] Bell (1990).

[13] In standard quantum theory, such a stuff-distribution is regarded as giving the probability density of detecting the particle at location $\mathbf{x}$ were a measurement of position to take place. We make no such connection here to a hypothetical measurement. Instead, the stuff is, as stated, just to be thought of as something that captures an aspect of reality.

How do we extract from this the number of nucleons *in* V? We give an example here. Suppose a "pointer," a macroscopic object containing N nucleons, is described by the normalized state vector at time t as being in a superposition of two quite separate places:

$$|\psi, t\rangle_N = c_1(t)|\psi_1\rangle + c_2(t)|\psi_2\rangle.$$

Let us consider a volume V that contains pointer 1, so $\hat{N}_V|\psi_1\rangle = N|\psi_1\rangle$ and $\hat{N}_V|\psi_2\rangle = 0$. Then,

$$\bar{n} \equiv {}_N\langle\psi, t|\hat{N}_V|\psi, t\rangle_N = |c_1(t)|^2 N,$$
$$\sigma^2 \equiv {}_N\langle\psi, t|[\hat{N}_V - \bar{n}]^2|\psi, t\rangle_N = |c_1(t)|^2 N^2 + \bar{n}^2 - 2|c_1(t)|^2 N\bar{n}$$
$$= N^2|c_1(t)|^2[1 - |c_1(t)|^2] = N^2|c_1(t)|^2|c_2(t)|^2$$

are the mean value of n and its variance in V respectively.

As the collapse proceeds under a particular random field, assume that field is one of those the probability rule says is of high probability. Then either $|c_1(t)|^2$ approaches 1 or 0. If $|c_1(t)|^2$ approaches 1, the pointer has ended up in region 1.If $|c_1(t)|^2$ approaches 0, the pointer has ended up in region 2. In either case, σ approaches zero

As σ then approaches zero, when it decreases beyond a certain threshold (to be made precise in Chapter 7), we say that $\bar{n}$ *is* the observable number of particles in V.

To repeat: for those special volumes V for which the standard deviation in the number of nucleons is sufficiently small, we go from the nucleons in V-stuff to a declaration of the observable number of nucleons in V. This is the *prediction of the theory,* to be compared with observation.

In summary, we have given a hierarchy of three entities: the state vector representing reality, arbitrary stuff distributions representing aspects of reality, and a subset of these, special stuff distributions representing observable reality. (Philosophers of physics use the terminology that the second supervenes on the first, the third supervenes on the second.) In subsequent sections, it is useful to keep this in mind as we see the details of how the collapse evolution works to narrow the standard deviation of particle number distributions, so that observability dynamically emerges.

1.3.1   Constraints on the Theory

The history of physics is replete with examples of a proposed "fundamental law" that, in time, is found to make predictions that conflict with experiment, thus has been found wanting and been replaced with another law. Although tremendously successful, why should quantum theory be exempt from this evolution?

Schrödinger's equation was a guessed law, something that (so far) apparently cannot be derived from something else, and it is applicable to a wide range of phenomena. More modest are guessed laws called phenomenological laws, invented to help resolve some problem with a current theory. That is the case for CSL, invented to give a description of the occurrence of events, which standard quantum theory does not do. However, CSL is expected to be applicable to a wide range of phenomena, which is unusual for phenomenological laws.

The constraints on such a dynamical collapse law are stringent. There must be the collapse behavior. The collapse outcome probabilities must obey the Born rule.

The new law has to work so that a superposition of macroscopically different states, such as occurs in a measurement situation, gets a rapid collapse treatment. On the other hand, a superposition of microscopically different states has to be negligibly affected by the collapse mechanism since the existence of such a superposition is experimentally verified, in interference of particles or atoms or molecules. (While "we" might be hard put to define with any precision what is meant by the words "macroscopic" and "microscopic," i.e., to say when the state vector is to be strongly or scarcely affected by the collapse dynamics, the law will decide for itself!)

Since it is a different law than standard quantum theory, some experimental predictions it makes will be different too. Chapter 6 considers a variety of these. It has been more than 30 years since the inception of CSL. Remarkably, these predictions do not disagree with the results of any relevant substantiated experiment performed so far. New and more stringent experiments are continually being proposed and executed to test it.

And, one would hope the new law could be described as simple or elegant (although aesthetic judgments are notoriously subjective). Dirac famously wrote:[14] "It seems that if one is working from the point of view of getting beauty in one's equations, and if one has really a sound insight, one is on a sure line of progress."

One can hope that the idea of dynamical collapse represents a sound insight. Certainly, it delivers a significant achievement. One is accustomed to thinking that standard quantum theory describes everything that occurs in Nature, because of its enormous predictive success. But, it does *not* describe everything that occurs in Nature. It does not describe that most obvious thing, the individual outcome actually seen in an experiment with multiple possible outcomes. And, that is what dynamical collapse does.

Also, I believe one may find that there is beauty, in the simplicity of the basic mechanism for collapse discussed in this book.

1.4   Requirements on State Vector Dynamics

Now we return to continuing the discussion begun in Sections 1.1 and 1.2, and start to look at how to make such a theory.

The thing to tackle first is the novel, crucial thing: how does one describe the collapse dynamics? In what follows, for simplicity, we specialize to two states (the discussion is, and will be, readily generalized to more states). Also, the discussion proceeds in terms of the normalized state vector $|\psi, t\rangle_N$ rather than the un-normalized $|\psi, t\rangle$. (Using the evolution equation, Eq. (1.1), we may find $|\psi, t\rangle$ and then just normalize it to get $|\psi, t\rangle_N$.)

We want to consider a state vector which, at time 0, is a superposition of macroscopically different states, and ask how it should evolve thereafter toward one of them. So, consider

$$|\psi, t\rangle_N = c_1(t)|a_1\rangle + c_2(t)|a_2\rangle \tag{1.2}$$

---

[14] Dirac (1963).

for $t \geq 0$. For example, this could be the state vector describing an experiment that took place prior to $t = 0$. Then, the states $|a_i\rangle$ correspond to different (macroscopic) measurement outcomes (sometimes called different "pointer positions," which evokes the days when a meter needle indicated the measurement result, as illustrated on the cover of this book).

These states are (orthogonal, normalized to 1) eigenstates of a Hermitian operator $\hat{A}$ we will call a *collapse-generating operator*, so

$$\hat{A}|a_i\rangle = a_i|a_i\rangle. \tag{1.3}$$

The eigenstates $|a_i\rangle$ are termed the "preferred basis."[15] As might be expected, $\hat{H}'$, the collapse-generating term added to the usual Hamiltonian, is a function of $\hat{A}$ (as well as a function of the random variable w(t)).

Now, under collapse dynamics, how is the state vector (1.2) to evolve? What is important for collapse behavior are the squared magnitudes, $x_i(t) \equiv |c_i(t)|^2$. They are to change in random fashion, since $\hat{H}'$, which depends on the random field $w(\mathbf{x}, t)$, will be a random function of time. They need to satisfy three necessary and sufficient conditions.[16]

First, since $1 = {}_N\langle \psi, t | \psi, t \rangle_N = |c_1(t)|^2 + |c_2(t)|^2$, there is the constraint

$$x_1(t) + x_2(t) = 1. \tag{1.4}$$

Second, collapse means, after a sufficiently long time, certainly by time ∞, that

$$either \ x_1(\infty) = 1, x_2(\infty) = 0, \ or \ x_1(\infty) = 0, x_2(\infty) = 1. \tag{1.5}$$

Then, the state vector (1.2) has evolved to either $|a_1\rangle$ or $|a_2\rangle$ (up to a possible phase factor). Collapse is then complete: we have obtained a result from the mathematics that is consistent with what is seen.

For the third condition, consider the collection of all possible evolutions (the ensemble of evolutions). The Born rule, which connects the initial state vector to the final results of the ensemble, must be obeyed. The Born rule says that $|\langle a_i | \psi, 0 \rangle|^2 = x_i(0)$, the *initial* value of the variable, is the probability that the *final* state of the collapse dynamics is $|a_i\rangle$. This means, for the *ensemble* of evolutions, that ^{C}

$$x_1(\infty) = 1, x_2(\infty) = 0, \text{ occurs with probability } x_1(0),$$
$$x_1(\infty) = 0, x_2(\infty) = 1, \text{ occurs with probability } x_2(0). \tag{1.6}$$

1.5 Gambler's Ruin Game

Thus chance has been admitted, not merely as a mathematical tool for physics, but as part of its warp and weft.[17]

---

[15]Pearle (1986).
[16]Pearle (1976).
[17]Wiener (1954).

Fig. 1.1 Gambler's Ruin Game: the probability of winning is equal to the initial percentage of the stakes.

Perhaps surprisingly, what appears to be a hodgepodge of requirements (1.4), (1.5), (1.6), occurs in a natural way, in what is called the "Gambler's Ruin game."[18] This seemingly off–the–wall game,[D] having nothing to do with quantum mechanics has, in its evolution, the very heart of the evolution in dynamical collapse.[19]

Consider two gamblers, Willie Oneka and Tillie Twosday (sketched in Fig. 1.1). They have between them \$100. Say Tillie starts with \$60 and Willie \$40. They toss a coin: if the result is heads, Tillie gives a dollar to Willie. If the result is tails, Willie gives a dollar to Tillie. Thus, the money each possesses fluctuates up and down but, of course, the total money in the game is constant. The game ceases when one of them ends up with all the money, which inevitably happens.[E]

They play the same game over and over. What is the probability of winning?

As shown below, Tillie wins 60% of the games, while Willie wins 40% of the games. The initial stakes determine the probability of winning, precisely as the initial squared amplitudes determine the Born rule probabilities of the collapse outcomes!

The analogy with the collapse requirements is *precise*, as will now be shown.

Denote Tillie's (Willie's) fraction of the total sum of \$100 at any time as $x_1(t)$ $(x_2(t))$. Thus, $(x_1(t), x_2(t))$ can take on the values $(0, 1), (.01, .99), (.02, .98) \ldots (.5, .5)$, $\ldots (.98, .02), (.99, .01), (1, 0)$. Of course, Eq. (1.4) is satisfied.

[18] Feller (1950).
[19] Pearle (1982).

And, (1.5) is satisfied: the game ends, with either $(0, 1)$ or $(1, 0)$ as the final result. So, it just remains to prove that Eq. (1.6) is satisfied.

To enable this, define $Q(x)$ as the following conditional probability: given that a gambler possesses the fraction x (possesses $\$100x$), that this gambler eventually wins. (This game is an example of a Markov process, which entails that the probability of the future depends only upon the present.) Since the game is completely fair, meaning that it looks the same to either participant, this conditional probability is the same function for either gambler.

Then, based upon the rules of probability theory, one can write down an equation relating $Q(x)$ to $Q(x + .01)$ and to $Q(x - .01)$.

For, there are only two possible ways that a gambler can possess the fraction x of the money at a particular step in the game, and then go on to win the game.

One way is that, on the previous step, the gambler had the fraction $x - 0.01$ (had $\$(100x-1)$) and won that toss, which is an event of probability $1/2$. So, the probability is $1/2$ that the gambler's probability of winning the game is $Q(x - 0.01)$.

The other way is that, on the previous step, the gambler had the fraction $x + 0.01$ (had $\$(100x + 1)$) and lost that toss, which is an event of probability $1/2$. So, the probability is $1/2$ that the gambler's probability of winning the game is $Q(x + 0.01)$.

Thus, according to the rules of probability theory, $Q(x)$ satisfies the equation

$$Q(x) = \frac{1}{2}Q(x - 0.01) + \frac{1}{2}Q(x + 0.01). \tag{1.7}$$

The general solution of this difference equation is $Q(x) = A + Bx$.

The constants A and B are determined by the boundary conditions. Since $Q(0) = 0$ (if the gambler has $\$0$, it is impossible to win), we see that $A = 0$. Since $Q(1) = 1$ (the gambler who has all the money has already won), we see that $B = 1$.

Thus, $Q(x) = x$. The probability of winning, of eventually achieving $x = 1$ if one starts with the initial fraction x, is equal to the initial fraction x, just as in the Born Rule (1.6)!

So, it is quite a helpful analogy to *think of the collapse process as a Gambler's Ruin competition among the possible outcomes in a superposition.* The squared quantum amplitudes in Eq.(1.2), $|c_1(t)|^2 = x_1(t)$, $|c_2(t)|^2 = x_2(t)$, play the Gambler's Ruin game against each other, fluctuating up and down until one "wins": the collapse is complete.

True, in the Gambler's Ruin game, there is a finite change in $x_1(t), x_2(t)$ which occurs in discrete time, whereas the collapse entails an infinitesimal change in $x_1(t), x_2(t)$ occurring in continuous time. However expressing the gambler's ruin game as a continuous process with infinitesimal exchanges of cash is readily achieved.[20] And, one can provide an analog to a superposition of N quantum states collapsing to one of them by imagining N gamblers playing the game in pairs.

In the Gambler's Ruin game we see the behavior needed for dynamical collapse. So, we must turn to constructing equations describing dynamical collapse that implement Gambler's Ruin behavior.

---

[20]Pearle (1982).

1.6   Synopsis

Here is an outline of what follows in the rest of this book.

The first order of business is to present the CSL *general framework*, describing the continuous evolution of a state vector in a superposition of (preferred) basis states toward one of them. That is, we will give a dynamical equation for the state vector such that $|\psi, 0\rangle = \sum_i c_i(0)|a_i\rangle$ evolves to $|\psi, t\rangle \to |a_j\rangle$ under a Hamiltonian $\hat{H}'$ that depends upon white noise. Moreover, the probability rule entails that the result $|a_j\rangle$ of these evolutions occurs with probability $|c_j(0)|^2$. This is discussed in the next two chapters.

Specifically, in Chapter 2, the two suggested equations defining dynamical collapse, the modified Schrödinger equation and the probability rule, are presented in their simplest form, and how they work is elucidated.

Also, the density matrix for this simplest of cases is presented. The density matrix is a very useful mathematical tool for analyzing the behavior of a collection of state vector evolutions. And, CSL provides such a collection (each element of the collection is a different evolution under a different random function $w(t)$). (An introduction to the density matrix is given in Appendix E.)

In Chapter 3, this simplest form of the CSL modified Schrdinger equation is gradually generalized. These additional features are needed for non-relativistic CSL. Also here is placed the calculation of the wave function evolution for a free particle moving in one dimension under the combined influence of the free particle Hamiltonian and the collapse Hamiltonian, where the collapse–generating operator is the particle's position operator.

Then, in Chapter 4, these features are used to construct the modified Schrödinger equation for non-relativistic CSL.

Also presented in Chapter 4 is the corresponding density matrix evolution equation, derived from the two CSL equations. This is actually *what is utilized in the remainder of the book* (since it describes the ensemble of evolutions which is what experiment tests).

Then, two important consequences of non-relativistic CSL are obtained.

One is the collapse rate for an object in a superposition of two places. This is important because a two–outcome experiment, which evolves according to the Hamiltonian evolution, entails such a superposition, so this quantitatively shows how the collapse evolution takes over, giving one or the other outcome..

The other is a "universal" expression for the energy increase of any collection of particles (which turns out to be independent of the interaction potential between the particles). The physical reason for the increase is the narrowing of particle wave functions due to the collapse dynamics. This is important because there are experimental techniques to measure energy increases, thereby enabling tests of this prediction of the theory.

Chapter 5 is a pause to present a different collapse theory, the SL (spontaneous localization) theory[21] (also called GRW theory after the authors' initials). Attention is given to displaying the features of that theory incorporated in CSL.

---

[21] Ghirardi, Rimini, and Weber (1986), Bell (1987).

Chapter 6 then considers various experiments, proposed and already performed, for which CSL makes different predictions than standard quantum theory.

Finally, Chapter 7 discusses the interpretation of the theory, a picture of reality engendered by CSL and how to extract observable consequences from it, in more detail than has already been discussed.

It is my aim, in this book, to provide an accessible introduction to the CSL theory of dynamical collapse.

Notes

[A] This is often called Schrödinger's "cat paradox" paper. Equally famously, Schrödinger introduced and analyzed the notion of entanglement in this essay. However, his main theme, as outlined here, was to castigate the collapse postulate.

[B] It is well to mention here that, according to the theory, complete collapse, $|\psi, t\rangle \sim |\text{outcome}\rangle$, is only achieved at infinite time, and if only the collapse mechanism operates, that is, if the usual Hamiltonian $\hat{H}$ is set equal to 0. In practice, for finite time and with $\hat{H} \neq 0$, one has $|\psi, t\rangle \sim |\text{outcome}\rangle + |\epsilon\rangle$, where $\langle \epsilon | \epsilon \rangle / \langle \text{outcome} | \text{outcome} \rangle \ll 1$. How to view this "tail," as it is called, is discussed in Chapter 7. The upshot is essentially that the tail may be disregarded because its particle number content is negligibly different from 0.

[C] As an aside, it is possible to deduce the gambler's ruin conditions Eqs. (1.5) (which is expressed in terms of the outcomes of an *individual* run of the game) and Eq. (1.6) (which gives the probability of *individual* outcomes), by two conditions that are statements about *averages* (over the complete set of possible runs).[22] It is interesting that statements about ensemble averages lead to results for individual systems. This is shown in Chapter 8, the Supplement for Chapter 1.

[D] This problem began in 1656 with correspondence between the founders of probability theory, Blaise Pascal and Pierre de Fermat, and was treated by Christiaan Huygens in a 1657 book. It is fun to think that work by these early venerable theoretical physicists might play a role in resolving the 'measurement problem' of quantum mechanics.

[E] Consider an infinite string of coin tosses. Any particular sequence of toss outcomes is sure to occur. At most a run of 99 heads or tails in a row will end the game, whatever the initial stakes.

[22] Pearle (1976).

2
Continuous Spontaneous Localization (CSL) Theory

We begin with the CSL modified Schrödinger equation for the evolution of the state vector under the influence of a random noise function, for the simplest possible case, a two-state Hilbert space. It is shown how it and the probability rule work to give Gambler's Ruin-type collapse behavior.

This is followed by a derivation and discussion of the associated density matrix evolution equation for the ensemble of state vectors evolving under all possible random noises.

Consider the initial state vector

$$|\psi, 0\rangle = c_1(0)|a_1\rangle + c_2(0)|a_2\rangle \text{ with } |c_1(0)|^2 + |c_2(0)|^2 = 1. \tag{2.1}$$

It is written in terms of the preferred basis eigenstates $|a_i\rangle$ of the collapse–generating operator $\hat{A}$ (Eq.(1.3)). (It is assumed that $a_1 \neq a_2$.)

In the Gambler's Ruin game, the money a gambler possesses fluctuates with time, undergoing a random walk until the game ends. A random walk, in the continuous time limit is (mathematical) Brownian motion.[A] (Random walk and Brownian motion are discussed in detail in Appendices B and C.) Therefore, we expect collapse dynamics to involve Brownian motion.

2.1 The Two Equations of CSL

It is simplest to start with the *solution* $|\psi, t\rangle$ of the modified Schrödinger equation and show how that works. (Of course, the time derivative of the solution gives the modified Schrödinger equation that gives rise to that solution.)

The usual Hamiltonian $\hat{H}$ can alter the collapse dynamics, so we will set it equal to 0 for now.

Here is the state vector evolution we propose:

$$|\psi, t\rangle = e^{-\frac{1}{4\lambda t}[B(t) - 2\lambda \hat{A} t]^2} |\psi, 0\rangle. \tag{2.2}$$

What is in this equation, what governs the state vector evolution, are the collapse–generating operator $\hat{A}$, a parameter λ, and the function $B(t)$.

We will see that λ, which has dimension time^{-1}, characterizes the rate of collapse to one or the other state. Since the exponent has to be dimensionless, we note that $\hat{A}$ and $B(t)$ are dimensionless quantities.

The function $B(t)$ is supposed to be chosen by Nature from the class of mathematical Brownian motion functions, which may be thought of as a continuous but non-differentiable random walk of a "point" (see Fig. 2.1).

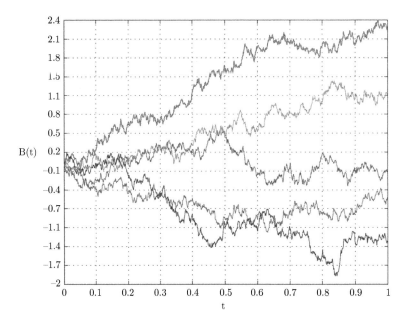

Fig. 2.1 $B(t)$ graphed as a function of t for five different Brownian motion trajectories.

There are many different Brownian motion trajectories of a point that starts from location $B(0) = 0$ and which happen to reach the same value of B at time t. Eq. (2.2) says that, as far as the state vector at time t is concerned, all that matters is $B(t)$. It is indifferent to the path taken to reach that value.[B]

For the *usual* Brownian motion, the probability for a location at time t lying in the interval $[B(t), B(t) + dB(t)]$ is (as discussed in Appendices B, C) given by the Gaussian probability rule

$$P(B(t))dB(t) = \frac{1}{\sqrt{2\pi\sigma^2 t}} e^{-\frac{B^2(t)}{2\sigma^2 t}} dB(t).$$

In this equation, t is a constant, a label of the particular time under consideration. $P(B(t))$ is called the probability density. The numerical factor in front of the Gaussian is chosen so that $\int_{-\infty}^{\infty} dB(t)P(B(t)) = 1$: the point's location $B(t)$ is sure to be somewhere.

The probability rule for $B(t)$ in CSL is *not* the simple standard one above. The probability of $B(t)$ depends upon the state vector that has evolved under this B until time t, and is

$$P(B(t))dB(t) \equiv \frac{1}{\sqrt{2\pi\lambda t}} \langle \psi, t|\psi, t\rangle dB(t). \tag{2.3}$$

So, the largest norm state vectors have the highest probability of occurring.

These two equations, (2.2), (2.3), a modified Schrödinger evolution and a probability rule, are all there is to this simplest application of CSL theory.[C1]

2.1.1 How CSL Works

Now, let us see how Eqs. (2.2), (2.3) work.

Insert the initial state vector (2.1) into both equations. Using $\hat{A}|a_i\rangle = a_i|a_i\rangle$ we obtain, for the state vector,

$$|\psi, t\rangle = c_1(0)e^{-\frac{1}{4\lambda t}[B(t)-2\lambda a_1 t]^2}|a_1\rangle + c_2(0)e^{-\frac{1}{4\lambda t}[B(t)-2\lambda a_2 t]^2}|a_2\rangle, \qquad (2.4a)$$

and for the probability density of $B(t)$

$$P(B(t)) = \frac{1}{\sqrt{2\pi\lambda t}}\left[|c_1(0)|^2 e^{-\frac{1}{2\lambda t}[B(t)-2\lambda a_1 t]^2} + |c_2(0)|^2 e^{-\frac{1}{2\lambda t}[B(t)-2\lambda a_2 t]^2}\right]. \quad (2.4b)$$

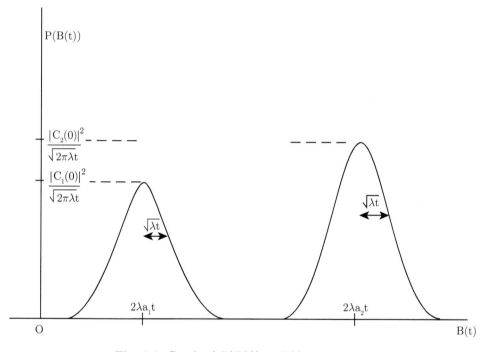

Fig. 2.2 Graph of $P(B(t))$ vs $B(t)$ at some time t.

Focus first upon the probability density expression (2.4b), the sum of two Gaussians (graphed in Fig. 2.2). Immediately, we confirm from (2.4b) that the total probability is 1 (using the Gaussian integral (A.3) in Appendix A): $\int_{-\infty}^{\infty} dB(t)P(B(t)) = |c_1(0)|^2 + |c_2(0)|^2 = 1$.

What are the high probability values of $B(t)$? We see that the *peaks* of the Gaussians move at different speeds, one at speed $2\lambda a_1$, the other at speed $2\lambda a_2$, and so

---

[1]It is shown in Pearle (2013) that CSL can be derived from two postulates, Gambler's Ruin behavior and a linear, real, Schrödinger equation.

separate linearly with time. However, the *widths* of the Gaussians are given by the standard deviation $\sqrt{\lambda t}$, and so spread much more slowly since $\sqrt{t}$ increases much more slowly than t.

Therefore, eventually, when the peak separation exceeds a few standard deviations, the probability is the sum of two Gaussians which overlap very little, as illustrated in Fig. (2.2). One therefore need only consider those B values that lie within Gaussian 1 or those B values that lie within Gaussian 2 (by "within" is meant a few standard deviations to either side of the peak). The remaining B values have very little probability of being realized.

Then, apply this to the state vector expression (2.4a). Consider a $B(t)$ value that lies within Gaussian 1. The magnitude of Gaussian 2 is negligibly small for that value of $B(t)$. Since the magnitude of Gaussian 2 multiplies $|a_2\rangle$, the contribution of $|a_2\rangle$ to the state vector (2.4a) is negligibly small. So, a $B(t)$ value that lies within Gaussian 1 is responsible for collapse to state $|a_1\rangle$.

(And, of course, this holds with the labels 1 and 2 reversed.)

So, to an excellent approximation (improving all the time and exact in the limit $t = \infty$),[2] we have one of two alternatives for state vectors where $B(t)$ has non–negligible probability:

$$|\psi, t\rangle \approx c_1(0)e^{-\frac{1}{4\lambda t}[B(t)-2\lambda a_1 t]^2}|a_1\rangle, \quad P(B(t)) \approx |c_1(0)|^2 \frac{1}{\sqrt{2\pi\lambda t}} e^{-\frac{1}{2\lambda t}[B(t)-2\lambda a_1 t]^2},$$

$$(2.5a)$$

$$|\psi, t\rangle \approx c_2(0)e^{-\frac{1}{4\lambda t}[B(t)-2\lambda a_2 t]^2}|a_2\rangle, \quad P(B(t)) \approx |c_2(0)|^2 \frac{1}{\sqrt{2\pi\lambda t}} e^{-\frac{1}{2\lambda t}[B(t)-2\lambda a_2 t]^2}.$$

$$(2.5b)$$

At $t = \infty$, $\approx$ is replaced by $=$ in these expressions.

In this way, the state vector collapses to one or the other possible eigenstates of $\hat{A}$ (albeit with a numerical multiplying factor).

And, when one integrates the probability expressions over all values of $B(t)$ that give rise to one or the other outcome, one gets $\int_{-\infty}^{\infty} dB(t) P(B(t)) \approx |c_i(0)|^2$, which are the Born probabilities.

A word about the numerical factor multiplying a collapsed state vectors. In order to explore the physical meaning of a state vector, one typically calculates expectation values of operators using the normalized state vector $|\psi, t\rangle_N$, which one can calculate from $|\psi, t\rangle$ at any time. This is a complicated expression during the initial state vector evolution (written in Eq. (2.6)), until one can make the approximations appropriate when the Gaussians are well-separated. Then, from (2.5a), (2.5b), $|\psi, t\rangle_N$ ends up simply as $|a_1\rangle$ or $|a_2\rangle$ to an excellent approximation.

To summarize, we have shown, for a collapse-generating operator $\hat{A}$ with a discrete spectrum, how the Eqs. (2.2) and (2.3) give rise to collapse, with the Born probabilities.[3]

---

[2] See Chapter 7 for discussion of the interpretational issue in the approximation of neglecting the Gaussian "tail" at a finite time.

[3] How does this work when there are more than two states? When $\hat{A}$ has N discrete eigenstates and eigenvalues, Fig. (2.2) is replaced by a similar picture, of N moving and separating Gaussians, with

2.2   CSL Density Matrix

We have considered above the state vector description of collapse. Eq.(2.4a) tells us that, at time t, for each $B(t)$ value, there is a corresponding state vector. Eq.(2.4b) tells us the probability of occurrence of that $B(t)$, and therefore the probability of occurrence of that corresponding state vector. We do not know which $B(t)$, and so which corresponding state vector, actually has been chosen by Nature. However, an ensemble of state vectors is generated by repeated evolutions that start out with the same initial state vector, but end up differently, as typically occurs in experiments. It therefore behooves us to consider the full ensemble of state vectors.

The most natural tool for this discussion is the density matrix (see Appendix F for an introduction). If we have an ensemble of state vectors labeled by a discrete index i, $|\psi_i, t\rangle$, and they are known to occur with respective probability $p_i(t)$ ($p_i(t) \geq 0$, $\sum_i p_i(t) = 1$), the density matrix (operator) is defined as

$$\hat{\rho}(t) \equiv \sum_i p_i(t)|\psi_i, t\rangle\langle\psi_i, t|$$

For our application, there is not a discrete sum labeled by i, there is a continuous integral labeled by $B(t)$.

Before applying the density matrix expression to the CSL situation, let us consider two useful notations, to denote two different kinds of averages that frequently occur.

The first, a stochastic average (in this case, an average over all values of $B(t)$) will be denoted by an overline. Thus the density matrix can be written as $\hat{\rho}(t) = \overline{|\psi, t\rangle\langle\psi, t|}$.

The second is a quantum average. What is called the "expectation value" of an operator $\hat{O}$, $\langle\psi, t|\hat{O}|\psi, t\rangle$, is often a useful thing to calculate. The reason is that it may be written as

$$\langle\psi, t|\hat{O}|\psi, t\rangle = \langle\psi, t|\hat{O}\Big[\sum_j |o_j\rangle\langle o_j|\Big]|\psi, t\rangle = \sum_j \langle o_j|\psi, t\rangle|^2 o_j.$$

(We have used $\sum_j |o_j\rangle\langle o_j| = \hat{1}$: the eigenstates $|o_i\rangle$ of $\hat{O}$ form a complete set.) If one has a system described by the state vector $|\psi, t\rangle$ and performs repeated measurements of the physical quantity corresponding to the operator $\hat{O}$, one obtains one or another o_i. The average result would be given by this expression, each eigenvalue multiplied by its probability of observation. So, we will write $\langle\hat{O}\rangle \equiv \langle\psi, t|\hat{O}|\psi, t\rangle$ for this average (mean, expectation) value.

The density matrix is so useful because the expectation value of an operator over all the state vectors in the ensemble can easily be found using it:

$$\overline{\langle\hat{O}\rangle} = \sum_i p_i(t)\langle\psi_i, t|\hat{O}|\psi_i, t\rangle = \sum_{ij} o_j p_i(t)|\langle o_j|\psi_i, t\rangle|^2$$

$$= \sum_j \langle o_j|\hat{O}\Big[\sum_i p_i(t)|\psi_i, t\rangle\langle\psi_i, t|\Big]|o_j\rangle \equiv \mathrm{Tr}\hat{O}\hat{\rho}(t),$$

accompanying equations just like those discussed. An example is given in Chapter 9, the Supplement for this chapter, where $\hat{A}$ has a continuous spectrum of eigenvalues. There, the collapse-generating operator is $\hat{A} = \hat{X}$, the position operator for a particle in one dimension. It is shown how the initial wave function undergoes collapse toward one or another eigenstate $|x\rangle$ with the Born probability.

(Tr denotes the trace operation, the sum of diagonal matrix elements of an operator: see Appendix E). This equation combines both notations. For example, the quantum expectation value of the energy for the ensemble of state vectors is $\langle \hat{H} \rangle$.

We now turn to construct the CSL density matrix for the ensemble of state vectors in (2.4a). The state vectors must be normalized to 1, which is then

$$|\psi, t\rangle_N = \frac{|\psi, t\rangle}{\sqrt{\langle \psi, t|\psi, t\rangle}} = \frac{c_1(0)e^{-\frac{1}{4\lambda t}[B(t) - 2\lambda a_1 t]^2}|a_1\rangle + c_2(0)e^{-\frac{1}{4\lambda t}[B(t) - 2\lambda a_2 t]^2}|a_2\rangle}{\sqrt{|c_1(0)|^2 e^{-\frac{1}{2\lambda t}[B(t) - 2\lambda a_1 t]^2} + |c_2(0)|^2 e^{-\frac{1}{2\lambda t}[B(t) - 2\lambda a_2 t]^2}}}.$$

$$(2.6)$$

Now, this looks very complicated, made so by the normalization factor in the denominator. However, what is neat is that the probability, given by the probability rule, Eq. (2.4b), cancels this normalization factor in the density matrix expression! Then, the integral over the probability, $\int dB(t)$, involves just the readily performed integral of a product of Gaussians.

In detail, starting with the definition of the density matrix,

$$\hat{\rho}(t) = \overline{|\psi, t\rangle_{NN}\langle \psi, t|} = \int_{-\infty}^{\infty} dB(t) P(B(t)) |\psi, t\rangle_N \, {}_N\langle \psi, t| \qquad (2.7)$$

we insert the probability expression (2.3) and the normalized state vector expression (2.7),

$$\hat{\rho}(t) = \int_{-\infty}^{\infty} dB(t) \frac{1}{\sqrt{2\pi\lambda t}} \langle \psi, t|\psi, t\rangle \frac{|\psi, t\rangle}{\sqrt{\langle \psi, t|\psi, t\rangle}} \frac{\langle \psi, t|}{\sqrt{\langle \psi, t|\psi, t\rangle}} \qquad (2.8)$$

which, following cancellation, is the very simple expression

$$\hat{\rho}(t) = \int_{-\infty}^{\infty} dB(t) \frac{1}{\sqrt{2\pi\lambda t}} |\psi, t\rangle\langle \psi, t|. \qquad (2.9)$$

Then, putting the state vector expression (2.4a) into (2.9), this is

$$\hat{\rho}(t) = \int_{-\infty}^{\infty} dB(t) \frac{1}{\sqrt{2\pi\lambda t}} \sum_{i,j=1}^{2} c_i(0) c_j^*(0) e^{-\frac{1}{4\lambda t}[B(t) - 2\lambda a_i t]^2} |a_i\rangle\langle a_j| e^{-\frac{1}{4\lambda t}[B(t) - 2\lambda a_j t]^2},$$

$$(2.10)$$

which leaves only a Gaussian integral of the form (A5) to be performed, with the result:

$$\hat{\rho}(t) = \sum_{i,j=1}^{2} c_i(0) c_j^*(0) |a_i\rangle\langle a_j| e^{-\frac{\lambda t}{2}[a_i - a_j]^2}. \qquad (2.11)$$

Thus we have obtained the density matrix for the ensemble of collapsing state vectors arising from the single initial state vector $|\psi, 0\rangle$ in (2.1).

Indeed, setting $t = 0$ in (2.11), we get $\hat{\rho}(0) = |\psi, 0\rangle\langle \psi, 0|$.

The asymptotic behavior in (2.11) is just what should be expected from dynamical collapse. The eventual density matrix should be

$$\hat{\rho}(\infty) = |c_1(0)|^2|a_1\rangle\langle a_1| + c_2(0)|^2|a_2\rangle\langle a_2|, \qquad (2.12)$$

where every state vector of non-negligible probability ends up as either $|a_1\rangle$ or $|a_2\rangle$, each with the the the Born Rule probability.

We see from (2.11) how the density matrix evolution accomplishes this. It keeps the diagonal elements constant at all times,[D]

$$\langle a_i|\hat{\rho}(t)|a_i\rangle = |c_i(0)|^2$$

while the off-diagonal elements of the density matrix decay:

$$\langle a_1|\hat{\rho}(t)|a_2\rangle = c_1(0)c_2^*(0)e^{-\frac{\lambda t}{2}[a_1-a_2]^2}. \qquad (2.13)$$

The collapse rate is given by the expression[t]

$$\text{collapse rate} = \frac{\lambda}{2}[a_1 - a_2]^2. \qquad (2.14)$$

It depends on two things. It is proportional to λ, the collapse rate parameter. And, it depends upon the square of the eigenvalue differences of the preferred basis eigenstates competing in the Gambler's Ruin collapse game. *In general, this is how CSL works*: λ sets the overall collapse rate, and the rate is larger when the squared difference between eigenvalues (of the collapse-generating operator $\hat{A}$) is larger.

2.3 Evolution Equation for the CSL Density Matrix

Now we derive this density matrix again, in a different way that is basis-independent. The technique illustrated here will be useful to apply in more complicated situations.

Starting from Eq. (2.9), instead of putting in the state vector expression (2.4a) which contains the eigenvalues of $\hat{A}$ as was done in (2.10), we put in the state vector expression (2.2) written in terms of $\hat{A}$,

$$\hat{\rho}(t) = \int_{-\infty}^{\infty} \frac{dB(t)}{\sqrt{2\pi\lambda t}} e^{-\frac{1}{4\lambda t}[B(t)-2\lambda\hat{A}t]^2} |\psi,0\rangle\langle\psi,0|e^{-\frac{1}{4\lambda t}[B(t)-2\lambda\hat{A}t]^2}. \qquad (2.15)$$

It is worth learning how to do the integral in (2.15), since this situation arises so frequently. It is an integral of the product of two Gaussians, just like Eq.(A5), just waiting to be evaluated. Although there is an operator $\hat{A}$ in place of the number in (A5), the operator can be treated like a number in all manipulations, in that it commutes with itself, as a number does. The problem is that there is a $\hat{A}$ on the left side of $|\psi,0\rangle\langle\psi,0|$ and a $\hat{A}$ on the right side. How does one deal with that?

There is a neat trick. On the left side, rewrite $\hat{A}$ as $\hat{A}_L$ and on the right side rewrite $\hat{A}$ as $\hat{A}_R$. This means that you can put the gaussians together on either side of $|\psi,0\rangle\langle\psi,0|$, with the understanding that, when all manipulations are completed, the $\hat{A}_L$'s are to be put to the left, the $\hat{A}_R$'s are to be put to the right, and then both changed back to $\hat{A}$ (and, of course, $[\hat{A}_L, \hat{A}_R] = 0$ for all manipulations).[E] In this notation, for example,

$$[\hat{A}_L - \hat{A}_R]^2 \hat{\rho}(0) = \hat{A}^2 \hat{\rho}(0) + \hat{\rho}(0)\hat{A}^2 - 2\hat{A}\hat{\rho}(0)\hat{A} = [\hat{A}, [\hat{A}, \hat{\rho}(0)]]. \tag{2.16}$$

So, we write Eq. (2.15) as

$$\hat{\rho}(t) = \int_{-\infty}^{\infty} \frac{dB(t)}{\sqrt{2\pi\lambda t}} e^{-\frac{1}{4\lambda t}[B(t) - 2\lambda A_L t]^2} e^{-\frac{1}{4\lambda t}[B(t) - 2\lambda \hat{A}_R t]^2} \hat{\rho}(0),$$

and use Eq.(A5) to perform the integral and thus obtain the density matrix:

$$\hat{\rho}(t) = e^{-\frac{\lambda t}{2}[\hat{A}_L - \hat{A}_R]^2} \hat{\rho}(0). \tag{2.17}$$

It is easy now to obtain the differential equation that the density matrix $\hat{\rho}(t)$ satisfies:

$$\frac{d}{dt}\hat{\rho}(t) = -\frac{\lambda}{2}[\hat{A}_L - \hat{A}_R]^2 e^{-\frac{\lambda t}{2}[\hat{A}_L - \hat{A}_R]^2} \hat{\rho}(0) = -\frac{\lambda}{2}[\hat{A}_L - \hat{A}_R]^2 \hat{\rho}(t) = -\frac{\lambda}{2}[\hat{A}, [\hat{A}, \hat{\rho}(t)]]. \tag{2.18}$$

This is an important equation. Appendix F discusses a general form of evolution equation for the density matrix, Eq. (F.7), called the Lindblad equation.[4] Eq. (2.18) is the simplest example of a Lindblad equation. Equations of Lindblad form appear in many different physical contexts, not just in the context of dynamical collapse.

The solution of Eq. (2.18) is readily seen to be (2.11). Taking the matrix element of (2.18),

$$\frac{d}{dt}\langle a_i|\hat{\rho}(t)|a_j\rangle = -\frac{\lambda}{2}[a_i - a_j]^2 \rho(t) \text{ so } \langle a_i|\hat{\rho}(t)|a_j\rangle = \langle a_i|\hat{\rho}(0)|a_j\rangle e^{-\frac{\lambda t}{2}[a_i - a_j]^2} \tag{2.19}$$

which is the matrix element of (2.11).

2.4   True Collapse and False Collapse

Starting from the ensemble of state vectors undergoing collapse, the density matrix just discussed behaves as one would expect. However, the converse is not true: a density matrix that behaves in this way does not necessarily indicate that the individual state vectors underwent collapse. For, a specific density matrix can be constructed in numerous ways, from different ensembles of state vectors with different probabilities.

It is possible, *with standard quantum theory* and proper choice of time-dependent Hamiltonian, starting with the same initial state vector (2.1), to have an ensemble of unitarily evolving state vectors that has *exactly* the same density matrix as (2.11) at *any* time t, but there is *no* collapse for *any* state vector in the ensemble.

---

[4]Lindblad (1976), Gorini, Kossakowski, and Sudarshan (1976), Pearle (2012).

Choose $\hat{H} = \frac{d}{dt}B(t)\hat{A}$, where $B(t)$ is a Brownian motion function. The Schrödinger equation is

$$\frac{d}{dt}|\psi, t\rangle = -i\hat{H}|\psi, t\rangle = -i\frac{d}{dt}B(t)\hat{A}|\psi, t\rangle,$$

with solution (using the initial state vector (2.1), and taking $B(0) = 0$),

$$|\psi, t\rangle = c_1(0)e^{-iB(t)a_1}|a_1\rangle + c_2(0)e^{-iB(t)a_2}|a_2\rangle \qquad (2.20)$$

(compare with CSL's (2.4a)). To calculate the density matrix, use the definition (2.7) with the standard Brownian motion probability density and the state vector (2.20):

$$\hat{\rho}(t) = \int_{-\infty}^{\infty} dB(t)\frac{1}{\sqrt{2\pi\lambda t}}e^{-\frac{1}{2\lambda t}B^2(t)}\sum_{i=1}^{2}c_i(0)e^{-iB(t)a_i}|a_i\rangle\sum_{j=1}^{2}\langle a_j|e^{iB(t)a_j}c_j^*(0),$$

which, upon performing the Gaussian integral using (A.2), results in

$$\hat{\rho}(t) = \sum_{i,j=1}^{2} c_i(0)c_j^*(0)|a_i\rangle\langle a_j|e^{-\frac{\lambda t}{2}[a_i-a_j]^2}. \qquad (2.21)$$

This is *exactly* the same as the density matrix (2.11) that describes CSL collapse dynamics. But, in this case, there is no collapse. There are an infinite number of state vectors (2.20) in the ensemble, each in a superposition of $|a_1\rangle, |a_2\rangle$ with the same amplitudes, but with random phases: the phases contrive to cancel to give this density matrix.

It is often said that the density matrix gives all information about an ensemble of state vectors. That is not so.

It is tempting to infer, from the asymptotic density matrix in (2.21) (identical to (2.13)), that, at time ∞, there is either one pointer position or the other, because the off-diagonal elements of the density matrix have vanished. Tempting, but incorrect.

On the other hand, the inference is correct if one has the additional knowledge that there is actual collapse going on at the state vector level.

This underscores that there are many ensembles of state vectors that can give the same density matrix.[F] Although the density matrix tells you everything that can *subsequently* be measured on an ensemble, it does not allow one to know how it got there, whether by "true collapse" or "false collapse."[5]

Notes

[A] Physical Brownian motion is the behavior of small bits of matter immersed in a liquid, arising from random impacts of the liquid molecules. Mathematical Brownian motion is an idealized description of this motion in one dimension. Because of his meticulous observations, physical Brownian motion is named after the botanist Robert Brown (1773–1858), the greatest microscopist of his time (he discovered the cell nucleus). Having immersed triangular-shaped pollen of the plant *Clarkia pulchella* in water, Brown observed the pollen burst at the corners, emitting round and oblong particles

[5] Pearle (1994).

(spherosomes or fat containers and amyloplasts or starch containers, first identified in the reference below) which danced about in the water. This phenomenon had been seen earlier, for the first time by Antonie van Leeuwenhoek (1632–1723), but Brown was the first to investigate it thoroughly, which is why it is named after him. He initially thought it could be biological behavior. He eventually concluded it was physical behavior, because he discovered that small enough non-biological particles, like a ground-up piece of the Sphinx, or the soot in the London air, behaved likewise. He concluded that the cause was a mystery to him. That it is due to collisions of the particles with water molecules was only elucidated by Einstein in one of his four 1905 *annus mirabilis* papers. I found this story fascinating: here is a web site (http://physerver.hamilton.edu/Research/Brownian/index.html) with a detailed description of its history, botany and physics, with videos! Also, there is a paper giving a truncated, physics oriented treatment.[6]

[B]That will not be the case when the usual Hamiltonian evolution is added, $\hat{H} \neq 0$. A non–zero Hamiltonian, of course, is needed for the complete physical description. If $[\hat{H}, \hat{A}] \neq 0$, the dependence of the collapse evolution on the whole Brownian trajectory is required, with both the Hamiltonian and the collapse acting simultaneously every dt of the evolution. Then, (2.2) has to be modified,. Instead of containing Brownian motion, the solution of the evolution equation contains its time-derivative, white noise. Chapter 3 explains how that is to be accomplished.

[C]I named this framework the continuous spontaneous localization (CSL) theory, in part to distinguish it from, and also to honor, the spontaneous localization (SL) theory[7] discussed in Chapter 5. (SL is a dynamical collapse theory where the Schrödinger equation is not modified by having a term added to its Hamiltonian. Instead, the state vector abruptly and randomly changes in a well-defined way.)

[D]In Chapter 8, the Supplement to Chapter 1, it is pointed out that this is equivalent to a necessary feature of the Gambler's Ruin game, that the game (i.e., the coin) is "fair", so no gambler has an advantage over the other.

[E]This procedure is justified by consideration of the expressions $\int_{-\infty}^{\infty} dB e^{-B^2} F(B, \hat{A}) \hat{\rho}(0) G(B, \hat{A})$ and $\int_{-\infty}^{\infty} dB e^{-B^2} F(B, \hat{A}_L) G(B, \hat{A}_R) \hat{\rho}(0)$ where F, G are expressed as Taylor series in B with coefficients that are arbitrary functions of $\hat{A}$. One obtains the same result from both expressions, term by term.

[F]If one is only interested in the density matrix, it is sometimes useful to construct, as in this example, an ensemble of unitary evolutions possessing the same density matrix as the ensemble of collapse evolutions. This can simplify a problem that combines a complicated Hamiltonian evolution with collapse. Qijia Fu[8] first used this construction. He combined quantum electrodynamics and collapse (see Chapter 6) to calculate the CSL-generated "extra" or "spontaneous" radiation of an isolated charged particle that undergoes acceleration when "shaken" by the field $w(\mathbf{x}, t)$. Experiments looking for such radiation provide what is currently the most stringent test of CSL, compared to standard quantum theory, for which there is no such radiation.

[6]Pearle et al. (2010).

[7]Ghirardi, Rimini, and Weber (1986,1987), Bell (1987)

[8]Fu (1997).

3

CSL Theory Refinements

We have now conveyed the idea of how the simplest example of CSL works. It is necessary to make additions to the formalism so far presented, in order to have an evolution equation with features rich enough to give the non–relativistic collapse dynamics to be presented in Chapter 4.

Section 3.1 shows how to replace the Brownian function $B(t)$ in the two equations of CSL, Eqs. (2.2), (2.3), by white noise $w(t) = \frac{dB(t)}{dt}$. This is necessary in order to properly add the Hamiltonian evolution to the collapse evolution, the topic of Section 3.2. Then, Section 3.3 briefly looks at the density matrix and its evolution equation with this addition.

In Section 3.4, as promised earlier, the Schrödinger equation satisfied by the state vector is obtained, by taking the time derivative of the state vector.

Section 3.5 takes a pause on the road of generalizing Eqs. (2.2),(2.3) to solve an interesting problem made possible by the discussion thus far. The description of the free particle moving in one dimension appears early in quantum texts. So, here we treat it early too, except we add collapse dynamics, with the position operator as the collapse–generating operator, $\hat{A} = \hat{X}$.

We consider the free particle problem here, because we can! Also, to show the reader, most likely familiar with this most elementary problem in standard quantum theory, how the Hamiltonian evolution and the collapse evolution interfere with each other to give a rich, interesting behavior. Moreover, these results can be applied to an experimental test of CSL, as shown in Chapter 6, since the dynamics here is a good approximation to the dynamics of the full–fledged non–relativistic CSL theory, in certain circumstances.

After consideration of this problem, we return to the main theme of this section, additions to the formalism. Section 3.6 considers how the Schrödinger equation, probability rule and density matrix are altered when there is not just one collapse-generating operator, but a countable number of mutually commuting collapse-generating operators $\hat{A}_i$. Finally, Section 3.7 generalizes to a continuum of collapse-generating operators, replacing the index i by a position label $\mathbf{x}$, so we have a field of mutually commuting collapse–generating operators $\hat{A}(\mathbf{x})$.

3.1 Collapse Equations with White Noise

Eq. (2.2) says that the state vector at time t depends only upon $B(t)$, the value at time t of the Brownian motion sample function provided by Nature. This is independent of the past history of the Brownian motion: it doesn't matter how the Brownian motion got to that value from its initial value $B(0) = 0$. We want to redress that, to rewrite Eq. (2.2) so that the state vector evolution depends upon the whole Brownian trajectory, because that is needed when the Hamiltonian contribution to the state vector evolution is included. We have to enlarge the formalism to allow this description. We are going to replace the Brownian motion by white noise.

Eq. (2.2) was written assuming the initial state vector is given at time 0, and assuming $B(0) = 0$. We generalize it to the case of initial time t_0, and non–zero initial value $D(t_0)$:

$$|\psi, t\rangle = e^{-\frac{1}{4\lambda(t-t_0)}[B(t)-B(t_0)-2\lambda\hat{A}(t-t_0)]^2}|\psi, t_0\rangle. \tag{3.1}$$

Now, let the time interval be infinitesimal: set $t - t_0 = \Delta t$. We express the Brownian motion difference $B(t) - B(t_0)$ in terms of the effective Brownian motion velocity over this interval[1]

$$w_{\Delta t}(t) \equiv \frac{B(t) - B(t - \Delta t)}{\Delta t} \tag{3.2}$$

(see Fig. 3.1). The limit of $w_\Delta(t)$ as Δt approaches 0 is $w(t)$, white noise. (A detailed introduction to white noise is provided in Appendix D.)

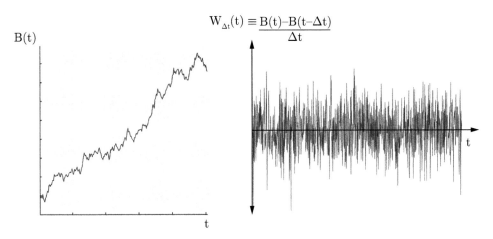

Fig. 3.1 Depiction of a Brownian motion trajectory $B(t)$ and of white noise, the "velocity" of $B(t)$ over a small time interval Δt.

It should be emphasized, as in Appendix C, that Brownian motion over any time interval is statistically independent of Brownian motion over any other non–overlapping time–interval, so $w(t), w(t + dt), w(t + 2dt), \dots$ are all statistically independent.

Substituting Eq. (3.2) into Eq. (3.1), and letting $\Delta t \to dt$ results in

$$|\psi, t\rangle = e^{-\frac{dt}{4\lambda}[w(t)-2\lambda\hat{A}]^2}|\psi, t - dt\rangle. \tag{3.3}$$

We can now express the evolution of the state vector over the finite time interval (t_0, t) as a succession of infinitesimal–time evolutions (3.3):

[1] Regarding Brownian motion as a random walk in the limit of infinitesimal steps, this "velocity" is the distance covered by "many" random walk steps, divided by the time they take.

$$|\psi, t\rangle = e^{-\frac{dt}{4\lambda}[w(t)-2\lambda\hat{A}]^2} e^{-\frac{dt}{4\lambda}[w(t-dt)-2\lambda\hat{A}]^2} ... e^{-\frac{dt}{4\lambda}[w(t_0+dt)-2\lambda\hat{A}]^2} |\psi, t_0\rangle$$

which can simply be written as

$$|\psi, t\rangle \equiv e^{-\int_{t_0}^{t} \frac{dt'}{4\lambda}[w(t')-2\lambda\hat{A}]^2} |\psi, t_0\rangle. \tag{3.4}$$

This is the result we have sought. The state vector is now expressed in terms of the values of the random function $w(t)$ at every time in the interval $t_0 \leq t$, not just in terms of the values of the random function $B(t)$ at the ends of the time interval.

The probability rule (2.3), expressed in terms of probability density, $P_t(w) = \langle\psi, t|\psi, t\rangle$, is unchanged. However, to convert it to a probability expression, instead of multiplying it by dB, we now multiply it by dw. Substituting for the state vector from (3.4), we may write it in the form

$$P_t(w)Dw \equiv Dw\left[\langle\psi, t_0|e^{-\int_{t_0}^{t} \frac{dt'}{4\lambda}[w(t')-2\lambda\hat{A}]^2}\right]\left[e^{-\int_{t_0}^{t} \frac{dt'}{4\lambda}[w(t')-2\lambda\hat{A}]^2}|\psi, t_0\rangle\right]$$

$$= Dw\langle\psi, t_0|e^{-\int_{t_0}^{t} \frac{dt'}{2\lambda}[w(t')-2\lambda\hat{A}]^2}|\psi, t_0\rangle. \tag{3.5}$$

In Eq. (3.5),

$$Dw \equiv \frac{dw(t)}{\sqrt{2\pi\lambda/dt}} \frac{dw(t-dt)}{\sqrt{2\pi\lambda/dt}} ... \frac{dw(t_0+dt)}{\sqrt{2\pi\lambda/dt}}.$$

The numerical factors ensure, when (3.5) is integrated over each dw, that each Gaussian contributes the factor 1 so the total probability is 1:

$$\int P_t(w)Dw \equiv \int Dw\langle\psi, t_0|e^{-\int_{t_0}^{t} \frac{dt'}{2\lambda}[w(t')-2\lambda\hat{A}]^2}|\psi, t_0\rangle$$

$$= \langle\psi, t_0|\prod_{t\geq t'\geq t_0} \frac{1}{\sqrt{2\pi\lambda/dt}} \int_{-\infty}^{\infty} dw(t')e^{-\frac{dt'}{2\lambda}[w(t')-2\lambda\hat{A}]^2}|\psi, t_0\rangle$$

$$= \langle\psi, t_0|1|\psi, t_0\rangle = 1.$$

3.2 Including the Hamiltonian: Time–Ordering

In standard quantum theory, when the Hamiltonian is time dependent, and possibly $\hat{H}$ does not commute with itself at different times, $[\hat{H}(t), \hat{H}(t')] \neq 0$ for $t' \neq t$, the solution of Schrödinger's equation,

$$\frac{d}{dt}|\psi, t\rangle = -i\hat{H}(t)|\psi, t\rangle, \tag{3.6}$$

can be written as follows. For an infinitesimal time interval dt, one can consider that $\hat{H}(t)$ *is* constant over the interval, and so

$$|\psi, t\rangle = e^{-idt\hat{H}(t)}|\psi, t - dt\rangle. \tag{3.7}$$

Continuing in this vein the solution of (3.6) is

$$|\psi, t\rangle = e^{-idt\hat{H}(t)}e^{-idt\hat{H}(t-dt)}...e^{-idt\hat{H}(2dt)}e^{-idt\hat{H}(dt)}|\psi, 0\rangle,$$

which we write as an integral by introducing a new notation,

$$|\psi, t\rangle \equiv Te^{-i\int_0^t dt'\hat{H}(t')}|\psi, 0\rangle. \tag{3.8}$$

T is called the time–ordering operator: (3.8) means nothing more or less than the expression preceding it.[A]

Often it is complicated to evaluate the product of the time-ordered exponential operators. But, one simple thing about it is that the time derivative of (3.8) is just (3.6).

Combining the Hamiltonian evolution (3.7) and the collapse evolution (3.3), the state vector evolution over an infinitesimal time interval can be written in various ways:

$$|\psi, t\rangle = e^{-idt\hat{H}(t)}e^{-\frac{dt}{4\lambda}[w(t)-2\lambda\hat{A}]^2}|\psi, t-dt\rangle \text{ or}$$

$$|\psi, t\rangle = e^{-\frac{dt}{4\lambda}[w(t)-2\lambda\hat{A}]^2}e^{-idt\hat{H}(t)}|\psi, t-dt\rangle \text{ or}$$

$$|\psi, t\rangle = e^{-idt\hat{H}(t)-\frac{dt}{4\lambda}[w(t)-2\lambda\hat{A}]^2}|\psi, t-dt\rangle,$$

since, even if $\hat{H}$ and $\hat{A}$ do not commute, the difference in the expressions is of order $(dt)^2$ and therefore negligible.

The state vector evolution over a finite time interval is therefore

$$|\psi, t\rangle = e^{-idt\hat{H}(t)-\frac{dt}{4\lambda}[w(t)-2\lambda\hat{A}]^2}e^{-idt\hat{H}(t-dt)-\frac{dt}{4\lambda}[w(t-dt)-2\lambda\hat{A}]^2}...$$

$$e^{-idt\hat{H}(0)-\frac{dt}{4\lambda}[w(dt)-2\lambda\hat{A}]^2}|\psi, 0\rangle$$

which can be written more compactly as

$$|\psi, t\rangle = Te^{-\int_0^t dt'\left[i\hat{H}(t')+\frac{1}{4\lambda}[w(t')-2\lambda\hat{A}]^2\right]}|\psi, 0\rangle. \tag{3.9}$$

It also may be written as

$$|\psi, t\rangle = Te^{-i\int_0^t dt'\hat{H}(t')}e^{\int_0^t dt''\frac{1}{4\lambda}[w(t')-2\lambda\hat{A}]^2}|\psi, 0\rangle$$

or

$$|\psi, t\rangle = Te^{\int_0^t dt''\frac{1}{4\lambda}[w(t')-2\lambda\hat{A}]^2}e^{-i\int_0^t dt'\hat{H}(t')}|\psi, 0\rangle$$

regardless of the commutators $[\hat{H}(t), \hat{A}]$, $[\hat{H}(t), \hat{H}(t')]$, because the time–ordering operation "sorts it out," putting the operators in the right order.

To summarize, in Eq. (3.9), we have the expression for the state vector when the Hamiltonian evolution and the collapse evolution both occur.

What happens to the probability rule? It remains the usual one, $P_t(w)Dw = \langle\psi, t|\psi, t\rangle Dw$. By integrating over each $w(t')$ consecutively, $t \geq t' \geq dt$, starting with $w(t)$, then $w(t-dt)$, etc., we may verify that the total probability is 1.[B]

While it is useful to allow the possibility that the Hamiltonian can depend upon time, for example, to describe an external time-dependent force, from now on we will assume $\hat{H}$ does not depend upon time, since that obtains for all our applications.

3.3 The Density Matrix and Its Evolution Equation

The density matrix is

$$\hat{\rho}(t) = \int Dw \langle \psi, t | \psi, t \rangle \frac{|\psi, t\rangle}{\sqrt{\langle \psi, t | \psi, t \rangle}} \frac{\langle \psi, t|}{\sqrt{\langle \psi, t | \psi, t \rangle}} = \int Dw |\psi, t\rangle \langle \psi, t|, \qquad (3.10)$$

which is beautifully simple. We want to evaluate this when the state vector is (3.9). According to Eq. (3.9), $|\psi, t\rangle$ is expressed as the product of exponentials $E(t)$ in time–ordered form, $|\psi, t\rangle = E(t)E(t - dt)....E(0)|\psi, 0\rangle$. Therefore, $\langle \psi, t|$ is expressed as the product of exponentials in time–reversed–ordered form, $\langle \psi, t| = \langle \psi, 0|E^{\dagger}(0)...E^{\dagger}(t - dt)E^{\dagger}(t)$. We will use the notation T_R for time–reversed ordering.

To perform the integral over each $w(t)$, we use the "trick" in Chapter 2, Section 2.3, of labeling the operator $\hat{A}$ to the left of $\hat{\rho}(0)$ as $\hat{A}_L$, and to the right as $\hat{A}_R$. With this notation, the operators can be written down in any order at all, because the left and right ordering and the time ordering will sort them out at the end. Inserting the state vector (3.9) into (3.10), we obtain the expression

$$\hat{\rho}(t) = \int Dw T e^{-i \int_0^t dt' \hat{H}} e^{-\int_0^t dt' \frac{1}{4\lambda}[w(t') - 2\lambda \hat{A}_L]^2} |\psi, 0\rangle \langle \psi, 0|$$
$$\cdot T_R e^{i \int_0^t dt' \hat{H}} e^{-\int_0^t dt' \frac{1}{4\lambda}[w(t') - 2\lambda \hat{A}_R]^2},$$

and then, following integration over the product of the two Gaussians for each $w(t')$, using

$$\int_{-\infty}^{\infty} \frac{dw(t')}{\sqrt{2\pi\lambda/dt}} e^{-dt' \frac{1}{4\lambda}[w(t') - 2\lambda \hat{A}_L]^2} e^{-dt' \frac{1}{4\lambda}[w(t') - 2\lambda \hat{A}_R]^2} = e^{-dt' \frac{\lambda}{2}[\hat{A}_L - \hat{A}_R]^2},$$

we have

$$\hat{\rho}(t) = TT_R e^{-i \int_0^t dt' [\hat{H}_L - \hat{H}_R]} e^{-\int_0^t dt' \frac{\lambda}{2}[\hat{A}_L - \hat{A}_R]^2} |\psi, 0\rangle \langle \psi, 0|, \qquad (3.11)$$

where T time–orders the operators with the subscript L, and T_R time–reverse–orders the operators with the subscript R.

Because of the time ordering, it is easy to take the time derivative, and so obtain the density matrix evolution equation

$$\frac{d}{dt} \hat{\rho}(t) = -i[\hat{H}_L - \hat{H}_R]\rho(t) - \frac{\lambda}{2}[\hat{A}_L - \hat{A}_R]^2 \hat{\rho}(t) \qquad (3.12)$$

which is the same as

$$\frac{d}{dt} \hat{\rho}(t) = -i[\hat{H}, \hat{\rho}(t)] - \frac{\lambda}{2}[\hat{A}, [\hat{A}, \hat{\rho}(t)]], \qquad (3.13)$$

generalizing (2.18) to the case where $\hat{H} \neq 0$,

3.4 The CSL Modified Schrödinger Equation

In this book we only need the solution (3.9) to the modified Schrödinger equation, not the modified Schrödinger equation itself. Of course we can take the time derivative of (3.9) (easily, because of its time–ordered form), and arrive at the modified Schrödinger equation it satisfies:

$$\frac{d}{dt}|\psi,t\rangle = \left[-i\hat{H} - \frac{1}{4\lambda}w^2(t) + w(t)\hat{A} - \lambda\hat{A}^2\right]|\psi,t\rangle. \tag{3.14}$$

However, there is a Slight Technical Problem. This is a Stochastic Differential Equation and Mathematicians Do Not Allow a term $\sim w^2(t)$ into an SDE. (Rather stuffy of them, but they will be rigorous, so there is nothing to be done about it.) So, it must be removed from Eq. (3.14) if it is to be a true SDE enabling one to apply SDE techniques to it. This is easily accomplished. Change the state vector to

$$|\psi',t\rangle \equiv e^{\int_{t_0}^t dt' \frac{1}{4\lambda}w^2(t')}|\psi,t\rangle,$$

obtaining as a true SDE the modified Schrödinger equation

$$\frac{d}{dt}|\psi',t\rangle = [-i\hat{H} + w(t)\hat{A} - \lambda\hat{A}^2]|\psi',t\rangle. \tag{3.15}$$

Removal of this time-dependent factor has changed the magnitude of the state vector, so the probability rule is changed to

$$P_t(w)Dw = Dw\langle\psi',t|\psi',t\rangle e^{-\int_{t_0}^t dt' \frac{1}{2\lambda}w^2(t')}.$$

Since the physical content of the state vector lies in its direction in Hilbert space and not in its norm, changing the norm from $|\psi,t\rangle$ to $|\psi',t\rangle$ has no physical consequences. Thus, the expectation value of an operator $\hat{O}$ is

$$_N\langle\psi,t|\hat{O}|\psi,t\rangle_N = \frac{\langle\psi,t|\hat{O}|\psi,t\rangle}{\langle\psi,t|\psi,t\rangle} = \frac{\langle\psi',t|\hat{O}|\psi',t\rangle}{\langle\psi',t|\psi',t\rangle}.$$

A second quote from Dirac's *Scientific American* article[2] (the first quote appears at the end of Section 1.2) is "It seems to be one of the fundamental features of nature that fundamental physical laws are described in terms of a mathematical theory of great beauty and power, needing quite a high standard of mathematics for one to understand it."

The CSL modified Schrödinger equation (3.15) is not a fundamental physical law. Rather it is phenomenological but, hopefully, an approximation to a fundamental physical law. However, it *is* readily described in terms of a mathematical theory (SDE's) which (in my opinion) possesses quite a bit of beauty and power. But, in keeping with its more modest status, it is not *so* mathematically demanding as all *that*!

---

[2]Dirac (1963).

3.5 Interregnum: Free Particle Example

At this point we pause in the development of the formalism to look at an example of how that formalism works. We consider that simplest of problems in standard quantum theory textbooks, the free particle moving in one dimension.[3]

We add to the free particle Hamiltonian the CSL Hamiltonian, where the collapse-generating operator is the particle position, $\hat{A} = \hat{X}$. The problem, while completely solvable, involves a lot of mathematical manipulations. For this reason, here we provide a sketch of the approach and then discuss the results. We refer a reader interested in the complete treatment to Chapter 10, the Supplement to this chapter.

We will later see, in the fully fleshed out non-relativistic CSL theory of Chapter 4, that the *center of mass* of a free visible object behaves the same way as the *particle* considered in this example, to an excellent approximation. Such behavior differs from behavior predicted by standard quantum theory, suggesting some experimental tests of CSL, as explained in Chapter 6.

3.5.1 Free Particle Wave Function

We suppose the free particle is initially in a Gaussian state, $\langle x|\psi',0\rangle = \frac{1}{[2\pi\sigma^2]^{1/4}}e^{-\frac{x^2}{4\sigma^2}}$. When there is no collapse, as shown in any textbook, the width of the wave packet obeying the Schrödinger equation spreads beyond the initial width σ as time increases, indeed, spreads to infinite width at $t \to \infty$.

What happens when CSL collapse dynamics is added? We ask for the solution of (3.15), expressed in the position basis,

$$\frac{\partial}{\partial t}\langle x|\psi',t\rangle = \left[\frac{i}{2m}\frac{\partial^2}{\partial x^2} + w(t)x - \lambda x^2\right]\langle x|\psi',t\rangle. \tag{3.16}$$

We guess (correctly) that the solution has the Gaussian form

$$\langle x|\psi',t\rangle = e^{-A(t)x^2+B(t)x+C(t)}. \tag{3.17}$$

We plug (3.17) into (3.16), obtaining three equations, the coefficients of x^2, of x and of 1:

$$\frac{d}{dt}A(t) - -\frac{2i}{m}A^2(t) + \lambda, \tag{3.18a}$$

$$\frac{d}{dt}B(t) = -\frac{2i}{m}A(t)B(t) + w(t), \tag{3.18b}$$

$$\frac{d}{dt}C(t) = \frac{i}{2m}[-2A(t) + B^2(t)], \tag{3.18c}$$

which may be solved.

---

[3]The other classic problem that is discussed early in every introductory textbook is the harmonic oscillator. When collapse on position is added, this gets quite complicated. But, we cannot leave out this interesting example. Therefore, the density matrix description is given in the Supplement to this chapter, showing how the initial ground state evolves into an ensemble of excited states. The discussion of the individual state vector evolutions appears in Chapter 18, since this is most easily solved using the stochastic differential equation formulation of CSL.

Eq. (3.16) implies that λ has dimensions $1/[L^2 T]$, so we may define a characteristic time $\tau \equiv \sqrt{\frac{m}{\lambda \hbar}}$ and characteristic squared length $s^2 \equiv \sqrt{\frac{\hbar}{m\lambda}}$ and express the results in terms of s, τ.

The first interesting result, which follows from solving Eq. (3.18a), is that, as time increases, the wave packet width grows from a small σ, or shrinks from a large σ, asymptotically approaching an *equilibrium value* $\sim s$ with time constant τ. Why the equilibrium size? The wave packet width tends to expand due to the usual Schrödinger evolution under the free particle Hamiltonian. The wave packet width tends to shrink due to the collapse dynamics, which attempts to localize the wave function to an eigenstate of the position operator (since $\hat{X}$ is the collapse-generating operator). At the size $\sim s$, one effect precisely counters the other.

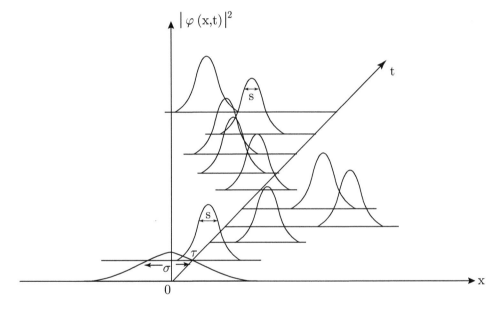

Fig. 3.2 One dimensional free particle undergoing collapse with position as the collapse-generating operator. Here the initial wave packet of width σ shrinks toward size s in characteristic time τ, and the peak undergoes a random walk.

The next interesting result, which follows from solving Eq. (3.18b), is movement of the wave packet center. We will see that the mean position and momentum moves randomly.

To simplify this investigation, instead of assuming the initial wave packet width is σ, we now assume that the initial wave packet width is the constant equilibrium width. We set $A(0) = \frac{1-i}{2s^2}$, which is the solution of (3.18a) when $\frac{d}{dt}A(t) = 0$. The wave packet's mean position and momentum, in terms of $A, B_R(t), B_I(t)$ ($B_R(t), B_I(t)$ are respectively the real and imaginary parts of $B(t)$), are given below.

It turns out that this solution for $B(t)$ is simplest when written, not in terms of $w(t)$, but rather in terms of another white noise function $v(t)$ whose associated

Brownian motion we call $b(t)$, that is, $v(t) = db(t)/dt$. The two white noise functions are related by

$$w(t) = v(t) + \frac{2}{\tau}b(t) + \frac{2}{\tau^2}\int_0^t dt' b(t').$$
(3.19)

The reason that $v(t)$ is simpler to use is that it turns out to be standard white noise, that is, the probability rule for it can be shown to be the standard Gaussian white noise probability distribution:

$$DwP(w) = Dw\langle\psi',t|\psi',t\rangle e^{-\int_{t_0}^t dt' \frac{1}{2\lambda}w^2(t')} = Dwe^{-\int_{t_0}^t dt' \frac{1}{2\lambda}w^2(t')}\int_{-\infty}^{\infty} dx\langle x|\psi',t\rangle|^2$$

$$= DvP(v) = Dve^{-\frac{1}{2\lambda}\int_0^t dt' v^2(t')},$$
(3.20)

and so the corresponding $b(t)$ is standard Brownian motion with variance $\overline{b^2(t)} = \lambda t$. The calculation in the Supplement then gives

$$B_R(t) = b(t) + \frac{1}{\tau}\int_0^t dt' b(t'), \ B_I(t) = -\frac{1}{\tau}\int_0^t dt' b(t').$$

Thus, we completely know the wave function.

We note that mean values of powers of position x and momentum p can be readily calculated using the wave function (3.17), for example

$$\overline{x} = \frac{\int dx x |\langle x|\psi',t\rangle|^2}{\int dx |\langle x|\psi',t\rangle|^2} = \frac{B_R}{2A_R}, \quad \overline{p} = \frac{\int dx \langle x|\psi',t\rangle^* \frac{1}{i}\frac{d}{dx}\langle x|\psi',t\rangle}{\int dx |\langle x|\psi',t\rangle|^2} = B_I - \frac{A_I B_R}{A_R},$$

In addition to these, other expectation values can be readily calculated from the wave function, using Gaussian integrals:

$$\langle \hat{X}\rangle = \overline{x} = s^2\left[b(t) + \frac{1}{\tau}\int_0^t dt' b(t'),\right]$$
(3.21a)

$$\langle \hat{X}^2\rangle - \langle \hat{X}\rangle^2 = s^2/2,$$
(3.21b)

$$\langle \hat{P}\rangle = \overline{p} = \hbar b(t),$$
(3.21c)

$$\langle \hat{P}^2\rangle - \langle \hat{P}\rangle^2 = \frac{\hbar^2}{s^2}.$$
(3.21d)

So, now we can extract physical information about the wave packet's motion (illustrated in Fig. 3.2).

Eq. (3.21b) confirms that the width of the wave function remains constant, at its equilibrium value.

Eq. (3.21d) says that, if we take the Fourier transform of the Gaussian wave function in position space to convert it to the Gaussian wave function in momentum space, the momentum–space width will be seen to be constant too.

Eq. (3.21c) says that the mean momentum undergoes Brownian motion (random walk).

Eq. (3.21a) says the mean position expression is the sum of Brownian motion plus an extra term. There is a classical analogy to this motion. The physical Brownian motion of a bit of matter immersed in a liquid is explained by its feeling two forces, the impulsive impacts of colliding liquid particles and the liquid viscous force. Under the joint influence of these forces, its momentum comes to thermal equilibrium with the liquid (the momenta of many such bits of matter obeys the Maxwell–Boltzmann probability distribution). But, if the viscous force was not there, the momentum of the bit of matter would instead undergo random walk as in (3.21c), leading to its position being described as in (3.21a).

Next, we can calculate the ensemble averages of the quantities in Eqs. (3.21a), (3.21b), (3.21c), (3.21d) since $b(t)$ is standard Brownian motion with variance λt.[4]

The ensemble mean energy increases linearly with time, as follows from Eqs.(3.21d), (3.21c):

$$\frac{1}{2m}\overline{\langle \hat{P}^2 \rangle} = \frac{\hbar^2}{2m}\left[\overline{b^2(t)} + \frac{1}{s^2}\right] = \frac{\hbar^2}{2m}\left[\lambda t + \frac{1}{s^2}\right], \tag{3.22}$$

using Appendix C's Eq. (C.3b). It is shown in Chapter 4, Section 4.5, that this *linear-in-time energy increase is ubiquitous, a property of any system of particles, under any potential whatsoever, undergoing non–relativistic collapse.*

Using $\bar{b}(t) = 0$, we find from (3.21a), (3.21c) that the ensemble mean position and mean momentum vanish. This is to be expected because, for any $w(t)$ driving a particle trajectory to the left, there is an equivalent $w(t)$ driving a particle trajectory to the right.

Lastly, we can find the mean squared position. It is rather fun to see that we can calculate this rather complicated expression using just the Brownian motion property

$$\overline{b(t)b(t')} = \lambda t' \text{ for } t \geq t' \tag{3.23}$$

(which was shown in Eq. (C.5), setting there $B(t) = b(t)$, $t_0 = 0$ and $B(0) = 0$). Starting with (3.21b),

$$\overline{\langle \hat{X}^2 \rangle} = \overline{\langle \hat{X} \rangle^2} + s^2/2,$$

we insert the expression (3.21a)

$$\overline{\langle \hat{X}^2 \rangle} = s^4\overline{\left[b(t) + \frac{1}{\tau}\int_0^t dt'b(t')\right]^2} + s^2/2$$

which, after multiplying out, is

$$\overline{\langle \hat{X}^2 \rangle} = s^4\left[\overline{b^2(t)} + \frac{1}{\tau}\int_0^t dt'\overline{2b(t)b(t')} + \frac{1}{\tau^2}\int_0^t dt'\int_0^t dt''\overline{b(t')b(t'')}\right] + s^2/2.$$

Now, using (3.23), we get

---

[4]The density matrix is the usual device for calculating ensemble averages. The density matrix approach gives the same results as obtained here, as shown in Chapter 10, this chapter's Supplement.

$$\overline{\langle \hat{X}^2 \rangle} = s^4 \left[\lambda t + \frac{2}{\tau} \int_0^t dt' \lambda t' + \frac{2}{\tau^2} \int_0^t dt' \int_0^{t'} dt'' \lambda t'' \right] + s^2/2,$$

which, upon performing the integrals, results in

$$\overline{\langle \hat{X}^2 \rangle} = s^2 \frac{t}{\tau} \left[1 + \frac{t}{\tau} + \frac{t^2}{3\tau^2} \right] + s^2/2. \tag{3.24}$$

This asymptotic $\sim t^3$ behavior of the position is characteristic of physical Brownian motion without viscous damping, a consequence of the momentum then undergoing random walk, as discussed earlier.

To summarize the free particle behavior, its wave function achieves an equilibrium size, its mean momentum undergoes random walk, its energy grows linearly with time, and its center undergoes random walk like a particle in a non-viscous fluid.

3.6 Adding More Collapse–Generating Operators

Now we return to the main task of this section, broadening the CSL formalism. In constructing non–relativistic CSL, a theory that describes behavior of particles in space, we need more than one collapse–generating operator, one for each spatial location $\mathbf{x}$.

We begin by generalizing to many collapse–generating operators Suppose we have a collection of mutually commuting operators $\hat{A}_i$. Such a collection of operators has joint eigenstates, $\hat{A}_i|a_1...a_i...\rangle = a_i|a_1...a_i...\rangle$. To use these as collapse-generating operators, we need to introduce more white noise functions, $w_i(t)$, one for each operator. Then, the generalization of (3.9) is simply

$$|\psi, t\rangle = T e^{-i \int_0^t dt' \hat{H}} e^{-\int_0^t dt' \frac{1}{4\lambda} \sum_i [w_i(t') - 2\lambda \hat{A}_i]^2} |\psi, 0\rangle. \tag{3.25}$$

When $\hat{H} = 0$, the time–ordering is unnecessary since all operators commute, so we can write the state vector as

$$|\psi, t\rangle = \prod_i e^{-\int_0^t dt' \frac{1}{4\lambda} [w_i(t') - 2\lambda \hat{A}_i]^2} |\psi, 0\rangle. \tag{3.26}$$

It is readily seen that the ith term in the product of exponentials acts independently of the other terms. Each individual term behaves as described in Chapter 2, and illustrated in Fig. 2.2. The ith term causes collapse to one of $\hat{A}_i$'s eigenvalues so, overall, the state vector collapses to one of the *joint* eigenvectors $|a_1...a_i...\rangle$.

The probability rule is unchanged,

$$P_t(w)Dw = Dw\langle \psi, t|\psi, t\rangle, \text{ where however } Dw = \prod_i Dw_i. \tag{3.27}$$

As for the density matrix, the multi-variable extension of the proof in (3.11) leads to the Lindblad equation which generalizes (3.12),

$$\frac{d}{dt}\hat{\rho}(t) = -i[\hat{H}, \hat{\rho}(t)] - \frac{\lambda}{2} \sum_i [\hat{A}_i, [\hat{A}_i, \hat{\rho}(t)]]. \tag{3.28}$$

3.7 Replacing $\hat{A}_i$ by $\hat{A}(\mathbf{x})$

Eqs. (3.25), (3.27), (3.28) define the CSL formalism when there are many (mutually commuting) collapse–generating operators $\hat{A}_i$. Now, we replace the discrete index i with a continuum index $\mathbf{x}$, a spatial position.

A classical field, like the components of the electric or magnetic or gravitational vector fields, is a variable that depends upon position. When one has an operator $\hat{A}(\mathbf{x})$ (it is customary to write the index as an argument of the operator, rather than as a subscript), so there is a different operator at every point of space, that operator is called a quantum field operator, or simply a quantum field.

How now do we write our state vector and density matrix evolutions?

Because Eqs. (3.25), (3.28) are written in terms of a sum of squares of the operator $\hat{A}_i$, and the sum has to be replaced by an integral, the replacement made in these equations is $\hat{A}_i \rightarrow \sqrt{d\mathbf{x}}\hat{A}(\mathbf{x})$. And, these operators, like their discrete counterparts, must mutually commute, $[\hat{A}(\mathbf{x}), \hat{A}(\mathbf{x})'] = 0$.

Moreover, since in Eq.(3.25) there is also the collection of independent white noise functions indexed by i, that index needs a continuum replacement too, so we must make the replacement $w_i(t) \rightarrow \sqrt{d\mathbf{x}}w(\mathbf{x}, t)$. Essentially, each point of space has an independent white noise function of time. Such a white noise classical field is discussed in Appendix D. So, Eqs.(3.25), (3.27), (3.28) are to be replaced as follows.

The evolution of the state vector is given by

$$|\psi, t\rangle = Te^{-i\int_0^t dt' \hat{H}} e^{-\int_0^t dt' d\mathbf{x}' \frac{1}{4\lambda}[w(\mathbf{x}',t')-2\lambda\hat{A}(\mathbf{x}')]^2} |\psi, 0\rangle. \tag{3.29}$$

The probability rule is unchanged,

$$P_t(w)Dw = Dw\langle\psi, t|\psi, t\rangle, \text{ where } Dw = \prod_{t'=0,\mathbf{x}}^t \frac{dw(\mathbf{x}, t)}{\sqrt{2\pi\lambda/dt'd\mathbf{x}'}}. \tag{3.30}$$

The density matrix satisfies the Lindblad equation

$$\frac{d}{dt}\hat{\rho}(t) = -i[\hat{H}, \hat{\rho}(t)] - \frac{\lambda}{2}\int d\mathbf{x}[\hat{A}(\mathbf{x}), [\hat{A}(\mathbf{x}), \hat{\rho}(t)]]. \tag{3.31}$$

So, in our proposal for a non–relativistic collapse theory, we posit that Nature supplies a random classical *field* $w(\mathbf{x}, t)$ to do the job. Reality is represented by two entities: the state vector and this field. Chapter 0 mentions speculations as to the physical nature of this field; maybe there could be a connection to something else already in the physics lexicon.[C] However, it could be that this field has no such connection, manifesting itself only in providing the essential probabilistic creation of events.

Notes

[A] Incidentally, other ways to write (3.8) are

$$|\psi, t\rangle = [1 - idt\hat{H}(t)][1 - idt\hat{H}(t - dt)]...[1 - idt\hat{H}(dt)]|\psi, 0\rangle$$

$$= \left[1 - i\int_0^t dt' \hat{H}(t') + (-i)^2 \int_0^t dt' \hat{H}(t') \int_0^{t'} dt'' \hat{H}(t'') + ...\right]|\psi, 0\rangle$$

$$= \left[1 - i\int_0^t dt' \hat{H}(t') + \frac{(-i)^2}{2!} \int_0^t dt' \int_0^t dt'' T\{\hat{H}(t')\hat{H}(t'')\} + ...\right]|\psi, 0\rangle$$

where $T\{\hat{A}(t)\hat{B}(t')\} \equiv \hat{A}(t)\hat{B}(t')$ if $t > t'$ and $\equiv \hat{B}(t)\hat{A}(t')$ if $t' > t$. In the first line, there is no need to include the second or higher order terms of the exponential in (3.8), since $[dt\hat{H}(t)]^2, \ldots$ make no contribution in the limit $dt \to 0$.

[B]To show that the total probability is 1, we have to perform the integrals

$$\int P_t(w)Dw = \int \frac{dw(t)}{\sqrt{2\pi\lambda/dt}} \frac{dw(t-dt)}{\sqrt{2\pi\lambda/dt}} \cdots \frac{dw(dt)}{\sqrt{2\pi\lambda/dt}} \langle \psi, t|\psi, t \rangle$$

We start by doing the first integral, integrating over $w(t)$, writing the state vector explicitly for an infinitesimal evolution from $t - dt$ to t:

$$\int P_t(w)Dw = \int \frac{dw(t-dt)}{\sqrt{2\pi\lambda/dt}} \cdots \frac{dw(dt)}{\sqrt{2\pi\lambda/dt}}$$

$$\cdot \int \frac{dw(t)}{\sqrt{2\pi\lambda/dt}} \langle \psi, t - dt | e^{idt'\hat{H}(t')} e^{-\frac{dt}{4\lambda}[w(t)-2\lambda\hat{A}]^2} e^{-\frac{dt}{4\lambda}[w(t)-2\lambda\hat{A}]^2} e^{-idt'\hat{H}(t')} |\psi, t - dt \rangle$$

The integral over $w(t)$ gives 1, and then $e^{idt'\hat{H}(t')}e^{-idt'\hat{H}(t')} = 1$, with the result

$$\int P_t(w) = Dw \int \frac{dw(t-dt)}{\sqrt{2\pi\lambda/dt}} \cdots \frac{dw(dt)}{\sqrt{2\pi\lambda/dt}} \langle \psi, t - dt|\psi, t - dt \rangle$$

Continuing in this manner, writing the state vector explicitly for an infinitesimal evolution from $t-2dt$ to $t - dt$, integrating over $w(t - dt)$, etc., we finally end up with

$$\int P_t(w) = \langle \psi, 0|\psi, 0 \rangle = 1.$$

[C] A white noise field has rather singular behavior. While it simplifies calculations, one may think it is physically implausible, likely an approximation to a more realistic random field whose frequency spectrum drops at high frequencies and where spatial and temporal correlations of nearby points are not zero.

4
Non–Relativistic CSL

At last we have arrived at the CSL structure (3.29), (3.30), (3.31), enabling a proposal for a non–relativistic specialization[1] that gives a reasonable collapse mechanism in physical situations. The choice simply comes down to what to choose for the collapse-generating field operator $\hat{A}(\mathbf{x})$.

4.1   Choosing $\hat{A}(\mathbf{x})$

The answer is fairly clear in its broad outlines. The point of a collapse theory is to have the state vector describe the world around us. A governing slogan might be, "What you see (in Nature) is what you get (from the theory)." What we see around us are objects localized in space. Therefore, the collapse must give us objects localized in space.

When a quantum experiment is performed, the usual Hamiltonian evolution of the microscopic system and the apparatus that observes the system produces a superposition: a state vector that is the sum of state vectors, each corresponding to a possible outcome of the experiment. The physical states of the apparatus described by these state vectors have to differ from each other in some way that is discernible by the experimenter. This means that some part of the apparatus has particles in different places for each outcome. The apparatus particle number spatial density distribution differs from outcome to outcome.

This suggests we consider choosing $\hat{A}(\mathbf{x}) \equiv \hat{D}(\mathbf{x})$, the operator that represents the particle number density at location $\mathbf{x}$. So, let us familiarize ourselves with this operator.

First consider the case of just one particle. We are familiar with the position operator $\hat{\mathbf{X}}$ whose eigenvalues are the possible locations of the particle. The particle number density operator is a function of this position operator:

$$\hat{D}(\mathbf{x}) \equiv \delta(\mathbf{x} - \hat{\mathbf{X}}).$$

If a particle is at location $\mathbf{x}_1$, that is, if the state vector describing it is the position eigenstate $|\mathbf{x}_1\rangle$, then it is also an eigenstate of the density operator:

$$\hat{D}(\mathbf{x})|\mathbf{x}_1\rangle = \delta(\mathbf{x} - \hat{\mathbf{X}})|\mathbf{x}_1\rangle = \delta(\mathbf{x} - \mathbf{x}_1)|\mathbf{x}_1\rangle.$$

This eigenvalue is what you would expect classically for the particle number density of a particle located at $\mathbf{x}_1$. The density is 0 everywhere except at $\mathbf{x}_1$ where it is infinite,

---

[1] Pearle (1989), Ghirardi, Pearle, and Rimini (1990a).

in such a way that the integral of the density over a volume V containing the particle is the number of particles, 1: $\int_V d\mathbf{x}\delta(\mathbf{x} - \mathbf{x}_1) = 1$.

For N particles, with position operators $\hat{\mathbf{X}}_1, \hat{\mathbf{X}}_2 \, ... \, \hat{\mathbf{X}}_N$, the particle number density operator is

$$\hat{D}(\mathbf{x}) \equiv \delta(\mathbf{x} - \hat{\mathbf{X}}_1) + \delta(\mathbf{x} - \hat{\mathbf{X}}_2) + ... \; \delta(\mathbf{x} - \hat{\mathbf{X}}_N)$$

The particle number density operators at two different points commute, $[\hat{D}(\mathbf{x}), \hat{D}(\mathbf{x}')] = 0$, since they are both functions of the same set of completely commuting operators.

The eigenbasis of the particle number density operator are the position basis states describing the particles as located at $\mathbf{x}_1, \mathbf{x}_2, \, ... \, \mathbf{x}_N$:

$$\hat{D}(\mathbf{x})|\mathbf{x}_1, \mathbf{x}_2, ...\mathbf{x}_N\rangle = [\delta(\mathbf{x} - \mathbf{x}_1) + \delta(\mathbf{x} - \mathbf{x}_2) + ... \; \delta(\mathbf{x} - \mathbf{x}_N)]|\mathbf{x}_1, \mathbf{x}_2, ...\mathbf{x}_N\rangle.$$

The eigenvalues of $\hat{D}(\mathbf{x})$ are what the classical particle number density is in this case. Since collapse is toward the eigenstates of $\hat{A}(\mathbf{x}) = \hat{D}(\mathbf{x})$, collapse is toward position eigenstates.

The particle number operator in a volume V, and the total particle number operator are respectively defined as

$$\hat{N}_V \equiv \int_V d\mathbf{x}\hat{D}(\mathbf{x}), \qquad \hat{N} \equiv \int d\mathbf{x}\hat{D}(\mathbf{x}),$$

where the second integral is over all space. When $\mathbf{x}$ in the eigenvalue equation above for $D(\mathbf{x})$ is integrated over V and over all of space respectively, we find that $|\mathbf{x}_1, \mathbf{x}_2, ...\mathbf{x}_N\rangle$ is an eigenstate of $\hat{N}_V$ whose eigenvalue is the number of particles in V and, similarly, $\hat{N}$'s eigenvalue is N.

There is more than just one kind of particle in Nature. When we consider different kinds of particles, we will distinguish the different number density operators by a subscript, for example, $\hat{D}_\alpha(\mathbf{x})$ with $\alpha = n$ corresponds to neutrons, $\alpha = p$ corresponds to protons, $\alpha = e$ corresponds to electrons, etc. We could make the correspondence that $\hat{A}(\mathbf{x})$ is the sum of all the particle number densities at $\mathbf{x}$, $\hat{A}(\mathbf{x}) \equiv \sum_\alpha \hat{D}_\alpha(\mathbf{x})$, or the sum could be over just certain types of particles.

And, there is no reason why the collapse rate λ for different particles should be the same: why not have a $\lambda_n, \lambda_p, \lambda_e, ...$? There are a lot of possible choices that could be made. But the same collapse rate for all massive particles is simplest. This was the proposal for the SL theory of dynamical collapse.[2] In my original proposal for CSL,[3] I too adopted that simplifying assumption.

But, there is actually experimental evidence against this choice. Briefly, the CSL collapse dynamics, in narrowing electron wave functions, excites atomic electrons, knocking them to higher (unoccupied) atomic states, and even out of the atom altogether. (This is called *anomalous* excitation, to make it clear that it is not a consequence of standard quantum theory.) Euan Squires and I soon realized[4] if the electron collapse rate is the same as the nucleon collapse rate, that the electron excitation out

---

[2]Ghirardi, Rimini, and Weber (1986): see Chapter 5 of this book.
[3]Pearle (1989).
[4]Pearle and Squires (1994).

of atoms would occur at too large a rate to agree with experimental results. So, λ_e has to be smaller than the nucleon collapse rate.

Also, one could put in the theory a random field for each particle type, $w_n(\mathbf{x}, t)$, $w_p(\mathbf{x}, t)$, $w_e(\mathbf{x}, t)$, ... But, it would be conceptually simpler if Nature supplied just one random field, $w(\mathbf{x}, t)$. We therefore proposed, rather than having the collapse-generating operator be the particle number density operator, instead it should be the mass density operator, $\hat{A}(\mathbf{x}) \equiv \hat{M}(\mathbf{x}) = \sum_\alpha m_\alpha \hat{D}_\alpha(\mathbf{x})$. This effectively makes the collapse rate for electrons quite a bit smaller than that for nucleons. So, this is the choice made for CSL.

And, there is experimental evidence that supports this choice, coming from an experimental limit on the anomalous electron excitation in a slab of Ge^5 (the first measurement to seriously constrain an alternative to quantum theory!), and from a later experiment,[6] coming from an experimental limit on the anomalous dissociation of deuterium.

A speculative remark: since gravity depends upon mass density, this experimentally mandated choice suggests that there might be a connection between collapse and gravity.

These topics are discussed in more detail in Chapter 6.

There is one last important consideration to our choosing the collapse-generating operators. After all this discussion, one finds that $\hat{A}(\mathbf{x}) \to \hat{D}(\mathbf{x})$ or $\hat{M}(\mathbf{x})$ has the consequence that the mean rate of energy increase is infinite!

This difficulty, first encountered and solved in SL, is readily overcome by choosing the collapse-generating operator to be a "smeared" mass density operator,

$$\hat{A}(\mathbf{x}) \to \sum_\alpha \frac{m_\alpha}{M} \int d\mathbf{z} f(|\mathbf{z} - \mathbf{x}|) \hat{D}_\alpha(\mathbf{z}). \tag{4.1}$$

M is some reference mass, which we will choose as of nucleon size, such as m_n, m_p, or $(m_n + m_p)/2$.

We wish $\hat{A}(\mathbf{x})$ to be proportional to the mass density in some spherically symmetric volume around $\mathbf{z}$, because of the isotropy of space. We will choose

$$f(|\mathbf{x} - \mathbf{z}|) \equiv \frac{1}{[\pi a^2]^{3/4}} e^{-\frac{(\mathbf{x}-\mathbf{z})^2}{2a^2}}. \tag{4.2}$$

(other functions would do as well). The "smearing volume" is determined by the length parameter a, *so this is an additional parameter fundamental to the theory* (in addition to λ).

With the choices of the factor M^{-1} and f's proportionality to $a^{-3/2}$, the dimension of $\hat{A}(\mathbf{x})$ is Length$^{-3/2}$. The $\pi^{-3/4}$ factor in f is chosen to simplify the density matrix evolution equation, as will be seen.

4.2   The Two Equations Defining Non–Relativistic CSL

To summarize, we choose the collapse-generating operator to be *smeared mass density*,

[5] Collett et al. (1995).
[6] Jones, Pearle, and Ring (2004).

$$\hat{A}(\mathbf{x}) = \sum_\alpha \frac{m_\alpha}{M} \frac{1}{[\pi a^2]^{3/4}} \int d\mathbf{z} e^{-\frac{(\mathbf{x}-\mathbf{z})^2}{2a^2}} \hat{D}_\alpha(\mathbf{z}). \tag{4.3}$$

We may crudely think of the Gaussian as having the value 1 for $|\mathbf{x} - \mathbf{z}| < a$ and the value 0 otherwise. Then, the integral in (4.3) may be crudely thought of as giving the number of particles of type α in a sphere of radius a centered on the point $\mathbf{x}$.

So, finally, the state vector evolution for non-relativistic CSL that we propose is given by (3.29) with the choice (4.3), that is

$$|\psi, t\rangle = Te^{-i\int_0^t dt' \hat{H}} e^{-\int_0^t dt' d\mathbf{x}' \frac{1}{4\lambda} \left[w(\mathbf{x}',t') - 2\lambda \sum_\alpha \frac{m_\alpha}{M} \frac{1}{[\pi a^2]^{3/4}} \int d\mathbf{z} e^{-\frac{(\mathbf{x}'-\mathbf{z})^2}{2a^2}} \hat{D}_\alpha(\mathbf{z}) \right]^2} |\psi, 0\rangle. \tag{4.4}$$

Eq.(4.4) and the probability rule (3.30) completely define non-relativistic CSL.

That this theory is indeed Galilean invariant is worth seeing. The proof is interesting, but a bit complicated and so is relegated to Chapter 11, the Supplement to this chapter.

4.3 Density Matrix Evolution Equation for Non–Relativistic CSL

Putting (4.3) into (3.31), and using the integral

$$\int d\mathbf{x} \frac{1}{[\pi a^2]^{3/4}} e^{-\frac{(\mathbf{x}-\mathbf{z})^2}{2a^2}} \frac{1}{[\pi a^2]^{3/4}} e^{-\frac{(\mathbf{x}-\mathbf{z}')^2}{2a^2}} = e^{-\frac{(\mathbf{z}-\mathbf{z}')^2}{4a^2}},$$

we obtain the evolution equation for the density matrix:

$$\frac{d}{dt}\hat{\rho}(t) = -i[\hat{H}, \hat{\rho}(t)] - \frac{\lambda}{2} \sum_{\alpha,\beta} \frac{m_\alpha}{M} \frac{m_\beta}{M} \int d\mathbf{z}' \int d\mathbf{z} e^{-\frac{(\mathbf{z}-\mathbf{z}')^2}{4a^2}} [\hat{D}_\beta(\mathbf{z}'), [\hat{D}_\alpha(\mathbf{z}), \hat{\rho}(t)]]. \tag{4.5}$$

Eq. (4.5) (and its restriction to just nucleons, Eq. (4.6) below) *is all we are going to use in the rest of this book.* In applications, one is most interested in the behavior of the ensemble of state vectors, and that is the purview of the density matrix.

4.3.1 Approximate Density Matrix Evolution Equation for Nucleons

A good approximation to (4.5), when describing the collapse rate of a superposition of states of macroscopic matter in different locations, such as in an apparatus, is to neglect the electrons because of their relatively small mass. Also a good approximation is to regard the neutron and proton as a single particle, a "nucleon," having mass M, since their masses are so close. Then, Eq.(4.5) becomes

$$\frac{d}{dt}\hat{\rho}(t) = -i[\hat{H}, \hat{\rho}(t)] - \frac{\lambda}{2} \int d\mathbf{z}' \int d\mathbf{z} e^{-\frac{(\mathbf{z}-\mathbf{z}')^2}{4a^2}} [\hat{D}(\mathbf{z}'), [\hat{D}(\mathbf{z}), \hat{\rho}(t)]]. \tag{4.6}$$

4.4 Collapse of a Superposition of an Object in Two Spatially Separated States

We have a new theory. Let's start to use it! In this and the next section we do two quite useful calculations.

Perhaps the hallmark of a dynamical collapse theory is its treatment of an object described by a superposition of two states, each with the object in a different place. If the object is macroscopic, this is the quintessential superposition of an apparatus "pointer." The dynamics should be collapse to one or the other of these states, since that is what we see.

To see how this works in some detail, we will consider the evolution of the initial state vector $|\psi, 0\rangle = \frac{1}{\sqrt{2}}[|L\rangle + |R\rangle]$ (initial density matrix $\rho(0) = \frac{1}{2}[|L\rangle + |R\rangle][\langle L| + \langle R|]$), where L, R refer to the object at the left and at the right, relatively translated by the vector $\vec{\ell}$, of length ℓ.

4.4.1   Collapse of a Superposition of a Nucleon in Two Spatially Separated States

To get our feet wet, let's first apply (4.6) to a single nucleon in such a superposition. Let $|L\rangle$ be the state of a nucleon located "to the left" at $\mathbf{x}_L$. We ignore its internal structure, taking as a good approximation that it is located at a point of space, an eigenstate of the density operator satisfying $D(\mathbf{x})|L\rangle = \delta(\mathbf{x} - \mathbf{x}_L)|L\rangle$. Similarly, for the nucleon "to the right," $D(\mathbf{x})|R\rangle = \delta(\mathbf{x} - \mathbf{x}_R)|R\rangle$.

To see how fast the collapse takes place on average, we calculate $\langle L|\rho(t)|R\rangle$. Neglecting the Hamiltonian term, for one nucleon, Eq. (4.6) yields

$$\frac{d}{dt}\langle L|\hat{\rho}(t)|R\rangle = -\frac{\lambda}{2}\int d\mathbf{z}'\int d\mathbf{z}\, e^{-\frac{(\mathbf{z}-\mathbf{z}')^2}{4a^2}}\langle L|[\hat{D}(\mathbf{z}'),[\hat{D}(\mathbf{z}),\hat{\rho}(t)]]|R\rangle.$$

With the double commutator written out, this becomes

$$\frac{d}{dt}\langle L|\hat{\rho}(t)|R\rangle = -\frac{\lambda}{2}\int d\mathbf{z}'\int d\mathbf{z}\, e^{-\frac{(\mathbf{z}-\mathbf{z}')^2}{4a^2}}\langle L|\Big[\hat{D}(\mathbf{z}')\hat{D}(\mathbf{z})\hat{\rho}(t) + \hat{\rho}(t)\hat{D}(\mathbf{z})\hat{D}(\mathbf{z}')$$

$$-\hat{D}(\mathbf{z})\hat{\rho}(t)\hat{D}(\mathbf{z}') - \hat{D}(\mathbf{z}')\hat{\rho}(t)\hat{D}(\mathbf{z})\Big]|R\rangle.$$

Letting the density operators act on $\langle L|$ and $|R\rangle$, we get

$$\frac{d}{dt}\langle L|\hat{\rho}(t)|R\rangle = -\frac{\lambda}{2}\langle L|\hat{\rho}(t)|R\rangle\int d\mathbf{z}'\int d\mathbf{z}\, e^{-\frac{(\mathbf{z}-\mathbf{z}')^2}{4a^2}}$$

$$\cdot\Big[\delta(\mathbf{z}' - \mathbf{x}_L)\delta(\mathbf{z} - \mathbf{x}_L) + \delta(\mathbf{z} - \mathbf{x}_R)\delta(\mathbf{z}' - \mathbf{x}_R)$$

$$-\delta(\mathbf{z} - \mathbf{x}_L)\delta(\mathbf{z}' - \mathbf{x}_R) - \delta(\mathbf{z}' - \mathbf{x}_L)\delta(\mathbf{z} - \mathbf{x}_R)\Big]$$

$$= -\lambda\langle L|\hat{\rho}(t)|R\rangle\Big[1 - e^{-\frac{\ell^2}{4a^2}}\Big] \tag{4.7}$$

with solution

$$\langle L|\hat{\rho}(t)|R\rangle = \langle L|\hat{\rho}(0)|R\rangle e^{-\lambda t\left[1 - e^{-\frac{\ell^2}{4a^2}}\right]}$$

$$= \frac{1}{2}e^{-\lambda t\left[1 - e^{-\frac{\ell^2}{4a^2}}\right]}$$

$$\approx \frac{1}{2}e^{-\lambda t} \text{ for } \ell \gg a. \tag{4.8}$$

Thus, the collapse rate for a single nucleon in widely separated locations is λ. *One may take that statement to be the meaning of λ.*

4.4.2 Collapse of a Superposition of an Extended Object in Two Spatially Separated States

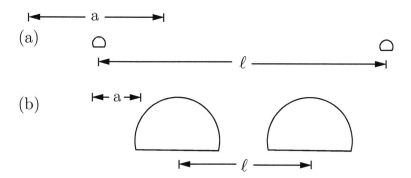

Fig. 4.1 Collapse of a superposition of an extended object in two places. a) Small object compared to a. b) Large object compared to a.

Next, we apply (4.6) to an extended homogeneous object that has volume V and contains N nucleons. As in the previous example, initially it is in a superposition of two states $|L\rangle$ and $|R\rangle$ with equal amplitudes, where there is no spatial overlap of the two physical states of the object.

To simplify the calculation, we model the object by supposing its mass is uniformly distributed over the object, so these states are eigenstates of the density operator satisfying $D(\mathbf{x})|L\rangle = \frac{N}{V}\Theta_L(\mathbf{x})|L\rangle$ and $D(\mathbf{x})|R\rangle = \frac{N}{V}\Theta_R(\mathbf{x})|R\rangle$. Here, $\Theta_L(\mathbf{x})$ is a step function, taking on the value 1 if $\mathbf{x}$ lies in the volume of the object at the left and having value 0 everywhere else (sometime called the characteristic function of the object), so $\int_\infty^\infty d\mathbf{x}\Theta_L(\mathbf{x})... = \int_{V_L} d\mathbf{x}...$ (and of course, similarly for R).

Putting this into (4.6), the calculation proceeds as in (4.7), except the delta functions are replaced by the step functions:

$$\frac{d}{dt}\langle L|\hat{\rho}(t)|R\rangle = -\frac{\lambda}{2}\left[\frac{N}{V}\right]^2 \langle L|\hat{\rho}(t)|R\rangle \int d\mathbf{z}' \int d\mathbf{z}\, e^{-\frac{(\mathbf{z}-\mathbf{z}')^2}{4a^2}}$$

$$\cdot \left[\Theta_L(\mathbf{z}')\Theta_L(\mathbf{z}) + \Theta_R(\mathbf{z})\Theta_R(\mathbf{z}') - \Theta_L(\mathbf{z})\Theta_R(\mathbf{z}') - \Theta_L(\mathbf{z}')\Theta_R(\mathbf{x})\right]$$

$$= -\lambda\left[\frac{N}{V}\right]^2 \langle L|\hat{\rho}(t)|R\rangle \left\{ \int_{V_L}\int_{V_L} - \int_{V_L}\int_{V_R} \right\} d\mathbf{z} d\mathbf{z}'\, e^{-\frac{(\mathbf{z}-\mathbf{z}')^2}{4a^2}}$$

$$\equiv -\lambda\left[\frac{N}{V}\right]^2 \langle L|\hat{\rho}(t)|R\rangle f(a). \tag{4.9}$$

In going from the first equation to the second, we have used the symmetry of the double integral under the exchanges $\mathbf{z} \leftrightarrow \mathbf{z}'$ ($\int_{V_L}\int_{V_L}$ could as well be $\int_{V_R}\int_{V_R}$). The last equation defines the function $f(a)$.

Therefore, we obtain

$$\langle L|\hat{\rho}(t)|R\rangle = \langle L|\hat{\rho}(0)|R\rangle e^{-\lambda t \left[\frac{N}{V}\right]^2 f(a)}. \tag{4.10}$$

For two cases it is easy to calculate $f(a)$ to a good approximation: a small object whose dimensions are $<< a \lesssim \ell$ and a large object whose dimensions are $>> a$. These are illustrated in Fig. (4.1).

Small object. In the first double integral in the expression for $f(a)$, since the separation $|\mathbf{z} - \mathbf{z}'|$ of any two points in V_L is much less than a, we make the approximation $e^{-\frac{(\mathbf{z}-\mathbf{z}')^2}{4a^2}} \approx 1$. In the second double integral in the expression for $f(a)$, all points $\mathbf{z}$ in V_L are very close to being equidistant from all points $\mathbf{z}'$ in V_R, so we make the approximation $e^{-\frac{(\mathbf{z}-\mathbf{z}')^2}{4a^2}} \approx e^{-\frac{\ell^2}{4a^2}}$. Therefore, from (4.10),

$$f(a) \approx \int_{V_L}\int_{V_L} d\mathbf{z}d\mathbf{z}' - \int_{V_L}\int_{V_R} d\mathbf{z}d\mathbf{z}' e^{-\frac{\ell^2}{4a^2}} = V^2\left[1 - e^{-\frac{\ell^2}{4a^2}}\right]$$

Putting this into (4.10), we find that the collapse rate is

$$\Gamma \approx \lambda N^2 \left[1 - e^{-\frac{\ell^2}{4a^2}}\right]. \tag{4.11}$$

In particular, it goes $\sim N^2$: the collapse goes faster with increasing number of particles *and as their square.*

Large object. Regarding the second double integral in the expression for $f(a)$, even if the two images abut, most points $\mathbf{z}$ within V_L are much farther than a from most points $\mathbf{z}'$ within V_R. Thus, we may neglect the second double integral in the curly bracket of (4.9), due to the smallness of the exponential,

For the first double integral in the curly bracket, for almost any point $\mathbf{z}$ within the object (except for points on or near its boundary), we can imagine a second point $\mathbf{z}'$ initially coinciding with $\mathbf{z}$ and then moving away in any direction, going much further away than a, but still lying in the object. Therefore, it is adding very little to extend the limits of the integral over $\mathbf{z}'$ to $(-\infty, \infty)$, with the resulting integral over the Gaussian equal to $(4\pi)^{3/2}a^3$. The integral over $\mathbf{z}$ then gives V. So, putting $f(a) \approx V(4\pi)^{3/2}a^3$ into (4.10), we find that the collapse rate is

$$\Gamma \approx (4\pi)^{3/2}\lambda\frac{N^2 a^3}{V} \equiv (4\pi)^{3/2}\lambda N n_a \tag{4.12}$$

where $n_a \equiv \frac{N}{V}a^3$. So, in this case, *the collapse rate is proportional to the product of the number of nucleons N in the object and the number of nucleons n_a contained in a cube of side a.*

This result can also be applied to the case where the two images of the object overlap. This is because the nucleons in the overlap region have identical mass density and so play no role in the collapse.[A] Thus, in the overlap case, the collapse rate is also given by Eq. (4.12) where, however, N is the number of nucleons in the non-overlap region.

In both cases, the large N value of a macroscopic object causes a large collapse rate, giving us the result we see. The small N value of a microscopic object means the object suffers negligible collapse, allowing it to behave as in standard quantum theory.

4.4.3   Lindblad Equation in Terms of Position Operators

Treating an object as having uniform density was a good approximation for the previous calculation,. However, matter actually consists of atoms separated by atomic distances. For some calculations, a more exact, but still approximate, model of matter is useful, still neglecting the electron contribution to collapse, but taking the nucleons as being point particles (not uniformly smeared out, as in the previous calculation). Then, one may express (4.5) in the nucleon position representation $|\mathbf{x}\rangle \equiv |\mathbf{x}_1, \mathbf{x}_2, ...\mathbf{x}_N\rangle$, in which case $D(\mathbf{x})|\mathbf{x}_1, \mathbf{x}_2, ...\mathbf{x}_N\rangle = \sum_{i=1}^{N} \delta(\mathbf{z} - \mathbf{x}_i)|\mathbf{x}_1, \mathbf{x}_2, ...\mathbf{x}_N\rangle$. Eq. (4.5) then implies

$$\frac{d}{dt}\langle\mathbf{x}|\hat{\rho}(t)|\mathbf{x}'\rangle = -i[\langle\mathbf{x}|\hat{H}, \hat{\rho}(t)]|\mathbf{x}'\rangle - \frac{\lambda}{2}\int d\mathbf{z}' \int d\mathbf{z} e^{-\frac{(\mathbf{z}-\mathbf{z}')^2}{4a^2}}$$

$$\cdot \sum_{i,j=1}^{N} [\delta(\mathbf{z}' - \mathbf{x}_i)\delta(\mathbf{z} - \mathbf{x}_j) + \delta(\mathbf{z}' - \mathbf{x}'_i)\delta(\mathbf{z} - \mathbf{x}'_j)$$

$$-\delta(\mathbf{z}' - \mathbf{x}_i)\delta(\mathbf{z} - \mathbf{x}'_j) - \delta(\mathbf{z}' - \mathbf{x}'_i)\delta(\mathbf{z} - \mathbf{x}_j]\langle\mathbf{x}|\hat{\rho}(t)|\mathbf{x}'\rangle$$

$$= -i\langle\mathbf{x}|[\hat{H}, \hat{\rho}(t)]|\mathbf{x}'\rangle$$

$$-\frac{\lambda}{2}\sum_{i,j=1}^{N}\left[e^{-\frac{(\mathbf{x}_i-\mathbf{x}_j)^2}{4a^2}} + e^{-\frac{(\mathbf{x}'_i-\mathbf{x}'_j)^2}{4a^2}} - 2e^{-\frac{(\mathbf{x}_i-\mathbf{x}'_j)^2}{4a^2}}\right]\langle\mathbf{x}|\hat{\rho}(t)|\mathbf{x}'\rangle.$$

Since $\hat{\mathbf{X}}_i|\mathbf{x}\rangle = \mathbf{x}_i|\mathbf{x}\rangle$, this can be written in a basis-independent way:

$$\frac{d}{dt}\hat{\rho}(t) = -i[\hat{H}, \hat{\rho}(t)]$$

$$-\frac{\lambda}{2}\sum_{i,j=1}^{N}\left[e^{-\frac{(\hat{\mathbf{X}}_i-\hat{\mathbf{X}}_j)^2}{4a^2}}\hat{\rho}(t) + \hat{\rho}(t)e^{-\frac{(\hat{\mathbf{X}}_i-\hat{\mathbf{X}}_j)^2}{4a^2}} - 2e^{-\frac{(\hat{\mathbf{X}}_{iL}-\hat{\mathbf{X}}_{jR})^2}{4a^2}}\hat{\rho}(t)\right].$$

$$(4.13)$$

where, in the last term, we have used the subscript notation L, R, as explained in Section 2.4. This form for the density matrix evolution equation will prove useful for calculations presented in Chapter 6

4.5   Mean Energy Increase

Our second calculation is to find the mean energy increase of a collection of particles. It turns out that the result is independent of the physical state of the particles!

Moreover, no approximations whatsoever are made to obtain the result. This is important because, as mentioned earlier, this anomalous heating provides experimental tests for the theory. The reason for the energy increase is that the collapse continuously tries to narrow wave functions, working toward achieving position eigenstates. If, in a short time interval, the wave packet width is decreased, then the momentum space width is increased (according to the uncertainty principle), which implies a larger kinetic energy.

Using (4.5), the mean energy rate of change for the ensemble of particles is

$$\frac{d}{dt}\overline{\langle \hat{H}\rangle} = \frac{d}{dt}\mathrm{Tr}\hat{H}\hat{\rho}(t) = -i\mathrm{Tr}\hat{\rho}(t)[\hat{H},\hat{H}]$$
$$-\mathrm{Tr}\hat{\rho}(t)\frac{\lambda}{2}\sum_{\alpha,\beta}\frac{m_\alpha}{M}\frac{m_\beta}{M}\int d\mathbf{z}'\int d\mathbf{z}e^{-\frac{(\mathbf{z}-\mathbf{z}')^2}{4a^2}}[\hat{D}_\beta(\mathbf{z}'),[\hat{D}_\alpha(\mathbf{z}),\hat{H}]], \quad (4.14)$$

using the cyclic permutivity of the trace operation, and the symmetry under $\mathbf{z} \leftrightarrow \mathbf{z}'$, to take $\hat{\rho}(t)$ out of the commutators.[7]

The energy operator is $\hat{H} = \hat{K} + \hat{V}$, where $\hat{K}$ is the kinetic energy operator

$$\hat{K} \equiv \sum_\alpha \hat{K}_\alpha = \sum_\alpha \frac{1}{2m_\alpha}\hat{P}_\alpha^2.$$

The potential energy operator is[B]

$$\hat{V} \equiv \frac{1}{2}\int d\mathbf{x}d\mathbf{x}' \sum_{\alpha,\beta} V(\mathbf{x}_\alpha - \mathbf{x}'_\beta)\hat{D}_\alpha(\mathbf{x})\hat{D}_\beta(\mathbf{x}').$$

$\hat{V}$ does not contribute since it depends upon the particle density operator and so gives a vanishing commutator in (4.14).

It is only the kinetic energy term which is relevant. Because the density operator for one type of particle has vanishing commutator with the kinetic energy operator of another particle, Eq.(4.14) becomes the sum of terms for each type of particle. Therefore, we only need to calculate the mean energy increase for one type of particle:

$$\frac{d}{dt}\mathrm{Tr}\hat{H}_\alpha\hat{\rho}(t) = -\mathrm{Tr}\hat{\rho}(t)\frac{\lambda}{2}\left[\frac{m_\alpha}{M}\right]^2\int d\mathbf{z}'\int d\mathbf{z}e^{-\frac{(\mathbf{z}-\mathbf{z}')^2}{4a^2}}[\hat{D}_\alpha(\mathbf{z}'),[\hat{D}_\alpha(\mathbf{z}),\hat{K}_\alpha]]. \quad (4.15)$$

This is a complicated calculation (done in Chapter 11, the Supplement to this chapter), but with a simple result (Eq. (11.23):

$$\frac{d}{dt}\mathrm{Tr}\hat{H}_\alpha\hat{\rho}(t) = \frac{3\lambda m_\alpha}{4}\left[\frac{\hbar}{Ma}\right]^2\mathrm{Tr}\hat{\rho}(t)\hat{N}_\alpha = \frac{3\lambda m_\alpha}{4}\left[\frac{\hbar}{Ma}\right]^2 N_\alpha, \quad (4.16)$$

$\hat{N}_\alpha$ is the α-type particle number operator and N_α is the number of particles of type α: the last step is valid because the number of particles of each type does not change with time.

Since the reduced Compton wavelength of the nucleon is $\bar{\lambda} = \frac{\hbar}{Mc}$, the energy increase rate per particle can also be expressed as $\frac{3\lambda}{4}[m_\alpha c^2][\frac{\bar{\lambda}}{a}]^2$, the product of the nucleon collapse rate λ, the particle's mass-energy, and the square of the ratio of neutron Compton wavelength to a.

---

[7] $\mathrm{Tr}\hat{H}[\hat{A},[\hat{B},\hat{\rho}]] = \mathrm{Tr}\hat{H}[\hat{A}\hat{B}\hat{\rho} - \hat{A}\hat{\rho}\hat{B} - \hat{B}\hat{\rho}\hat{A} + \hat{\rho}\hat{B}\hat{A}]$ which, by cyclic permutation allowed by the trace operation equals $\mathrm{Tr}\hat{\rho}[\hat{H}\hat{A}\hat{B} - \hat{B}\hat{H}\hat{A} - \hat{A}\hat{H}\hat{B} + \hat{B}\hat{A}\hat{H}] = \mathrm{Tr}\hat{\rho}[\hat{B}[\hat{A},\hat{H}]]$

Summing (4.16) over the contribution from all particles, and integrating with respect to time, we obtain the exact expression for the average of the quantum expectation value of the energy, $\langle \psi, t | \hat{H} | \psi, t \rangle$, for the complete ensemble of state vectors $|\psi, t\rangle$ undergoing collapse:

$$\overline{\langle \hat{H} \rangle(t)} - \langle \hat{H} \rangle(0) = \lambda t \sum_\alpha \frac{N_\alpha m_\alpha}{M} \frac{3\hbar^2}{4Ma^2} = \lambda t \frac{M}{M} \frac{3}{4} \left[\frac{\hbar^2}{Ma^2} \right], \quad (4.17)$$

where M is the total mass of all the particles.

The energy increase is proportional to t. The factor 3 is there because of three dimensions. $\frac{\hbar^2}{2Ma^2}$ is the kinetic energy of a nucleon with momentum $\hbar k = \frac{\hbar}{a}$. And, neglecting the contribution of the small electron mass, $\frac{M}{M}$ is the number of nucleons.

What happens to the mean momentum? Although we have seen that particles get random shoves, for the ensemble there is an equal amount of shoving in each direction, so the total momentum for the ensemble is a conserved quantity.[8]

Eq. (4.17) gives the *total* energy increase for all particles in an object, regardless of the forces that hold it together. The total energy increase may be written as the sum of the center of mass energy increase of the object and the internal energy increase (for example, the "spontaneous" excitation of atoms, of nucleii, of lattice vibrations within the object). Just how the energy increase splits between the center of mass and internal degrees of freedom requires further calculation. The energy increase is subject to some experimental tests of non-relativistic CSL, which is the subject of Chapter 6.

But, before turning to these applications, it is incumbent to consider another dynamical collapse model.

Notes

[A]In (4.10), write $\int_{V_L} = \int_{V_{Lno}} + \int_{V_o}, \int_{V_R} = \int_{V_{Rno}} + \int_{V_o}$, where *no* means the integral is over the region of no overlap and *o* means the integral is over the overlap region. The integrals add up to $\int_{V_{Lno}} \int_{V_{Lno}} - \int_{V_{Lno}} \int_{V_{Rno}} + I$, where $I \equiv 2[\int_{V_{Lno}} - \int_{V_{Rno}}] \int_{V_o}$. But, because the Gaussian in the integrals of (4.10) is approximately 0 for one of the points $\mathbf{z}, \mathbf{z}'$ in the overlap volume and one of the points outside the overlap volume (the only, relatively small, contribution coming from the neighborhood of the boundary of the overlap region), I makes a negligible contribution.

[B]To see that this gives the correct potential term in the Schrödinger equation, for simplicity for one kind of particle, consider

$$\langle \mathbf{x} | \hat{V} | \psi \rangle = \frac{1}{2} \int d\mathbf{z} d\mathbf{z}' V(\mathbf{z} - \mathbf{z}') \langle \mathbf{x} | \hat{D}(\mathbf{z}) \hat{D}(\mathbf{z}') | \psi \rangle$$

$$= \frac{1}{2} \int d\mathbf{z} d\mathbf{z}' V(\mathbf{z} - \mathbf{z}') \sum_{i,j=1}^{N} \delta(\mathbf{z} - \mathbf{x}_i) \delta(\mathbf{z}' - \mathbf{x}_j) \langle \mathbf{x} | \psi \rangle$$

$$= \frac{1}{2} \sum_{i,j=1}^{N} V(\mathbf{x_i} - \mathbf{x_j}) \langle \mathbf{x} | \psi \rangle.$$

[8]Replacing $\hat{H}$ by $\hat{\mathbf{P}}$ on the left side of Eq. (4.14), the commutators vanish: $[\hat{\mathbf{P}}, \hat{H}] = 0$ (total momentum is conserved in the usual Schrödinger dynamics) and, for each particle type, $[D(\mathbf{z}), \hat{\mathbf{P}}] \sim \nabla D(\mathbf{z})$, and $[D(\mathbf{z}'), \nabla D(\mathbf{z})] = 0$.

5

Spontaneous Localization (SL) Theory

"A hit, a very palpable hit"[1]

In the collapse models I made in the years before finding CSL, there were two problems I did not know how to overcome.

In my first paper[2] on collapse models I wrote:

> The rate cannot be too fast, or else the usual quantum predictions will be interfered with. It also cannot be too slow, or else it will predict that a system can be observed in a superposition of macroscopically distinguishable states. Perhaps the matrix element magnitudes should be small for a microscopic system, but large for a macroscopic system.

I came to call this the "trigger problem": how to turn off the collapse term for a superposition of microscopically different states, but turn it on for a superposition of macroscopically different states. In these models I did it "by hand", but I knew that was not acceptable, the equations should do it.

The other I called the "preferred basis problem". For the collapse to give what we see around us, I knew the collapse should to be toward position eigenstates, but position eigenstates have infinite energy, so how could one handle that?

We will see how both those problems were solved by GianCarlo Ghirardi, Alberto Rimini and Tullio Weber (GRW) in their non-relativistic SL theory, which was enthusiastically promulgated by John Bell[3] and whose resolution is incorporated in CSL.

5.1 General Structure of SL

As is the case for CSL, SL has a general mathematical structure which can then be specialized to a non-relativistic proposal.

Consider a state vector $|\psi, t\rangle$, evolving according to the standard Schrödinger equation. It is supposed that there is a probability λdt (λ is a parameter of the theory with dimension time^{-1}) at any time t that, in the time interval dt, the state vector makes a sudden dramatic change to a new state vector.

---

[1] Shakespeare (1623).
[2] Pearle (1976).
[3] Ghirardi, Rimini, and Weber (1986, 1987), Bell (1987).

The new state vector is to be one of a continuum of possible state vectors, characterized by the parameter s ($-\infty < s < \infty$). The change that it undergoes, which GRW called a "spontaneous localization", and which I dubbed a "hit[4]" (going from nine syllables to one syllable simplified conversation enormously!) is multiplication by an operator. This results in a new state vector $\hat{O}(s)|\psi, t\rangle$ where, since s is a continuum variable, $\hat{O}(s)$ is a continuum of Hermitian operators. The state vector before the hit was normalized to 1. The new possible state vector normalized to 1 is:

$$|\psi, t\rangle_s \equiv \frac{1}{\sqrt{\langle\psi, t|\hat{O}^2(s)|\psi, t\rangle}} \hat{O}(s)|\psi, t\rangle \tag{5.1}$$

The probability of a hit in the time interval $(t, t+dt)$, with parameter in the interval $(s, s+ds)$, is chosen to be

$$P(t, s)dtds \equiv \lambda dtds\langle\psi, t|\hat{O}^2(s)|\psi, t\rangle. \tag{5.2}$$

Since, as mentioned above, the probability of a hit with any s is to be λdt, according to (5.2) this requires

$$\lambda dt = \int ds P(t, s) = \lambda dt \int_{-\infty}^{\infty} ds\langle\psi, t|\hat{O}^2(s)|\psi, t\rangle,$$

where $|\psi, t\rangle$ can be any state vector. Therefore, there must be the following constraint on the operators:

$$\int_{-\infty}^{\infty} ds\hat{O}^2(s) = \hat{1}. \tag{5.3}$$

Eqs. (5.1), (5.2) and (5.3) define the structure of the SL theory.

5.1.1   Two-State Collapse Example

An example is given here showing how, with proper choice of $\hat{O}(s)$, this structure can be used to describe state vector collapse.

Consider, as we did in Section II for CSL, the simplest example, a Hilbert space spanned by the two "preferred basis" states $|a_1\rangle, |a_2\rangle$. Choose

$$\hat{O}(s) = f_1(s)|a_1\rangle\langle a_1| + f_2(s)|a_2\rangle\langle a_2|,$$

where $f_1(s) > 0$ for $s < 0$ and $f_1(s) = 0$ for $s > 0$ and $\int_{-\infty}^{0} ds f_1^2(s) = 1$. Similarly, $f_2(s) > 0$ for $s > 0$ and $f_2(s) = 0$ for $s < 0$ and $\int_{0}^{\infty} ds f_2^2(s) = 1$. Then, this $\hat{O}(s)$ satisfies Eq. (5.3).[5]

Let the state vector just before the hit be $|\psi\rangle = c_1|a_1\rangle + c_2|a_2\rangle$. If s is negative then, after the hit but before normalization, the state vector is $c_1 f_1(s)|a_1\rangle$. After normalization, according to (5.1), it is $e^{i\theta_1}|a_1\rangle$ (where $c_1/|c_1| \equiv e^{i\theta_1}$). If s is positive, the state

---

[4]Ghirardi and Pearle (1990).
[5]$\hat{O}^2(s) = f_1^2(s)|a_1\rangle\langle a_1| + f_2^2(s)|a_2\rangle\langle a_2|$ so $\int_{-\infty}^{\infty} ds\hat{O}^2(s) = |a_1\rangle\langle a_1| + |a_2\rangle\langle a_2| = \hat{1}$.

vector is $e^{i\theta_2}|a_2\rangle$ (where $c_2/|c_1| \equiv e^{i\theta_2}$). Thus, the state vector has instantaneously collapsed to one of the preferred basis states.

Since the probability that there is a hit in the time interval dt is λdt, the conditional probability, given the hit takes place with parameter in range $(s, s+ds)$, is $P(t,s)dtds/\lambda dt$. Therefore, according to (5.2), the probability that the hit results in $|a_1\rangle$ is the probability $\int_{-\infty}^{0} dsP(t,s)/\lambda = |c_1|^2$, the Born rule prediction (and similarly for the hit resulting in $|a_2\rangle$).

5.2 SL Density Matrix

In the time interval dt, if the state vector is hit by operator $\hat{O}(s)$ with probability λdt, the resulting state vector is given by (5.1). The other possibility is that the state vector is *not* hit, which occurs then with probability $1 - \lambda dt$. In that case, over the time interval dt, the state vector evolves according to Schrödinger's equation, that is,

$$|\psi, t+dt\rangle = e^{-i\hat{H}dt}|\psi, t\rangle \xrightarrow[dt\to 0]{} [1 - i\hat{H}dt]|\psi, t\rangle.$$

Now that we know all the ways (two!) that the state vector can change over the interval dt, and the probabilities of each, we can see how an initial pure density matrix at time t, $\hat{\rho}(t) = |\psi, t\rangle\langle\psi, t|$, evolves to the density matrix $\hat{\rho}(t+dt)$. This is the sum of terms, according to the definition of the density matrix, each term |new state vector⟩⟨new state vector| multiplied by the probability of that state vector. Utilizing (5.1), (5.2), we have

$$\hat{\rho}(t+dt) = [1 - \lambda dt]\{[1 - i\hat{H}dt]|\psi, t\rangle\}\{\langle\psi, t|[1 + i\hat{H}dt]\}$$
$$+\lambda dt \int ds\langle\psi, t|\hat{O}^2(s)|\psi, t\rangle$$
$$\cdot\frac{1}{\sqrt{\langle\psi, t|\hat{O}^2(s)|\psi, t\rangle}}\hat{O}(s)|\psi, t\rangle\langle\psi, t|\hat{O}(s)\frac{1}{\sqrt{\langle\psi, t|\hat{O}^2(s)|\psi, t\rangle}}$$

(what is so neat is that the probability cancels the normalization factors) which, to first order in dt, is

$$\hat{\rho}(t+dt) = [1 - \lambda dt]|\psi, t\rangle\langle\psi, t| - dt\imath\lfloor\hat{H}, |\psi, t\rangle\langle\psi, t|\rfloor + \lambda dt\int ds\hat{O}(s)|\psi, t\rangle\langle\psi, t|\hat{O}(s).$$

(5.4)

Replacing $|\psi, t\rangle\langle\psi, t|$ by $\hat{\rho}(t)$ and writing $\frac{d}{dt}\hat{\rho}(t)$ as the limit of $[\hat{\rho}(t+dt) - \hat{\rho}(t)]/dt$, the SL density matrix evolution equation is

$$\frac{d}{dt}\hat{\rho}(t) = -i[\hat{H}, \hat{\rho}(t)] - \lambda\left[\hat{\rho}(t) - \int ds\hat{O}(s)\hat{\rho}(t)\hat{O}(s)\right].$$

(5.5)

Although, for simplicity, (5.5) was derived under the assumption that $\hat{\rho}(t) = |\psi, t\rangle\langle\psi, t|$ is a pure density matrix at time t, it holds if $\hat{\rho}(t)$ is a general density matrix (constructed as the sum of pure density matrices multiplied by associated probabilities).

Eq. (5.5) is an equation of Lindblad form (Appendix F, Eq. (F.7)) since, on account of (5.3), it may be written

$$\frac{d}{dt}\hat{\rho}(t) = -i[\hat{H}, \hat{\rho}(t)] - \frac{\lambda}{2} \int ds \left[\hat{O}^2(s)\hat{\rho}(t) + \hat{\rho}(t)\hat{O}^2(s) - 2\hat{O}(s)\hat{\rho}(t)\hat{O}(s) \right]. \quad (5.6)$$

For the two state collapse example above, with $\hat{H} = 0$, Eq. (5.5) is

$$\frac{d}{dt}\hat{\rho}(t) = -\lambda \left[\rho(t) - \sum_{i=1}^{2} |a_i\rangle\langle a_i|\rho(t)|a_i\rangle\langle a_i| \right],$$

so the diagonal matrix elements satisfying $\frac{d}{dt}\langle a_i|\hat{\rho}(t)|a_i\rangle = 0$ remain constant, while the off-diagonal matrix element satisfying $\frac{d}{dt}\langle a_1|\hat{\rho}(t)|a_2\rangle = -\lambda\langle a_1|\hat{\rho}(t)|a_2\rangle$ decays exponentially at collapse rate λ,

$$\langle a_1|\hat{\rho}(t)|a_2\rangle = c_1 c_2^* e^{-\lambda t}.$$

5.3 Non–Relativistic SL

Now we apply this general structure to construct a non-relativistic collapse theory. The SL proposal applies only to a collection of distinguishable particles. That is, unfortunately, SL is flawed, because it does not work for identical particles, for fermions and bosons which are, of course, the particles in our world. These are described by an antisymmetric or symmetric wave function, and a hit destroys the antisymmetry or symmetry. (This, as we have seen, is not the case with CSL).

If there are N indistinguishable particles, with position operators $\hat{\mathbf{X}}_j$, it is proposed that there be N independent hitting processes, one for each particle. We consider operators $\hat{O}_j(\mathbf{s})$, with $j = 1 \dots N$ where the one-dimensional parameter space s is replaced by a three dimensional space $\mathbf{s}$. Eqs. (5.1)-(5.6) are all to be modified by replacing $\hat{O}(s)$ by $\hat{O}_j(\mathbf{s})$ and, in Eqs. (5.4)-(5.6), the collapse term is to have $\sum_{j=1}^{N}$ in front of it.

The specific choice made is

$$\hat{O}_j(\mathbf{s}_j) \equiv \frac{1}{[\pi a^2]^{3/4}} e^{-\frac{(\hat{\mathbf{X}}_j - \mathbf{s}_j)^2}{2a^2}}. \quad (5.7)$$

We note that

$$\int d\mathbf{s}_j \frac{1}{[\pi a^2]^{3/2}} e^{-\frac{(\hat{\mathbf{X}}_j - \mathbf{s}_j)^2}{a^2}} = 1,$$

so (the generalization of) (5.3) is satisfied.

Thus, a hit by the operator corresponding to particle j is multiplication of the whole wave function by a Gaussian, which narrows the wave function's dependence on the jth particle's position coordinate.

This was called a "spontaneous localization" for the following reasons,.

First, a hit is certainly spontaneous, an unpredictable act of nature.

Second, suppose the particle is described by a wave function in a superposition of two far apart wave packets, "here"+"there". The hit is specified by the location of the "center" of the hit s_j, and a second parameter, a, the "width" of the hit, which has thereby been added to the theory (in addition to λ). If the hit is, say, "here", then the tail of the wave function multiplies the "there" part of the wave function. If the two packets are far apart, then the "there'"packet following the hit is of negligible amplitude. The wave function which was spread out in space has been localized to the neighborhood of "here".

If the wave function before a hit is $\langle \mathbf{x}_1 \dots \mathbf{x}_N | \psi, t \rangle \equiv \psi(\mathbf{x}_1 \dots \mathbf{x}_N, t)$, after suffering a hit on the jth particle, the wave function becomes (according to (5.1)),

$$\psi_{s,j}(\mathbf{x}_1 \dots \mathbf{x}_N, t) = \frac{\frac{1}{[\pi a^2]^{3/4}} e^{-\frac{(\mathbf{x}_j - \mathbf{s}_j)^2}{2a^2}} \psi(\mathbf{x}_1 \dots \mathbf{x}_N, t)}{\sqrt{\int d\mathbf{x}'_1 \dots d\mathbf{x}'_N \frac{1}{[\pi a^2]^{3/2}} e^{-\frac{(\mathbf{x}'_j - \mathbf{s}_j)^2}{a^2}} |\psi(\mathbf{x}'_1 \dots \mathbf{x}'_N, t)|^2}}. \tag{5.8}$$

The probability of such a hit in the time interval dt and in the volume ds_j about s_j, according to (5.2), is

$$P_j(t, \mathbf{s}_j) dt ds_j = \lambda dt ds_j \int d\mathbf{x}'_1 \dots d\mathbf{x}'_N \frac{1}{[\pi a^2]^{3/2}} e^{-\frac{(\mathbf{x}'_j - \mathbf{s}_j)^2}{a^2}} |\psi(\mathbf{x}'_1 \dots \mathbf{x}'_N, t)|^2. \tag{5.9}$$

To understand the import of (5.9), define

$$|\psi(\mathbf{x}_j)|^2 \equiv \int d\mathbf{x}'_1 \dots d\mathbf{x}'_{j-1} d\mathbf{x}'_{j+1} \dots d\mathbf{x}'_N |\psi(\mathbf{x}'_1 \dots \mathbf{x}_j \dots \mathbf{x}'_N, t)|^2.$$

Then, (5.9) becomes

$$P_j(t, \mathbf{s}_j) dt ds_j = \lambda dt ds_j \int d\mathbf{x}_j \frac{1}{[\pi a^2]^{3/2}} e^{-\frac{(\mathbf{x}_j - \mathbf{s}_j)^2}{a^2}} |\psi(\mathbf{x}_j)|^2.$$

Thus, roughly speaking, the probability of a hit on the jth particle is largest when the value of the parameter $\mathbf{s}_j$ is a location where $|\psi(\mathbf{x}_j)|^2$ is largest.

Eqs. (5.8), (5.9) comprise non-relativistic SL.

This solved the preferred basis problem. Because of the Gaussian, the collapse is not to a position eigenstate. Rather, it is to a position range. In CSL, I made use of this idea to smear the mass density to a range a

Now, how does the collapse work, to be slow for a microscopic superposition, but fast for a macroscopic superposition?

Consider the superposition of an object $|\psi, 0\rangle = \frac{1}{\sqrt{2}}[|L\rangle + |R\rangle]$, as in Section 6.3, in the case where the object's two images are far apart, $|\ell| \gg a$.

Suppose first that the object is as microscopic as possible, a single, localized particle, so $\langle \mathbf{x} | \psi, 0 \rangle = \frac{1}{\sqrt{2}}[\langle \mathbf{x} | L \rangle + \langle \mathbf{x} | R \rangle]$, with each wave function well localized, its extent very small compared to a. According to (5.9), the probability is largest that the hit occurs with center located where either wave function $\langle \mathbf{x} | L \rangle$ or $\langle \mathbf{x} | R \rangle$ is largest. Such a hit scarcely alters the hit wave function, since the Gaussian width a is much broader than the wave function. However, the Gaussian multiplies the other wave function with the tail of the Gaussian, diminishing its magnitude by a great amount. The wave function now is no longer a sum of equal amplitude terms, it is one term plus a tiny "tail": the wave function has instantaneously collapsed.

However, the hit occurs on the time scale λ^{-1}. If that is a very long time, then the Hamiltonian evolution of the superposition is the behavior most of the time, the collapse process only seldom interfering.

Now, suppose the object is macroscopic, so the wave function for each of $|L\rangle$, $|R\rangle$ describes many particles. A hit centered on particle j in $|L\rangle$ say, means the tiny Gaussian tail multiplies the particle j part of the wave function of $|R\rangle$, making the magnitude of the state $|R\rangle$ extremely small. *A hit on any one particle collapses the whole wave function*: that is perhaps the most important part of the SL process.

And, since there are N possible particles that can be independently hit in the interval dt, each with probability λdt, then the probability of one or another hit leading to a collapse is $N\lambda dt$.[A] The hits occur on the fast time scale of $(N\lambda)^{-1}$.

This suggested to me how to resolve my "trigger problem". The collapse need not be turned on "by hand" when there is a macroscopic superposition, and turned off by hand when there is a microscopic superposition. The collapse mechanism can be *always on*, since works at a slow rate for a single particle and at a fast rate for many particles.

Similarly, CSL's collapse dynamics is "always on." The collapse rate is slow for a single particle and fast when many particles are involved because mass density is the collapse-generating operator: the greater the mass density difference between the states that "compete in the gambler's ruin game," the faster goes the collapse.

5.4   Non–Relativistic SL Density Matrix

The density matrix evolution equation for N distinguishable particles in non–relativistic SL follows by inserting (5.7) into (5.5) (with $\hat{O} \to \hat{O}_j$):

$$\frac{d}{dt}\hat{\rho}(t) = -i[\hat{H},\hat{\rho}(t)] - \lambda \sum_{j=1}^{N}\left[\hat{\rho}(t) - \int ds_j \frac{1}{[\pi a^2]^{3/4}}e^{-\frac{(\hat{\mathbf{x}}_j-\mathbf{s}_j)^2}{2a^2}}\hat{\rho}(t)\frac{1}{[\pi a^2]^{3/4}}e^{-\frac{(\hat{\mathbf{x}}_j-\mathbf{s}_j)^2}{2a^2}}\right]$$

and, performing the integral,

$$\frac{d}{dt}\hat{\rho}(t) = -i[\hat{H},\hat{\rho}(t)] - \lambda \sum_{j=1}^{N}\left[\hat{\rho}(t) - e^{-\frac{(\hat{\mathbf{x}}_{jL}-\hat{\mathbf{x}}_{jR})^2}{4a^2}}\hat{\rho}(t)\right]. \tag{5.10}$$

Let us apply this to the superposition $\frac{1}{\sqrt{2}}[|L\rangle + |R\rangle]$ of a macroscopic object discussed in Section 4.4.2 (setting $\hat{H} = 0$) where, however, here the particles are distinguishable nucleons. Since $|\ell| >> a$, for the off-diagonal matrix element, the Gaussian term in (5.10) is negligibly small, resulting in

$$\frac{d}{dt}\langle L|\hat{\rho}(t)|R\rangle = -\lambda N \langle L|\hat{\rho}(t)|R\rangle \text{ or}$$
$$\langle L|\hat{\rho}(t)|R\rangle = \langle L|\hat{\rho}(0)|R\rangle e^{-\lambda N t}. \tag{5.11}$$

Thus we see that a microscopic object with small N has a slow collapse rate, and a macroscopic object with large N has a fast collapse rate.

We note that this collapse rate λN is slower than the CSL rate $\sim \lambda N n_a$ obtained in Eq. (4.9) (assuming the same parameter values in CSL as in SL).

5.5   Values of λ and a

The authors of SL gave rough arguments for tentatively choosing values $\lambda \approx 10^{-16}$ sec$^{-1} \approx 1/(300$ million yrs) and $a \approx 10^{-5}$ cm.

They chose a to be mesoscopic, enough larger than atomic size to minimize the disruption of an atom by a hit on an electron, yet smaller than visible light wavelengths.

This value of λ^{-1}, around 1/40th the age of the universe, is certainly small enough so that a single particle is seldom hit, but large enough so that a collection of particles making up a visible clump would have a reasonably rapid collapse rate in SL.

I decided to choose the same tentative values of λ, a for CSL as this helps make comparison of the predictions of the two theories easier. If CSL dynamics is actually chosen by nature, if its effects show up in experiments, then these tentatively chosen parameter values will be replaced by measured ones.

CSL has some preferable features compared to SL.

One is that CSL's collapse rate is faster (making SL more vulnerable to refutation by experiments that place an upper limit on the collapse rate).

Another is that CSL can be expressed as a linear modified Schrödinger equation. (The abrupt change of the state vector in SL does not entail a modified Schrödinger equation. Rather, it can be thought of as a sophisticated, well-defined, collapse postulate.)

Yet another: there is greater generality in the Lindblad equation of CSL (which enables other applications of the CSL formalism).

Finally, unfortunately, SL only applies to particles that are not identical which are, of course, not present in Nature. (If one starts with a properly antisymmetrized fermion wave function, one hit destroys the antisymmetry.) One reason for my choice of mass density as the collapse-generating operator in CSL was to correct this deficiency. Later, however, a symmetry-sustaining hitting process was discovered[6] and rediscovered.[7] This is quite a bit more complicated than the original SL proposal.

In the next section, we return to CSL, for which predictions of the results of various experiments are discussed.

Notes

[A] The hits are independent, so the probability of no hits in dt is the product of the N individual probabilities of no hits, $(1 - \lambda dt)^N \approx (1 - N\lambda dt)$. The probabiliity of just one hit on one of the particles is N times the probability λdt of one hit multiplied by the probability of no hits on the other particles, or $N(1 - \lambda dt)^{N-1}\lambda dt \approx N\lambda dt$. The probability of two hits in dt is $\sim (dt)^2$, and this (as well as the probability of any higher number of hits in dt) makes no contribution in the limit $dt \to 0$.

[6] Dove and Squires (1995).
[7] Tumulka (2006b, 2018).

6
Some Experiments Testing CSL

Non–relativistic CSL has now been completely specified.

Because this is a different theory than standard quantum theory, it makes different predictions about a range of phenomena. This has been pursued both theoretically and experimentally. Exploring some of this effort is the subject of most of this chapter. Our sole tool will be the density matrix evolution Eq.(4.5). It is interesting to see the richness of the consequences of this one equation.

The experiments provide constraints on the parameters λ, a. For simplicity, for most of this Chapter, the SL value $a = 10^{-5}$cm will be assumed and the end result of each discussion will be a constraint placed upon λ.

Additionally, other applications of the mathematical structure of CSL have been proposed: to quantum field theory, to gravity, to cosmology, and some will also be discussed here.

There have been some quite comprehensive reviews of the experimental tests of CSL: here I cite a recent one.[1] The material presented here should make such reviews more comprehensible..

6.1 Theoretical Constraint on the Parameters by Collapse of a Superposition

Even without any experimental constraint, one can see that there must be a lower limit to λ. If $\lambda = 0$, then a macroscopic superposition does not collapse at all, so the state vector cannot represent reality. But the same conclusion obtains if λ is so small that such a superposition exists for what might be considered an unacceptably long time.

The purpose of a dynamical collapse theory is to describe what we see around us. To make that qualitative statement into a more precise statement, take it to mean, even if no one is looking, that any object that *could* be seen, that is placed in a spatial superposition, should collapse to one or the other state of the object faster than human perception time.[A] I called this a "theoretical constraint"[2] on the parameters.

What is meant by "collapse to one or the other state of the object?" As we have seen, in CSL, collapse is not "complete" in a finite time. Suppose there is an object in a superposition of two relatively displaced states, described by the initial state vector $|\psi, 0\rangle = \frac{1}{\sqrt{2}}[|L\rangle + |R\rangle]$. After time t large enough to bring about significant

---

[1] Carlesso et al. (2022).

[2] Collett et al. (1995). Feldman and Tumulka (2012) call it a "philosophical constraint" in their discussion of the issue.

collapse, one expects the ensemble of (normalized) state vectors to consist of half of the form $|\psi, t\rangle = \sqrt{1 - \epsilon^2}|L\rangle + \epsilon|R\rangle \approx |L\rangle + \epsilon|R\rangle$ and half looking like $|\psi, t\rangle \approx \epsilon|L\rangle + |R\rangle$. (We are assuming that the states $|L\rangle, |R\rangle$ are negligibly affected by the collapse dynamics.) Here ϵ, different for the different collapsed state vector,s is very, very small with overwhelming likelihood (and most likely decreasing at a rapid rate), but it is not zero.

This small amplitude remnant of the original state vector is called the "tail." Chapter 7 argues that a tail can be ignored when it would have no observable consequences. That point of view is adopted here.

The density matrix for large t describes half the state vectors collapsing to $|L\rangle$ and half to $|R\rangle$ so, denoting by p_{Li}, p_{Ri} the probabilities associated to these state vectors. we expect

$$\hat{\rho}(t) \approx \sum_i p_{Li}\Big[(|L\rangle + \epsilon_i|R\rangle)(\langle L| + \epsilon_i\langle R|)\Big] + \sum_i p_{Ri}\Big[(|R\rangle + \epsilon_i|L\rangle)(\langle R| + \epsilon_i\langle L|)\Big]$$

$$\approx \frac{1}{2}\Big[|R\rangle\langle R| + |L\rangle\langle L|\Big] + \bar{\epsilon}\Big[|R\rangle\langle L| + |L\rangle\langle R|\Big].$$

Here, we have used $\sum_i p_{Li} = \sum_i p_{Ri} = 1$, taken the probabilities and ϵ_i to be the same for both terms in this expression because of physical symmetry under exchange of L, R, neglected ϵ_i^2 terms, and written the ensemble average $\bar{\epsilon} \equiv \sum_i p_i \epsilon_i$.

The off-diagonal element of the density matrix is therefore $\langle L|\hat{\rho}(t)|R\rangle = \bar{\epsilon}$.

According to Eq. (4.10), the off-diagonal element of the density matrix for such a superposition has the form $\langle L|\rho(t)|R\rangle = \frac{1}{2}e^{-\Gamma t}$ so, at time t, the ensemble average of the tail amplitude is

$$\bar{\epsilon} = \frac{1}{2}e^{-\Gamma t}.$$

In order to obtain the smallest possible collapse rate for our considerations, we want the object described by the superposition to be as small as possible. But, the object has to be visible to the eye. We will consider the object to be a sphere, the smallest that can be seen through an optical microscope, which is about 4×10^{-5}cm in diameter, the wavelength of blue light,

In the superposition, the centers of the L and R spheres are to be separated by a distance ℓ. Here, ℓ has to be larger than the sphere diameter in order that, when one or the other outcome occurs, the eye can distinguish which one.

Also, the states $|L\rangle, |R\rangle$ are to include the light source illuminating the sphere, as well as the light bouncing off, heading, among other places, toward an eye. However, as mentioned in endnote A, we do not consider the states to include a human being actually involved in optical perception.

How does the calculation proceed? We choose a value of a and choose a value of λ, and calculate the associated Γ and therefore $\bar{\epsilon}$. With this value of $\bar{\epsilon}$, we calculate the flux of photons that would come to the eye within human perception time (taken to be ≈ 0.1 s). For sufficiently large λ, if this flux is too small to be perceived, the theory is doing its job, predicting that we perceive one or the other image.

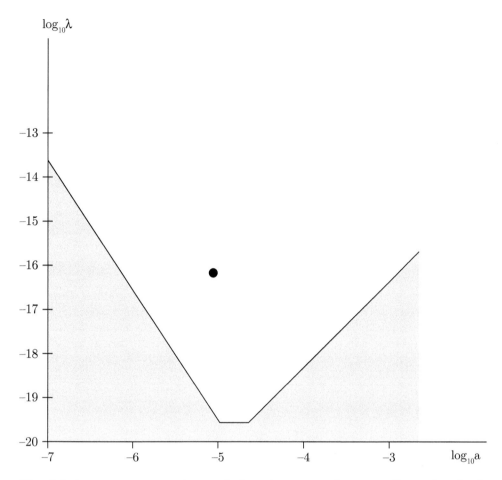

Fig. 6.1 A parameter constraint graph, $\log_{10} \lambda_{s^{-1}}$ versus $\log_{10} a_{cm}$, illustrating the theoretical constraint. The shaded region is outlawed. The solid lines, from left to right, are the boundaries of the three constraints 1), 2), 3). Also indicated is $\lambda = 10^{-16} \mathrm{s}^{-1}$, $a = 10^{-5} \mathrm{cm}$, a working point suggested by SL.

But, as one decreases λ, eventually $\bar{\epsilon}$ becomes so large that the number of photons from the collapsed sphere could be seen: approximately 9 photons at the 0.1 s perception time[3] Therefore, this and smaller values of λ are forbidden.

The reader can see that this is a fairly complicated argument: the details are relegated to Appendix F. In Appendix F are considered three different configurations encompassing a range of a from very small to very large: (1) $a \ll$ sphere diameter, (2) sphere diameter $< a < \ell$ and (3) sphere diameter $< \ell < a$. The results are: (1) $\lambda a^3 \lesssim 2 \times 10^{-35}$, (2) $\lambda \lesssim 2 \times 10^{-20}$ and (3) $\frac{\lambda}{a^2} \lesssim 5 \times 10^{-11}$ (see Fig. 6.1).

---

[3] See https://math.ucr.edu/home/baez/physics/Quantum/see_a_photon.html, "Can a Human See a Single Photon?" which says: "The researchers found that about 90 photons had to enter the eye for a 60% success rate in responding. Since only about 10% of photons arriving at the eye actually reach the retina, this means that about 9 photons were actually required at the receptors."

These considerations provide a *lower* bound on the allowed values of λ in the parameter constraint graph.[4] The result of experiments that test CSL are often presented in such a graph. They typically provide an *upper* bound to λ, drawn as a function of a. Fig. 6.1 is the only such graph appearing here. As mentioned earlier, for simplicity, in this chapter we instead assume the value $a = 10^{-5}$cm, and compare the upper bounds on λ provided by various experiments.

The reader can see, in order to obtain quantitative results, that some fairly arbitrary choices are being made. Indeed, a number of other authors have considered the same issue with different physical setups. Not surprisingly, the limitations displayed in a graph of the lower limit forbidden region in λ, a space, while qualitatively similar, are quantitatively different.

While the idea behind the theoretical constraint is important, because of this arbitrariness, one should consider anyone's results, including the results given here and illustrated in Fig. 6.1, as a rough guideline, rather than as a precise limitation. Moreover, as pointed out in endnote A, the criteria that a visible superposition not last longer than perception time, strictly speaking, is incomplete without inclusion of a perceiver in the state vector.

When that is done, the lower limit in the parameter constraint graph is decreased due to the increased mass distribution difference in the "human apparatus," corresponding to the different perceptions. These occur in the eye's rods and cones in the retina, in the optic nerve, in the brain's neurons transmitting and storing the different information and giving rise to different perceptions, and possibly to resultant different physical behavior. The quantitative effect of this on the parameter exclusion diagram is, of course, very difficult to estimate. One must conclude that this lower limit should be taken with a big grain of salt!

6.2   Mass Dependence of the Collapse Rate

As mentioned, collapse gives energy to particles. We "piggy–back" on experiments designed to detect something else, look at the data from the experiment, subtract the predicted excitation and take the remaining data to be due to collapse.

We consider two such experiments (the processes involved are illustrated in Fig. 6.2).

CSL predicts that electrons can be knocked out of atoms, so we work with an experiment designed to detect electrons knocked out of Ge atoms

CSL predicts that nucleons can be knocked out of nucleii, so we work with an experiment designed to observe deuterium nucleii broken apart, into a free proton and a free neutron.

What will be seen is that the experimental results strongly suggest the mass-proportional coupling of the collapse rate we have already adopted in Eq. (4.3), $\hat{A}(\mathbf{x}) = \sum_\alpha \frac{m_\alpha}{M} \cdots$.

In order to see how the experiments impact on this issue, we will not suppose the collapse rate is mass–proportional. Instead, we suppose a general form $\hat{A}(\mathbf{x}) = \sum_\alpha g_\alpha \cdots$ with arbitrary constants g_α. Then we will see these experiments put on the constraint,

---

[4]Collett et al. (1995) introduced the use of the parameter constraint graph.

for electrons, of $11\frac{m_e}{M} \gtrsim g_e$ and, for nucleons, g_n/g_p is equal to m_n/m_p to better than 1%. This is certainly consistent with, and suggestive of, our choice $g_\alpha = \frac{m_\alpha}{M}$.

6.2.1   Excitation of Bound States: General Analysis

We begin by calculating the ensemble average probability/sec, Γ, of a transition from the ground state of an atom at time 0, to an excited state at time dt, where the electrons and the nucleons can be excited.

The Hamiltonian for all the particles in the atom can be written as the sum of kinetic energy for the center of mass, with the remainder of the Hamiltonian (kinetic and potential energies) expressed in coordinates relative to the center of mass coordinates.

A center–of–mass (cm) state is denoted $|\chi_k\rangle$. A state relative to the center of mass is denoted $|\phi_n\rangle$. These states are energy eigenstates. Discrete indices are used for notational convenience: the cm states are actually continuum states, as are the relative states when the electron is freed from the atom. The relative eigenstates are orthonormal, $\langle\phi_m|\phi_n\rangle = \delta_{mn}$. The cm states are to be normalized, but do not need to be orthogonal The ground state is $|\psi_0\rangle \equiv |\phi_0\rangle|\chi_0\rangle$, the excited state is $|\psi_{nk}\rangle \equiv |\phi_n\rangle|\chi_k\rangle$.

At time 0, the density matrix is $|\psi_0\rangle\langle\psi_0|$. In time dt, most atoms in the ensemble are still in the ground state, but some are excited. The fraction of excited atoms in the ensemble that are in state $|\psi_{nk}\rangle$, divided by dt, can be expressed in terms of the density matrix at time dt:

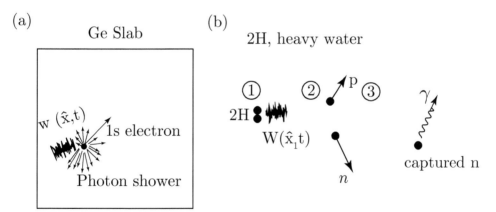

Fig. 6.2 Two experiments looking for spontaneous excitation due to collapse dynamics. Experiment a) A 1s electron in an atom in a slab of Ge is spontaneously knocked out of the atom. Within the atom, the remaining electrons readjust, emitting a photon shower. These photons and the ejected electron travel through the slab, collide with and knock out electrons from atoms in their paths, causing a current pulse that is detected. Experiment b) 1) In a pool of heavy water with some salt added, the proton and neutron in a deuterium nucleus are 2) spontaneously knocked apart. 3) The neutron slows down and gets captured (making tritium or a heavier chlorine), with emission of a gamma. The gamma collides with electrons in the water, giving them a speed faster than the speed of light in water, so these electrons emit Cherenkov (shock–wave) radiation detected by phototubes.

$$\Gamma \equiv \frac{\langle \chi_k | \langle \phi_n | \hat{\rho}(dt) | \phi_n \rangle | \chi_k \rangle}{dt}$$

$$= \frac{\langle \chi_k | \langle \phi_n | [\hat{\rho}(dt) - \hat{\rho}(0)] | \phi_n \rangle | \chi_k \rangle}{dt} = \langle \chi_k | \langle \phi_n | \frac{d\rho(t)}{dt} | \phi_n \rangle | \chi_k \rangle_{t=0}, \qquad (6.1)$$

where, in going to the second line, we have used

$$\langle \chi_k | \langle \phi_n | \hat{\rho}(0) | \phi_n \rangle | \chi_k \rangle = \langle \chi_k | \langle \phi_n | \phi_0 \rangle | \chi_0 \rangle \langle \chi_0 | \langle \phi_0 | \phi_n \rangle | \chi_k \rangle = 0$$

since, regardless of the value of $\langle \chi_0 | \chi_k \rangle$, we have $\langle \phi_0 | \phi_n \rangle = 0$ (which will be responsible for terms vanishing in expressions below).

We treat all particles with a unified notation. The position basis $|\mathbf{x}\rangle \equiv |\mathbf{x}_1...\mathbf{x}_N\rangle$ describes all the particles in the atom: particles 1 to Z are electrons, the next Z are protons and the last are $A - Z$ neutrons (the atomic number is A) so $N = Z + N_p + N_n = Z + A$. The center of mass operator $\hat{\mathbf{X}}_{\text{cm}}$ and relative coordinate operators are respectively

$$\hat{\mathbf{X}}_{\text{cm}} \equiv \frac{1}{m_e Z + m_p N_p + m_n N_n} \sum_{i=1}^{N} m_i \hat{\mathbf{X}}_i$$

$$\hat{\mathbf{R}}_i \equiv \hat{\mathbf{X}}_i - \hat{\mathbf{X}}_{\text{cm}}, \text{ so } \sum_{i=1}^{N} m_i \hat{\mathbf{R}}_i = 0. \qquad (6.2)$$

The density matrix evolution equation, applied to this most general case, is Eq. (4.13) with factors $g_i g_j$ inserted:

$$\frac{d}{dt}\hat{\rho}(t) = -i[\hat{H}, \hat{\rho}(t)]$$

$$-\frac{\lambda}{2} \sum_{i,j=1}^{N} g_i g_j \left[e^{-\frac{(\hat{\mathbf{X}}_i - \hat{\mathbf{X}}_j)^2}{4a^2}} \hat{\rho}(t) + \hat{\rho}(t) e^{-\frac{(\hat{\mathbf{X}}_i - \hat{\mathbf{X}}_j)^2}{4a^2}} - 2e^{-\frac{(\hat{\mathbf{X}}_{iL} - \hat{\mathbf{X}}_{jR})^2}{4a^2}} \hat{\rho}(t) \right].$$

$$(6.3)$$

To find (6.1), using (6.3) we calculate $\langle \chi_k | \langle \phi_n | \frac{d\rho(t)}{dt} | \phi_n \rangle | \chi_k \rangle_{t=0}$. Setting

$$\rho(0) = |\phi_0\rangle |\chi_0\rangle \langle \phi_0 | \langle \chi_0 |,$$

the matrix elements of the first terms in the bracket vanish, leaving

$$\Gamma = -\frac{\lambda}{2} \sum_{i,j=1}^{N} g_i g_j \left[-2\langle \chi_k | \langle \phi_n | e^{-\frac{(\hat{\mathbf{X}}_{iL} - \hat{\mathbf{X}}_{jR})^2}{4a^2}} \hat{\rho}(0) | \phi_n \rangle | \chi_k \rangle \right]. \qquad (6.4)$$

Now, we expand the exponential in (6.4). This gives a series in powers of order $(\ell/a)^2$, where ℓ is either the atomic size or nuclear size, depending on which excited state we are considering. For atoms $(\ell/a)^2 \approx (10^{-8}/10^{-5})^2 = 10^{-6}$, and for nucleons

$(\ell/a)^2 \approx (10^{-12}/10^{-5})^2 = 10^{-14}$. So, in either case, to high accuracy we may retain just the leading term:

$$\Gamma \approx \lambda \sum_{i,j=1}^{N} g_i g_j \langle \chi_k | \langle \phi_n | \left[1 - \frac{(\hat{\mathbf{X}}_{iL} - \hat{\mathbf{X}}_{jR})^2}{4a^2} \right] \left[|\phi_0\rangle |\chi_0\rangle \langle \chi_0 | \langle \phi_0 | \right] |\phi_n\rangle |\chi_k\rangle.$$

The 1 in the bracket gives a vanishing contribution. Letting the $(\hat{\mathbf{X}}_{iL}, \hat{\mathbf{X}}_{jR})$ operators act to the left and right of the density matrix, only the cross term does not vanish:

$$\Gamma = \frac{\lambda}{4a^2} \sum_{i,j=1}^{N} g_i g_j \Big[- \langle \chi_k | \langle \phi_n | \hat{\mathbf{X}}_i^2 | \phi_0\rangle |\chi_0\rangle \langle \chi_0 | \langle \phi_0 | \phi_n\rangle |\chi_k\rangle$$

$$- \langle \chi_k | \langle \phi_n | \phi_0\rangle |\chi_0\rangle \langle \chi_0 | \langle \phi_0 | \hat{\mathbf{X}}_j^2 | \phi_n\rangle |\chi_k\rangle + 2\langle \chi_k | \langle \phi_n | \hat{\mathbf{X}}_i | \phi_0\rangle |\chi_0\rangle \langle \chi_0 | \langle \phi_0 | \hat{\mathbf{X}}_j | \phi_n\rangle |\chi_k\rangle \Big]$$

$$= \frac{\lambda}{2a^2} \langle \chi_k | \langle \phi_n | \sum_{i=1}^{N} g_i \hat{\mathbf{X}}_i | \phi_0\rangle |\chi_0\rangle \langle \chi_0 | \langle \phi_0 | \sum_{j=1}^{N} g_j \hat{\mathbf{X}}_j | \phi_n\rangle |\chi_k\rangle. \tag{6.5}$$

We now observe that, *if there is mass-proportionality, so* $g_i \sim m_i$*, the expression* (6.5) *vanishes.*[5] For, if $\sum_{i=1}^{N} g_i \hat{\mathbf{X}}_i \sim \sum_{i=1}^{N} m_i \hat{\mathbf{X}}_i$, then by (6.2), $\sum_{i=1}^{N} m_i \hat{\mathbf{X}}_i \sim \hat{\mathbf{X}}_{\mathrm{cm}}$, and the resulting matrix element is:

$$\langle \chi_k | \langle \phi_n | \sum_{i=1}^{N} m_i \hat{\mathbf{X}}_i | \phi_0\rangle |\chi_0\rangle \sim \langle \chi_k | \hat{\mathbf{X}}_{\mathrm{cm}} | \chi_0\rangle \langle \phi_n | \phi_0\rangle = 0.$$

In that case, to order $(\ell/a)^2$, there is no excitation of any bound state!

To get a non–zero excitation rate, it is necessary to expand the exponential in (6.4) to the next order, obtaining a rate $\sim \ell^4/a^4$, so small that it is presently far from being measurable.

This is why mass-proportionality is suggested by experiment. The smaller the experimental upper bound on the excitation rate Γ, the closer g_i has to be to m_i.

To finish evaluation of Eq.(6.5), we use (6.2) to write particle coordinates in terms of relative and cm coordinates, $\hat{\mathbf{X}}_i \equiv \hat{\mathbf{R}}_i + \hat{\mathbf{X}}_{\mathrm{cm}}$.

$$\Gamma = \frac{\lambda}{2a^2} \sum_{i,j=1}^{N} g_i g_j |\langle \chi_k | \chi_0\rangle|^2 \langle \phi_n | \hat{\mathbf{R}}_i | \phi_0\rangle \langle \phi_0 | \hat{\mathbf{R}}_j | \phi_n\rangle \tag{6.6}$$

The cm operator has disappeared from the equation, since its matrix elements are multiplied by $\langle \phi_n | \phi_0\rangle = 0$.

Eq.(6.6) is our general result for the excitation rate of a bound state. We now apply it to the two experiments.

---

[5] Pearle and Squires (1994).

6.2.2 Electron Excitation: Spontaneous Radiation in Germanium

What is expected to happen if an atom within a chunk of Ge spontaneously *ejects* a 1s (ground state) electron? The other electrons in the atom readjust, jumping down levels, emitting photons with total energy 11.1 keV, essentially the ionization energy of the 1s electron. These photons travel through the Ge slab, knocking electrons out of encountered atoms. Moreover, the emergent 1s electron, with energy E, does the same. The net result is that, within the chunk of Ge, there is a bunch of ejected electrons providing a current pulse of energy $(11.1 + E)$ keV.

For quite a few years, experiments have been looking for such sudden pulses of electrons, due to X-rays suddenly appearing within a chunk of Ge in a very low background setting over a long time interval.[6] The most recent data[7] useful to us presents a graph of observations of 2 kg of ^{76}Ge over 40 days, including the region of interest to us, 10-20 keV. The data shows counts of the order of 0.1 ± 0.05 counts per keV per kg per day in that energy region. The experimenters expected that the only counts would be due to experimental noise. Take the most conservative position and assume what is observed is all due to CSL excitation. If we take $\lambda = 10^{-16} s^{-1}$, an upper limit on g_e is obtained as follows.

We assume mass–proportionality for the nucleons, $g_n = \frac{m_n}{M}, g_p = \frac{m_p}{M}$, but not for the electrons.

It follows from Eq. (6.2) that

$$m_e \sum_{j=1}^{Z} \hat{\mathbf{R}}_j + \sum_{j=Z+1}^{N} m_j \hat{\mathbf{R}}_j = 0.$$

Using this, the quantity $\sum_{j=1}^{N} g_j \hat{\mathbf{R}}_j$ in Eq. (6.6) may be written as

$$\sum_{j=1}^{N} g_j \hat{\mathbf{R}}_j = \sum_{j=1}^{Z} g_j \hat{\mathbf{R}}_j + \sum_{j=Z+1}^{N} \frac{m_j}{M} \hat{\mathbf{R}}_j = g_e \sum_{j=1}^{Z} \hat{\mathbf{R}}_j - \frac{m_e}{M} \sum_{j=1}^{Z} \hat{\mathbf{R}}_j$$

$$= \left[g_e - \frac{m_e}{M} \right] \sum_{j=1}^{Z} \hat{\mathbf{R}}_j,$$

so dependence on $\hat{\mathbf{R}}_i$ for nucleons has been eliminated! The electron excitation rate (6.6) is therefore expressed solely in terms of the matrix element of the electron position operator between ground and excited state:

$$\Gamma = \frac{\lambda}{2a^2} \left[g_e - \frac{m_e}{M} \right]^2 \sum_{i,j=1}^{Z} |\langle \chi_k | \chi_0 \rangle|^2 \langle \phi_n | \hat{\mathbf{R}}_i | \phi_0 \rangle \langle \phi_0 | \hat{\mathbf{R}}_j | \phi_n \rangle,$$

where $|\phi_n\rangle$ describes a free particle of energy E.

---

[6]These experiments (e.g., Miley et al. (1990), Garcia et al. (1995), Arnquist et al. (2022)) have been looking for double beta-decay of a Ge nucleus, or for a dark matter Wimp collision with a Ge nucleus.

[7]Aalseth et al. (2002), Irastorza et al. (2002).

The remainder of this calculation is complicated, so further analysis is relegated to Chapter 13, the Supplement to this chapter. In broad outline, one integrates over all possible states of the cm and over all directions in which the ejected 1s electron can emerge with energy between $E, E + dE$. Since this is the probability/sec of excitation in an infinitesimal energy range dE about E, it is appropriate to rename it $d\Gamma(E)$. Finally, evaluation of the matrix elements results in[8]

$$\frac{d\Gamma(E)}{dE} = \frac{\lambda}{2}\left(\frac{a_0/Z}{a}\right)^2\left(g_e - \frac{m_e}{M}\right)^2\gamma(E). \tag{6.7}$$

In Eq. (6.7), $a_0 \approx 0.53 \times 10^{-8}$cm is the Bohr radius and $\frac{a_0}{Z}$ is the mean radius of the Ge atom's 1s electron wave function. $\gamma(E)$ is calculated to have the values in keV^{-1} of $\gamma(0) \approx 0.55, \gamma(3) \approx 0.4, \gamma(5) \approx 0.15$ and $\gamma(7) \approx 0.1$, corresponding respectively to X-ray pulses in keV of $11.1, 14.1, 16.1$ and 18.1.

Eq.(6.7) gives the counts/sec for one atom. In order to compare with experiment, we must multiply by the number of atoms per kg of Ge, $\approx 8 \times 10^{24}$ and by the number of seconds per day, $\approx 8.6 \times 10^4$. Thus (6.7) yields

$$\frac{d\Gamma(E)}{dE}\text{counts/keV/kg/day} \approx \frac{10^{-16}}{2}\left(\frac{0.53 \times 10^{-8}/32}{10^{-5}}\right)^2\left(g_e - \frac{m_e}{M}\right)^2\gamma(E)$$
$$\cdot 8 \times 10^{24} \times 8.6 \times 10^4$$
$$\approx 10^4\left(g_e - \frac{m_e}{M}\right)^2\gamma(E).$$

Setting $\gamma(E) \approx 0.5$ and the experimental value at 0.1 counts per keV per kg per day, we obtain the constraint:

$$0.1 \gtrsim 10^4\left(g_e - \frac{m_e}{M}\right)^2 0.5 \text{ or } 5 \times 10^{-3} \gtrsim \left|g_e - \frac{m_e}{M}\right| \text{ or } 11\frac{m_e}{M} \gtrsim g_e. \tag{6.8}$$

(The last step uses $\frac{m_e}{M} \approx 5 \times 10^{-4}$ so $5 \times 10^{-3} = 10\frac{m_e}{M}$).

The inequality (6.8) is certainly consistent with mass-proportionality, $\frac{m_e}{M} = g_e$, and far from the assumption that the electron has the same collapse rate as a nucleon, $1 = g_e$.[9]

6.2.3 Nuclear Excitation: Dissociation of Deuterium

We next apply (6.5), with $Z = 1$ and $N = 2$, (ignoring the electron) to CSL breaking apart deuterium.[10] $|\phi_0\rangle$ is the ground state of the deuterium nucleus and $|\phi_n\rangle$ is a dissociated state, the neutron and the proton moving freely, so the complete set of states is normed as $|\phi_0\rangle\langle\phi_0| + \sum_n |\phi_n\rangle\langle\phi_n| = \hat{1}$ (again, the discrete index n is for notational convenience only, since these are continuum eigenstates of energy).

---

[8]Collett et al. (1995), Pearle et al. (1999b).

[9]It would require greatly increased experimental accuracy to do much better, which is very difficult to achieve. But, were the experiment 10 or 100 times more accurate, so the upper limit would be 0 .01 or 0.001count/keV/kg/day, then the constraint would respectively be $2.1 \frac{m_e}{m_n} \gtrsim g_e \gtrsim 0$.

[10]Jones, Pearle, and Ring (2004).

$\hat{\mathbf{X}}_p$, $\hat{\mathbf{X}}_n$ are the proton and neutron position operators. The cm operator and the relative position operator are then respectively

$$\hat{\mathbf{Q}}_{\text{cm}} = \frac{1}{m_p + m_n}[m_p\hat{\mathbf{X}}_p + m_n\hat{\mathbf{X}}_n]$$

and $\hat{\mathbf{R}} \equiv \hat{\mathbf{X}}_p - \hat{\mathbf{X}}_n$. The proton and neutron coordinates relative to the center of mass can be written in terms of $\hat{\mathbf{R}}$:

$$\hat{\mathbf{R}}_p \equiv \hat{\mathbf{X}}_p - \hat{\mathbf{Q}}_{\text{cm}} = \frac{1}{m_p + m_n}m_n\hat{\mathbf{R}},$$

$$\hat{\mathbf{R}}_n \equiv \hat{\mathbf{X}}_n - \hat{\mathbf{Q}}_{\text{cm}} = -\frac{1}{m_p + m_n}m_p\hat{\mathbf{R}}.$$

Using these relations, the operator that appears in the expression (6.5) for the excitation rate Γ can be written in terms of $\hat{\mathbf{R}}$ and $\hat{\mathbf{Q}}_{\text{cm}}$:

$$g_p\hat{\mathbf{X}}_p + g_n\hat{\mathbf{X}}_n = g_p\hat{\mathbf{R}}_p + g_n\hat{\mathbf{R}}_n + (g_p + g_n)\hat{\mathbf{Q}}_{\text{cm}}$$

$$= \frac{1}{m_p + m_n}(g_p m_n - g_n m_p)\hat{\mathbf{R}} + (g_p + g_n)\hat{\mathbf{Q}}_{\text{cm}}.$$

As before, because the excited states are orthogonal, the matrix element of $\hat{\mathbf{Q}}_{\text{cm}}$ vanishes:

$$\langle\chi_k|\langle\phi_n|\hat{\mathbf{Q}}_{\text{cm}}|\phi_0\rangle|\chi_0\rangle = \langle\chi_k|\hat{\mathbf{Q}}_{\text{cm}}|\chi_0\rangle\langle\phi_n|\phi_0\rangle = 0.$$

The resulting transition rate (6.5) of the deuterium nucleus to a dissociated state $|n\rangle$, summed over all possible final cm states, is therefore

$$\Gamma_n = \frac{\lambda}{2a^2}|\langle\phi_n|\left[\frac{1}{m_p + m_n}(g_p m_n - g_n m_p)\hat{\mathbf{R}}\right]|\phi_0\rangle|^2 \sum_k |\langle\chi_k|\chi_0\rangle|^2$$

$$= \frac{\lambda}{2a^2}\left[\frac{1}{m_p + m_n}(g_p m_n - g_n m_p)\right]^2 |\langle\phi_n|\hat{\mathbf{R}}|\phi_0\rangle|^2$$

$$= \frac{\lambda g_n^2}{2a^2}\left[\frac{\frac{g_p}{g_n} - \frac{m_n}{m_n}}{1 + \frac{m_p}{m_n}}\right]^2 |\langle\phi_n|\hat{\mathbf{R}}|\phi_0\rangle|^2$$

where, in going from the first step to the second step, we have used

$$\sum_k |\langle\chi_k|\chi_0\rangle|^2 = \langle\chi_0|\sum_k\left[|\chi_k\rangle\langle\chi_k|\right]|\chi_0\rangle = \langle\chi_0|\hat{1}|\chi_0\rangle = 1.$$

In Section 4.1, it was shown that λ is the collapse rate for a nucleon whose wave function describes it as in a superposition two widely separated places. There mass proportionality was assumed. Had we used there g_n instead of $m_n/M = 1$, we would have gotten the result that the neucleon collapse rate is λg_n^2. Since we wish the collapse rate for such a neutron state to be $\lambda = 10^{-16}\text{cm}^{-1}$, we will replace λg_n^2 by λ in the equation above:

$$\Gamma_n = \frac{\lambda}{2a^2}\left[\frac{\frac{g_p}{g_n} - \frac{m_p}{m_n}}{1 + \frac{m_p}{m_n}}\right]^2 |\langle\phi_n|\hat{\mathbf{R}}|\phi_0\rangle|^2. \tag{6.9}$$

A dissociated state is a continuum state of relative momentum $\mathbf{k}$ so we make the replacement $|\phi_n\rangle \to \sqrt{d\mathbf{k}}|\mathbf{k}\rangle$, thereby preserving the norm of the states:

$$\hat{1} = \sum_n |\phi_n\rangle\langle\phi_n| \to |\phi_0\rangle\langle\phi_0| + \int d\mathbf{k}|\mathbf{k}\rangle\langle\mathbf{k}|.$$

To obtain the dissociation rate regardless of the final state, we have to sum (6.9) over the set of free-particle states labeled by n, which now becomes an integral over $\mathbf{k}$;

$$\int d\mathbf{k}|\langle\mathbf{k}|\hat{\mathbf{R}}|\phi_0\rangle|^2 = \langle\phi_0|\hat{\mathbf{R}}\int d\mathbf{k}\Big[|\mathbf{k}\rangle\langle\mathbf{k}|\Big]\hat{\mathbf{R}}|\phi_0\rangle = \langle\phi_0|\hat{\mathbf{R}}\Big[1 - |\phi_0\rangle\langle\phi_0|\Big]|\hat{\mathbf{R}}|\phi_0\rangle$$

$$= \langle\phi_0|\hat{\mathbf{R}}^2|\phi_0\rangle - \langle\phi_0|\hat{\mathbf{R}}|\phi_0\rangle^2 = \langle\phi_0|\hat{\mathbf{R}}^2|\phi_0\rangle$$

since $\langle\phi_0|\hat{\mathbf{R}}|\phi_0\rangle = 0$ (the wave function of the deuterium nucleus is centered at $\mathbf{R} = 0$). Thus, the probability/sec (6.9) of dissociating the deuterium regardless of outgoing relative momentum is

$$\Gamma = \frac{\lambda}{2a^2}\left[\frac{\frac{g_p}{g_n} - \frac{m_p}{m_n}}{1 + \frac{m_p}{m_n}}\right]^2 \langle\phi_0|\hat{\mathbf{R}}^2|\phi_0\rangle \tag{6.10}$$

Nuclear models give $\langle\phi_0|\hat{\mathbf{R}}^2|\phi_0\rangle \approx (3 \times 10^{-13}\text{cm})^2$, the square of a typical nucleus size. If a volume of V cubic meters of heavy water, D_2O, whose density is $\approx \frac{2}{3} \times 10^{29} D_2$ atoms/m^3, is observed for Tyr ($\approx 3 \cdot 10^7$sec/yr), the predicted number of observed dissociations N is the rate (6.10) multiplied by the number of atoms and the seconds observed:

$$N = \frac{10^{-16}}{2 \times (10^{-5})^2}\left[\frac{\frac{g_p}{g_n} - \frac{m_p}{m_n}}{1 + \frac{m_p}{m_n}}\right]^2 [3 \times 10^{-13}]^2 \times \frac{2}{3} \times 10^{29} V(3 \times 10^7)T$$

$$\approx 10^5 \left[\frac{\frac{g_p}{g_n} - \frac{m_p}{m_n}}{1 + \frac{m_p}{m_n}}\right]^2 VT. \tag{6.11}$$

We use the data of an experiment performed at the Sudbery Neutrino Observatory.[11] A 12 m diameter shell contained heavy water and was surrounded by photo-tubes. The experiment was designed to observe solar neutrinos (and detect neutrino oscillations), which occasionally would collide with a deuterium atom and dissociate it. We assune that the standard solar model correctly predicted the events where the neutrinos dissociated deuterium, and that the CSL dissociation occurred alongside.

It was the ejected neutrons that were observed. They collided with atoms and were slowed down (were thermalized) as they travelled through the bath, until they were

---

[11]SNO Collaboration (2004).

either captured by a deuterium nucleus to make tritium, with attendant emission of a γ, or captured by a ^{35}Cl nucleus (the water was salted), also with attendant emission of a gamma pulse. The gammas then Compton scattered off electrons in the bath whose Cherenkov radiation was detected by the phototubes.

The number of detected ejected neutrons, over $T = 254.2\text{days} = 0.70\text{yr}$, multiplied by an efficiency factor (more events occurred than were counted) were observed to be $N_{\text{expt}} = 3360 \pm 300$. The theoretical prediction of the standard solar model is $N_{\text{ssm}} \sim 3300 \pm 600$. Therefore, the CSL contribution could have been as much as (adding the experimental and theoretical errors in quadrature as is usually done), $N_{\text{expt}} - N_{\text{ssm}} \approx 3360 - 3300 + \sqrt{300^2 + 600^2} \approx 700$. Putting this into (6.11), with the volume $V = \frac{4\pi}{3}6^3 \approx 900\text{m}^3$, we obtain the constraint

$$700 \gtrsim 10^5 \left[\frac{\frac{g_p}{g_n} - \frac{m_p}{m_n}}{2}\right]^2 900 \times 0.7 \text{ or } 0.007 \gtrsim \left|\frac{g_p}{g_n} - \frac{m_p}{m_n}\right| \text{ or}$$

$$\frac{g_p}{g_n} = \frac{m_p}{m_n} \pm 0.007 = 0.9986 \pm 0.007.$$

$$(6.12)$$

Thus the experiment constrains the collapse rate for the proton to be within 1% of mass proportionality.

We have gone over the calculations in the last two sections in such detail because of their importance. For their SL theory, GRW chose the collapse rate to be the same for all particles, regardless of their mass (Chapter 5). Following their lead, I did the same initially for CSL. But, here we have experimental evidence that this cannot be, strongly suggesting that collapse depend upon mass density, not upon particle number density. I know of no other theory meant to resolve the "measurement problem" that has been so decisively affected by experiment.

6.3   Gravity?

A mass–proportional interaction is suggestive of a link to gravity. So, let us pause in our discussion of experimental tests to take a brief theoretical excursion into various proposals of a relation between wave function collapse and gravity.

In an early paper on collapse dynamics, I concluded[12]

> But, ultimately a theory such as this needs to be legitimized by being a consequence of a larger theory that has more ties to established physics...the juncture between general relativity, which describes events but does not describe microscopic behavior, and quantum theory which describes microscopic behavior but does not describe events, might be an attractive place to look.

For some time[13] Roger Penrose has argued that

> a macroscopic quantum superposition of two differing mass distributions is *unstable* (analogous to an unstable particle). Accordingly, such a state would decay, after a characteristic lifetime T, into one or the other of the two states.

---

[12]Pearle (1979).
[13]Penrose (1994a, 1994b, 1996), Fuentes and Penrose (2018).

His idea is that, in the context of an imagined framework that combines quantum theory with general relativity, each term in the superposition has a different space-time metric. He proposes that the actual metric is chosen by collapse to one or the other metric. This is to occur at a rate determined by the energy–time uncertainty relation. Let T^{-1} be the collapse rate, and ΔU be the gravitational potential self–energy of a mass distribution equal to the difference of the two mass-distributions:

$$\Delta U = -G \int d\mathbf{x}d\mathbf{x}'[\rho_1(\mathbf{x}) - \rho_2(\mathbf{x})][\rho_1(\mathbf{x}') - \rho_2(\mathbf{x}')]/|\mathbf{x} - \mathbf{x}'|.$$

The energy time uncertainty relation is $T\Delta U = \frac{\hbar}{4\pi}$, so the proposed time it takes to collapse is

$$T = \frac{\hbar}{4\pi|\Delta U|}.$$

Thus, the bigger the mass placed in superposition, the smaller the duration of the collapse T. This is qualitatively the same behavior as in CSL, but it has the virtue of there being just one, known, parameter. However, Penrose did not suggest how his idea might be applied to more general situations, or propose any altered dynamics of the state vector due to this mechanism.

Let us apply this to the superposed state of the barely visible carbon sphere (mass $m \approx 6 \times 10^{-14}$gm, radius $R \approx 10^{-5}$cm) used to illustrate the theoretical constraint in Section 6.1. If we consider a large displacement $>> R$, then ΔU is twice the gravitational self-energy of such a sphere, $\Delta U = -3Gm^2/R$, which gives the result $T \approx 1s$. While T is ten times larger than the perception time 0.1 s we used in our discussion of the theoretical constraint, since that is a fairly rough criterion we may give this a pass. The collapse time is so long because the collapse time is proportional to the inverse of the gravitational constant, which is small.

The earliest consideration of a connection between gravity and collapse was made in the thesis of Frigyes Károlyházy.[14] He argued that the center of mass wave function of a freely moving object alternates between Schrödinger equation expansion and collapse brought about by an uncertainty in the metric. While not a systematic approach to collapse, it predicted random walk of the object's center of mass, a behavior shared with the CSL description. Interestingly contemporaneously, the first systematic dynamical collapse model was offered by David Bohm and Jeffrey Bub,[15] with however no gravitational connection nor random behavior: the state vector evolution is deterministic, the outcomes depending upon "Wiener-Siegal" hidden-variables.

A concrete gravitationally related dynamical proposal was made by Lajos Diosi.[16] He introduced a stochastic differential equation for the state vector similar to that of non–relativistic CSL, whose density matrix evolution equation has the form of (4.5), with the Gaussian replaced by $\frac{1}{|\mathbf{z}-\mathbf{z}'|}$, densities replaced by particle densities modeled as spheres of their Compton wavelength size and λ replaced by the gravitational constant G. He called it "quantum mechanics with universal density localizations"

[14]Károlyházy(1966), Károlyházy, Frenkel, and Lukács (1986).
[15]Bohm and Bub (1966).
[16]Diosi (1989).

(QMUDL). Whereas the value of the CSL collapse rate λ is unspecified, he wrote that the "strength is proportional to the gravitational constant. This choice makes QMUDL a parameter free unification of micro and macrodynamics."

But, it was pointed out,[17] because the particle sizes are so small in this model, that the energy produced by collapse is too much. The CSL result (4.17), applied to a nucleon $\frac{dE}{dt} \sim \lambda \frac{\hbar^2}{Ma^2} \approx 10^{-36}$erg/s, in this model is replaced by $\frac{dE}{dt} \sim \frac{GM^2}{\hbar r_c} \frac{\hbar^2}{Mr_c^2} \approx 10^{-19}$erg/s ($r_c \approx 10^{-13}$cm is the Compton wavelength). That is enough to heat a cup of water by $1°$K in about 7 minutes.[18] They went on to point out a cure, smearing the mass density with a Gaussian of width a as in SL and CSL. But, that removes the "parameter free" nature of the model Diosi was hoping to achieve.

However, applied as in Section 4.4, to a homogeneous object of N particles and volume V, of dimensions larger than a, initially in the superposition $|\psi, 0\rangle = \frac{1}{\sqrt{2}}[|L\rangle + |R\rangle]$, the density matrix dependence on a essentially disappears as Diosi wished, with the result (4.9), (4.10) replaced by

$$\langle L|\rho(t)|R\rangle = \frac{1}{2}e^{-\Gamma t}, \text{ where } \Gamma \approx \frac{G}{\hbar}\left(\frac{NM}{V}\right)^2\left\{\int_{V_L}\int_{V_L} - \int_{V_L}\int_{V_R}\right\}d\mathbf{z}d\mathbf{z}'\frac{1}{|\mathbf{z}-\mathbf{z}'|}.$$

$\Gamma\hbar$ is just the change in gravitational potential energy, ΔU, of two initially overlapping copies of the object, moved to their new locations at L and R. So, this is a CSL-type embodiment of Penrose's proposal.

Euan Squires and I suggested[19] a relation between CSL and gravity, namely a gravitational cause of the random field $w(\mathbf{x}, t)$. We hypothesized that the source of the gravitational potential is not only the mass of particles, but also mass fluctuations arising from the vacuum. We modeled these fluctuations as randomly appearing and disappearing Planck masses, which exert a random gravitational force on a freely moving object.

As mentioned before Eq. (3.22), a random force without viscous damping results in the object's energy growing linearly with time. By equating that energy increase to the CSL energy increase, and relating the fluctuating gravitational potential to $w(\mathbf{x}, t)$, we obtained equations for λ and a in terms of two model parameters. These parameters, specifying the random motion of the Planck masses, are a characteristic time to appear and disappear and a characteristic length, their mean separation. These were respectively chosen to be the Planck time and the reduced Compton length $\lambda_C = h/Mc$ of a nucleon.

Using these, we obtained expressions for the CSL parameters (m_P is the Planck mass):

$$a = (3/\pi^2)^{1/4}(\lambda_C/4)(m_P/M)^{1/2} \approx 1.4 \times 10^{-5}\text{cm}$$

(essentially the CSL choice for a) and

$$\lambda = (3\pi)^{-1/2}(GM^2/a\hbar) \approx 2 \times 10^{-24}$$

[17]Ghirardi, Grassi, and Rimini (1990b), Bahrami, Smirne, and Bassi (2014).

[18]A gram of water contains around 10^{24} nucleons, so their collapse dynamics supplies $\approx 10^{-19} \times 10^{24} = 10^5$erg/s. The specific heat of water is $\approx 4 \times 10^7$erg/gram/degree, so it takes ≈ 400s for 10^5erg/sec to heat a gram of water by 1 degree.

[19]Pearle and Squires (1996).

(up to a numerical factor, the Diosi collapse rate for a nucleon). So, this is a parameter–free model which suggests expressions for the CSL parameters in terms of known physical constants.

Stephen Adler has recently suggested that the randomly fluctuating part of the metric be imaginary, in order that the anti-Hermitian collapse term of the Schrödinger equation be obtained, and this idea involving complex general relativity has received some further development.[20]

Daniel Sudarsky[21] and coworkers have argued that a number of problems involved in combining general relativity and quantum theory could be solved by invoking collapse dynamics. For example, in a quantum theory of the early universe (involving the evolution of a quantum field, called the inflaton field, which ultimately decays, generating the particles with which we are familiar) the state vector evolves into a superposition of possible universes. The addition of CSL collapse dynamics provides a mechanism that chooses one such universe.[22]

Recently, an approach[23] borrowed the non-relativistic CSL collapse mechanism for a model describing collapse in the early universe. The authors argued that their model's prediction of the cosmological microwave radiation conflicted with experiment, and that this defect implies a defect in non-relativistic CSL. The validity of this argument has been contested.[24]

These are some examples of ideas that connect gravity and collapse.

Penrose has expressed qualms about the energy increase in CSL, and thinks it would not occur when a deeper understanding of the relation between gravity and collapse is obtained. He writes about his own efforts:

> It should be made clear, however, that this proposal does not provide a theory of quantum state reduction. It merely indicates the level at which deviations from standard linear Schrödinger (unitary) evolution are to be expected owing to gravitational effects. Indeed, it is this author's personal opinion that the correct theory uniting general relativity with quantum mechanics will involve a major change in our physical world–view — of a magnitude at least comparable with that involved in the shift from Newtonian to Einsteinian gravitational physics.

This suggests that looking for a deeper, gravitational, basis for the phenomenological dynamical collapse models may provide a path toward a quantum theory of gravity.

6.4 Special Relativity?

There is a vast array of special–relativity affirming experiments: particle physics is rife with them. But, the macroscopic parts of the apparatuses placed in a superposition by the experiments are non–relativistic devices. Thus far, combining non–relativistic CSL with the correct relativistic Hamiltonian has not led to an experimental conflict. The Hamiltonian mostly governs the behavior of the relativistic particles (the collapse effect

---

[20] Adler (2016), Gasbarri et al. (2017).
[21] Okon and Sudarsky (2014), Modak et al. (2015), Josset, Perez, and Sudarsky (1917).
[22] Canate, Pearle, and Sudarsky (2013).
[23] Martin and Vennin (2020), (2021).
[24] Bengochea et al. (2020), Gundhi et al. (2021).

is negligible for such low mass-density objects) while the apparatus's non–collapse dynamics utilizes only the Hamiltonian's non-relativistic limit.

Nonetheless, after arriving at non–relativistic CSL, it was natural to try to make a special–relativistic CSL too.[25] And, it is possible, but a serious defect is encountered.

Before discussing this, it is worth taking a brief pause to point out that collapse can look differently from different Lorentz frames. Take as an example a wave packet for a particle that meets a "beam splitter," with the result that half the packet goes to meet a detector to the right, and half goes to meet a detector to the left, both detectors at equal distance from the beam splitter in the rest frame of all these pieces of apparatus.. A superposition results, with the left packet state multiplied by a state where the left apparatus indicates detection and the right apparatus does not (and similarly for the right packet state). The mass density differences in the different apparatus states "trigger" the collapse.

In this reference frame, both packets impact their respective detectors at the same time, and so both trigger the collapse. But time is frame–dependent: from other frames, the left packet impacts its detector first and so triggers the collapse, while from the rest of the frames, it is the right packet that impacts its detector first and triggers the collapse. The "cause" of the collapse is frame–dependent, as are the amplitudes of the packets in the space–time region between impacts. Of course, the final result (left packet detected OR right packet detected) has to be, and is, the same in all frames.

Now we return to the problem of making a relativistic collapse model. The simplest way to apply the CSL formalism to make a special relativistic theory is to choose the collapse-generating operator to be a scalar quantum field associated with particles of mass m, obeying the Klein–Gordon equation. Then, one indeed has a perfectly good relativistic theory, insofar as the collapse toward eigenstates of the field works very well.[26] Unfortunately, along with the collapse, comes production of particles out of the vacuum.

Indeed, in a special relativistic theory, production of just one particle out of the vacuum implies that particles of all momenta must come out of the vacuum.[27] For, if a particle of any momentum is excited out of the vacuum, in another Lorentz frame it looks like a different momentum. But, since the dynamics is relativistically invariant, all reference frames are equivalent, so that different momentum must have also been excited in the original frame. Since all possible momenta appear in all possible reference frames, then all possible momenta must appear in the original frame. What happens is that there is a finite production of particles of each momentum, but all momenta, right on up to infinite magnitude momenta, are produced in any reference frame, which adds up to production of infinite energy per second per unit volume. The calculation that shows this appears in Chapter 13, the Supplement to this chapter. Obviously, this conflicts with observation.

The only way to avoid this is to have a relativistic theory in which the collapse generates no particle production out of the vacuum whatsoever. This was first achieved

---

[25] Pearle (1990), Ghirardi, Grassi, and Pearle (1990c), Pearle (1999), Nicrosini and Rimini (2003).
[26] Bedingham and Pearle (2019).
[27] Pearle (1999, 2015), Myrvold (2017).

in a model by Daniel Bedingham, and I later found a different model.[28] However, these models are somewhat unusual.

Using usual ideas, it is possible to generalize non–relativistic CSL "part way" toward a relativistic dynamical collapse theory for quantum fields.[29] It is possible to find a collapse–generating operator with some good features: no particle production out of the vacuum, the collapse dynamics is valid in a special Lorentz frame in which the particles have relativistic energy–momentum expressions, the non-relativistic limit is non-relativistic CSL. However, invariance under Lorentz transformations is at best only *approximately* obeyed, that is, for "nearby" (not too rapidly moving) reference frames.

Indeed, one might take the point of view that this problem with making a relativistically invariant theory is a message that collapse is not meant to be special-relativistically invariant, it is meant to work in a preferred frame, perhaps the co-moving frame of the universe.

An approach to partial employment of relativistic ideas is to replace the particle mass-density–proportionality in non–relativistic CSL by relativistic particle energy-density–proportionality (divided by c^2). This reduces to non–relativistic CSL when the particles move non-relativistically, but gives a larger collapse contribution when the particles move relativistically. In particular, in non–relativistic CSL, since photons have no mass, they do not participate in collapse dynamics. But, with collapse that is relativistic energy-density–proportional, dynamical collapse due to differences in photon energy densities occurs.[30] Then, there are new predicted effects, such as alteration of photon motion or energy in an intense laser beam or in the cosmic background radiation.

6.5   Motion of an Isolated Object

Now we return to considering experimental consequences of non-relativistic CSL.

We have seen that a particle, moving in one dimension, whose wave function evolves under the free particle Hamiltonian and CSL collapse with $\hat{A} \equiv \hat{X}$, undergoes random walk (Section 3.5, with details in Chapter 10, the Supplement to Chapter 3, and a SDE treatment appears in Chapter 17). The density matrix evolution equation for this is Eq. (3.13) (or Eq. (10.38a)):

$$\frac{d}{dt}\hat{\rho}(t) = -i\frac{1}{2m}[\hat{P}^2, \hat{\rho}(t)] - \frac{\lambda}{2}[\hat{X}, [\hat{X}, \hat{\rho}(t)]]. \tag{6.13}$$

Here we show that not only a particle, but the cm of *any* freely moving object subject to CSL dynamics undergoes "anomalous" random walk.[31] This behavior has the potential to provide an experimental test of CSL because an ensemble of such identical objects "spreads out" over a significantly greater distance in a given time interval than is predicted by the wave function spread of standard quantum theory.

---

[28] Bedingham (2010), Bedingham et al. (2014), Pearle (2015).
[29] Pearle (2005).
[30] Pearle (2018).
[31] Collett and Pearle (2003), Bassi (2005), Pearle (2013).

In non–relativistic CSL, the density matrix evolution (4.6), applied to the cm of a freely moving object, turns out to be a three dimensional version of (6.13). This is shown in Chapter 13, the Supplement to this chapter, for an object assumed homogeneous and composed of N nucleons. If the object is smaller in size than a, the result is Eq. (13.13):

$$\frac{d}{dt}\hat{\rho}_{cm}(t) = \sum_{j=1}^{3}\left\{-i\frac{1}{2NM}[\hat{P}_{cm}^{j2},\hat{\rho}_{cm}(t)] - \frac{\lambda N^2}{4a^2}[\hat{X}_{cm}^{j},[\hat{X}_{cm}^{j},\hat{\rho}_{cm}(t)]]\right\}. \quad (6.14)$$

where $\hat{X}_{cm}^1, \hat{X}_{cm}^2, \hat{X}_{cm}^3$ are the cm position coordinate operators. Thus, all the results obtained for the one-dimensional particle problem are results for each direction of motion of the center of mass for the object, provided we make substitutions for the mass and collapse rate of the particle:

$$m \to NM, \qquad \lambda \to \lambda\frac{N^2}{2a^2}. \quad (6.15)$$

When the object is not limited in size, the generalization of (6.14) is (13.18):

$$\frac{d}{dt}\hat{\rho}_{cm}(t) = \sum_{j=1}^{3}\left\{-i\frac{1}{2Nm}[\hat{P}_{cm}^{j2},\hat{\rho}(t)] - \frac{\lambda N^2 f^j}{4a^2}[\hat{X}_{cm}^{j},[\hat{X}_{cm}^{j},\hat{\rho}(t)]]\right\},$$

where

$$f^j \equiv \frac{1}{V^2}\int_V d\mathbf{z}\int_V d\mathbf{z'}e^{-\frac{(\mathbf{z}-\mathbf{z'})^2}{4a^2}}\left[1 - \frac{1}{2a^2}[z^j - z'^j]^2\right], \quad (6.16)$$

where V is the volume of the object. If the object is a sphere of radius R, one finds from (6.16):

$$f^j\left(\frac{R}{a}\right) = 6\left(\frac{a}{R}\right)^4\left[1 - 2\left(\frac{a}{R}\right)^2 + \left(1 + 2\left(\frac{a}{R}\right)^2\right)e^{-\left(\frac{R}{a}\right)^2}\right]. \quad (6.17)$$

f^j is a monotonically decreasing function of its argument, with $f^j(0) = 1$ (the result (6.14), obtained by expanding the exponential in (6.17) to order $(R/a)^6$), $f^j(1) \approx 0.62$, $f^j(2) \approx 0.2$, $f^j(\frac{R}{a}) \to 6\left(\frac{a}{R}\right)^4$. So, the smaller the object, the larger the effect.

To get a feel for the magnitudes involved, consider the random walk of a blue–wavelength–sized sphere just visible under a microscope, of radius $R = 2 \times 10^{-5}$cm$=$ $2u$, as in Section 0.1. For concreteness, suppose it is made of carbon, from whose density we find it contains $N \approx 4 \times 10^{10}$ nucleons and has mass $NM \approx 6 \times 10^{-14}$gm. The one-dimensional results in Section 3.5 depend upon two constants: s, which characterizes the wave packet equilibrium width and τ, which characterizes the time needed to achieve that width. For this sphere, using the CSL parameter values for λ, a, the equilibrium size of the center-of-mass wave packet and the time to reach that size are given by

$$s = \left[\frac{\hbar}{m\lambda}\right]^{1/4} \to \left[\frac{\hbar}{NM\frac{\lambda N^2}{2a^2}f^j}\right]^{1/4} \approx 10^{-7}\text{cm}, \quad (6.18a)$$

$$\tau = \frac{ms^2}{\hbar} \to \frac{NMs^2}{\hbar} \approx \frac{2}{3}\text{ sec.} \quad (6.18b)$$

Consider an ensemble of such identical spheres (or, one sphere observed multiple times), where the initial center of mass location of each is known, and then each is observed after time T min has elapsed. A measure of the distance travelled, the root mean square displacement in one direction of the spheres, is predicted by (3.24) to be

$$\sqrt{\langle \hat{X}^2 \rangle} = s\sqrt{\frac{t}{\tau}\left[1 + \frac{t}{\tau} + \frac{t^2}{3\tau^2}\right] + 1/2} \approx 5 \times 10^{-5}T^{3/2}\text{cm}, \qquad (6.19)$$

slightly larger than the sphere diameter in one minute. By contrast, the standard quantum mechanical wave function spread of a packet initially of size s in T minutes is $\frac{\hbar t}{2NMs} \approx \frac{10^{-27}60T}{2\times6\cdot10^{-14}\times10^{-7}} = 5 \times 10^{-6}T\text{cm}$.

After one minute, the CSL spread is larger than the quantum mechanical spread by a factor of 10, and as time goes on, the discrepancy gets larger because of the difference in the power of T. However, the longer T is, the more likely it is that ambient gas, even in a good laboratory vacuum, will collide with the sphere and provide an unacceptably large error.

Therefore, it was suggested[32] that a more easily observed behavior is the rotation of a small disc "on edge". Under the CSL dynamics, a disc of thickness 0.5×10^{-5}cm and radius 2×10^{-5}cm was found to randomly walk through 2π radians in about a minute. For comparison, in the same time, according to standard quantum theory, the rotation would be through about 0.06 radians.

Technical issues challenging the experimenter are injection of the object into the viewing area, method of suspension of the object, method of viewing the object, limiting the effect of collisions of the object with ambient gas molecules or photons by keeping the pressure and temperature low and keeping the apparatus vibration low.

Following fairly recent proposals,[33] such an experiment, at room temperature, looking at rotation of a polyethylene glycol 1 micron radius sphere has been performed,[34] with an upper limit obtained, $\lambda < 10^{-6.5}\text{s}^{-1}$.

6.6   Heating

CSL heats up objects. We obtained the beautifully simple expression, Eq. (4.17),

$$\overline{\langle \hat{H} \rangle(t)} = \lambda t M \frac{3}{4}\left[\frac{\hbar}{Ma}\right]^2$$

for the mean energy increase of an object of mass $\mathcal{M}$, entailing no approximations. How much of this is due to center of mass kinetic energy increase and how much is due to internal heating?

We can answer this, at least for a freely moving object constrained not to rotate, because we studied the one-dimensional free particle and found its mean kinetic energy increase in Eq. (3.22) to be $E = \lambda t\frac{\hbar^2}{2m}$. As explained in Section 6.3, this applies to an

---

[32]Collett and Pearle (2003).
[33]Bera et al. (2015), Schrinski, Stickler, and Hornberger (2017), Carlesso et al. (2017).
[34]Zheng et al. (2020).

extended object's cm mean kinetic energy increase, provided we use (6.15) to replace m, λ with their appropriate equivalents (and take the three dimensions into account):

$$KE_{\rm cm} = 3\frac{\lambda N^2}{2a^2} f^j t \frac{\hbar^2}{2NM} = \lambda t \mathcal{M}\frac{3}{4}\left[\frac{\hbar}{Ma}\right]^2 f^j. \qquad (6.20)$$

We see this is smaller than the total mean energy increase by the factor f^j given in Eq. (6.17). Given the monotonically decreasing behavior of f^j with increasing object size (Eq.(6.17) et seq.), we see that the cm of a large ($>> a$) object scarcely budges, so essentially all the mean energy increase is due to internal heating. (If the object's size is less than a, then $f^j \approx 1$ and most of the mean energy increase is due to cm mean kinetic energy increase.)

These statements are also true for a harmonically bound extended object, since the mean energy increase of the one-dimensional harmonically oscillating particle is the same as that for the free particle, and the mean energy increase cited above for the free particle, derived in Section 4.5, holds also for the harmonically bound particle (just replace the free particle Hamiltonian with the harmonic oscillator Hamiltonian).

Experiments measuring the kinetic energy increase of a free or harmonically bound small object have been proposed.[35] The CSL-generated rigid body motion of the very large test masses in gravitational wave detectors have been analyzed, with the observed noise providing the upper limit $\lambda \lesssim 10^{-8}{\rm s}^{-1}$ (at $a = 10^{-5}{\rm cm}$).[36] And, experiments have been performed measuring the energy increase of an oscillating small cantilever with the most recent result[37] $\lambda < 2 \times 10^{-10}{\rm s}^{-1}$.

6.6.1   Temperature Increase of a Large (Dimensions $>> a$) Object

If energy ΔE is pumped into an object, the temperature increase is $\Delta T = \frac{\Delta E}{\mathcal{M}c}$, where $\mathcal{M}$ is its mass, and c is its specific heat. The smallest specific heat readily available (making ΔT as large as possible) is that of gold or lead, slightly larger than $c \approx 0.01{\rm erg/gm°K}$. Eq.(4.17) (Eq.(6.20) with f^j replaced by 1) gives the internal energy increase ΔE for a large (dimensions $>> a$) object over a time interval t. Therefore, the CSL-induced temperature increase in T minutes of a piece of gold or lead of any size is

$$\Delta T = \lambda t \mathcal{M}\frac{3}{4}\left[\frac{\hbar}{Ma}\right]^2 \frac{1}{\mathcal{M}c}$$

$$\approx \lambda(60T)\frac{3}{4}\left[\frac{10^{-27}}{(5/3) \times 10^{-24} \cdot 10^{-5}}\right]^2 \frac{1}{0.01} = 1.6\left[\frac{\lambda}{10^{-16}}\right] T \text{ nK}. \qquad (6.21)$$

(nanoKelvin=10^{-9}K). For example, a measurement placing an upper limit on the temperature increase to less than 1.6 mK (1.6 milliKelvin= 1.6×10^6 nK) per minute would put an upper limit on λ of $\approx 10^{-10}{\rm s}^{-1}$.

---

[35]Adler (2005), Goldwater, Paternostro, and Barker (2016), Li et al. (2017).
[36]Carlesso et al. (2016).
[37]Vinante et al. (2016), (2020).

Such a measurement, perhaps with a resistance temperature detector (RTD) such as platinum (accuracy perhaps 1mK), has to contend with environmental energy transfer (including heating by the current used to measure the detector resistance) and stability. A recently proposal suggests looking for the elevated temperature at the center of a sphere, above the temperature maintained at the boundary, due to this internally generated CSL source of heat.[38]

6.6.2 Heating of a Bose–Einstein Condensate

Bose and Einstein pointed out that a collection of integer angular momentum atoms (such as zero angular momentum ^{133}Cs or ^{87}Rb) should have wave functions symmetric in the particle coordinates, so that any number of them can occupy the same state. In particular, at 0 degrees Kelvin, they will all be in the ground state. Such a state is called a "Bose–Einstein condensate" (BEC).

Experimenters are able to achieve temperatures of the order of a nanoKelvin, and provide a potential to trap a substantial number of atoms in the ground state. (Typically, this is a magnetic or optical trap. The potential is however not solely that of the trap, as a dense BEC creates a substantial self-potential). Since this is not 0 degrees Kelvin, at thermal equilibrium atoms also occupy higher energy, bound states of the potential, giving rise to a "thermal cloud" surrounding the ground state atoms.

CSL heating, along with other competing mechanisms (collision of the atoms in the trap with particles in the environment), causes particle loss from the BEC, as well as from the cloud, with particles escaping the trap.

As the number of trapped atoms decreases, those remaining in the BEC and cloud may be observed at any time by suddenly removing the trap so that the atoms fly off, revealing to detectors their number densities and speeds. If there were no competing heating mechanisms, measurement of the decay of the number of particles in the BEC with time gives a measure of the CSL heating of the BEC atoms.

The calculation, given in Chapter 13, the Supplement to this chapter, proceeds in two steps. The first is to show that, as far as the density matrix is concerned, the atoms behave like point particles. That is, starting with the density matrix evolution equation (4.13) for all the nucleons in all the atoms, each atom effectively acts like a point particle of mass MA, where the atomic number is A (number of protons plus neutrons). The density matrix evolution equation for N_A atoms is shown to be

$$\frac{d}{dt}\langle \mathbf{x}_{cm}|\hat{\rho}_{cm}(t)|\mathbf{x}'_{cm}\rangle = -i\langle \mathbf{x}_{cm}|[\hat{H}_{cm},\hat{\rho}_{cm}(t)]|\mathbf{x}'_{cm}\rangle$$

$$-\frac{\lambda A^2}{2}\sum_{i,j}^{N_A}\left[e^{-\frac{(\mathbf{x}_{cmi}-\mathbf{x}_{cmj})^2}{4a^2}}+e^{-\frac{(\mathbf{x}'_{cmi}-\mathbf{x}'_{cmj})^2}{4a^2}}-2e^{-\frac{(\mathbf{x}_{cmi}-\mathbf{x}'_{cmj})^2}{4a^2}}\right]$$

$$\cdot\langle \mathbf{x}_{cm}|\hat{\rho}_{cm}(t)|\mathbf{x}'_{cm}\rangle.$$

$$(6.22)$$

---

[38] Adler (2017, 2018), Adler and Vinante (2018), Bahrami (2018).

where $|\mathbf{x}_{cm}\rangle \equiv |\mathbf{x}_{cm1}, \mathbf{x}_{cm2}, ... \mathbf{x}_{cmN_A}\rangle$ is the joint position eigenstate for the cm operators of all the atoms. In Eq. (6.22), $A^2 = [\frac{MA}{M}]^2$ replaces the factor $[\frac{m}{M}]^2$ in the usual CSL density matrix evolution equation for identical particles of mass m.

The second step is to show it follows from (6.22), if one starts with all atoms in the ground state, that the number of atoms in that state drops as CSL heating knocks them out according to simply

$$N_0(t) - N_0(0)o^{-\lambda A^2 t} \tag{6.23}$$

(the subscript 0 refers to the ground state) assuming a BEC wave function dimension $>> a$, as is usually the case. So, if the number of atoms in the BEC is observed to decay with time constant τ_{expt}, then $\tau_{expt} \leq \frac{1}{\lambda A^2}$, the equality sign holding if there is no other mechanism but collapse to knock atoms out of the BEC. For example, using (6.23), for ^{133}Cs and $\tau_{expt} = 10s$, the limit $\lambda \leq 6 \times 10^{-6} s^{-1}$ is obtained. A more complex analysis, of experiments taking into account known mechanisms for particle loss, involving the decrease of ^{133}Cs atoms in a BEC plus cloud, resulted in $\lambda \lesssim 10^{-7} s^{-1}$.[39]

A more recent experiment[40] observed the spreading of ^{87}Rb atoms released from a BEC trap, where the CSL heating is expected to increase the speeds of the atoms and so their spread: the authors reported $\lambda \lesssim 5 \times 10^{-8} s^{-1}$.

6.7 Spontaneous Radiation of a Free Charged Particle

When a charged particle accelerates, it radiates. Since, under CSL dynamics, a free particle undergoes random walk, and random walk entails acceleration, a free charged particle should emit "spontaneous" radiation. That radiation rate was calculated by Qijia Fu[41] to be

$$\frac{d\Gamma(E)}{dE} = \frac{1}{\pi} \lambda \frac{e^2 \hbar}{a^2 M^2 c^3} \frac{1}{E}$$
$$= \frac{1}{\pi} \lambda \left[\frac{e^2}{\hbar c}\right] \left[\frac{\hbar^2/Ma^2}{Mc^2}\right] \frac{1}{E} \approx 10^{-20} \lambda \frac{1}{E_{keV}} \text{photons/sec/keV}, \tag{6.24}$$

where $\Gamma(E)$ is the number of emitted photons/sec of energy E,[B] M is the nucleon mass, as usual. Without mass proportionality of the collapse rate, this radiation rate is proportional to m^{-2} where m is the mass of the charged particle (the smaller the mass, under a specified force, the larger the acceleration and therefore the radiation rate) but this is cancelled by the mass-proportional collapse rate $\sim [\frac{m}{M}]^2$.

Fu applied this to the experiment (discussed in Section 6.2.2) that placed an upper limit on X-ray radiation in a chunk of Ge, arguing that the four outermost electrons of each atom are so weakly bound they can be considered as free. With the data then available, the limit $\lambda \lesssim 8 \times 10^{-10} s^{-1}$ was obtained.

[39]Laloé, Mullin, and Pearle (2014).
[40]Billardello et al. (2016).
[41]Fu (1997).

At first Adler argued that application of this to the outermost electrons in Ge should also take into account the spontaneous radiation of the positively charged remainder of the atom, whose phase, he said, would severely cancel the electron radiation. However, he soon realized[42] that this cancellation was negligible for the short wavelengths of the ≈ 10 keV X-rays considered by Fu. In a current embodiment of this experiment,[43] a recent analysis[44] considers that not just 4, but 30 electrons in Ge, excluding the innermost 1s electrons, are free enough for this application for $E > 15$ keV, and they obtain the limit $\lambda < 7 \times 10^{-12}\mathrm{s}^{-1}$.

We can roughly see how this limit arises, using the experimentally obtained upper limit, $R_{\mathrm{expt}} \approx 0.1\mathrm{counts/keV}$-kg-day at $E \approx 10$keV. Eq. (6.24) gives the photon production number per electron per second. We need the number per kg per day, so we must multiply it by the $\approx 8.3 \times 10^{24}$ atoms in a kg of Ge, and then by the 30 contributing electrons, and then by the 8.6×10^4 seconds per day, obtaining the rate

$$R_{\mathrm{theor}} = 2 \times 10^{11}\lambda \frac{1}{E_{\mathrm{keV}}}\mathrm{photons/keV}\text{-kg-day} < R_{\mathrm{expt}} \text{ or } \lambda < 5 \times 10^{-12}\mathrm{s}^{-1}. \quad (6.25)$$

There have been three recent papers concerning this. In one,[45] the authors consider not the radiation due to the Ge electrons, but rather the radiation due to the protons in the Ge nucleus, and take the nucleus to accelerate as a single particle of charge $+32$, so one multiplies Fu's expression ($\sim e^2$ for a single electron) by 32^2. For measured X-ray emission in the range 1000–3,500 keV, they arrive at the limit $\lambda < 5 \times 10^{-13}\mathrm{s}^{-1}$.

In a second paper,[46] with 32.5 kg-year observations in the energy range 19–1000 keV, if the 30 electrons in the Ge atom are considered to be generating the radiation, they arrive at the limit $\lambda < 2 \times 10^{-13}\mathrm{s}^{-1}$ If it is the protons in the Ge nucleus that are considered to be generating the radiation, they arrive at the limit $\lambda < 5 \times 10^{-15}\mathrm{s}^{-1}$.

In a third paper,[47] data obtained in a dark matter search using around 8,000 kg of liquid xenon, data taken in the energy ranges 12–27 keV and 90–140 keV is claimed to yield a limit $\lambda \leq 2 \times 10^{-17}\mathrm{s}^{-1}$.

Since Fu's 1997 work, these considerations have provided the best upper limit on λ.

6.8   Interference

Consider the prediction, in standard quantum theory, of a two-slit interference pattern produced by a beam of identical objects, identically prepared. The object could be a particle or a structure composed of many particles (for example, a molecule or a miniscule metal sphere). How is this prediction altered by CSL dynamics of such an object?

The standard quantum theory description is as follows. With a centered, symmetric beam (assumed overlapping both slits), with a fairly well-defined cm momentum,

---

[42] Adler and Ramazanoglu (2007).
[43] Morales et al. (2002).
[44] Piscicchia et al. (2017).
[45] Donadi et al. (2021).
[46] Arnquist et. al. (2022)
[47] Kim (2002)

the wave function of a single object immediately emerging from the slits (assumed wide enough to allow the object to pass through) will be the sum of equal amplitude wave packets. These packets will spread and eventually overlap and so interfere. The interference pattern will have minima, located where the path length from one slit differs from the path length from the other slit by an integer-plus-a-half wavelength, and maxima where the difference is an integer number of wavelengths.

The squared sum of the two amplitudes at any point gives the probability of detection there, were a detector to be present. Suppose the beam hits a screen which is filled with many detectors. Their initial state is a direct product of states, each detector in the "unenergized" or "not-detected" state. As the state vector evolves, the detectors state becomes entangled with the single object's wave function, producing a superposition of states. Each state describes the object hitting one detector on the screen which is energized, while the rest renain unenergized. Finally, the collapse postulate says the superposition is to be replaced by one of the states in the superposition, occurring with the Born probability. This array of results displays the classic two-slit interference pattern.

The CSL description begins as follows. The two wave packets immediately start playing the Gambler's Ruin game upon emerging from the two slits. Therefore, for each different random field, the amplitudes of the two packets will generally be different by the time the packets hit the screen. Indeed, if the collapse rate is sufficiently large, so that collapse is essentially completed before the packets even overlap, then there is just a single packet hitting the screen, producing no interference pattern at all!

But, even if the Gambler's Ruin game between the packets only provides a small difference in the packet amplitudes before they reach the detector screen, the interference pattern resulting upon detection will not be the same as the classic equal–amplitude two–slit pattern. In the different evolutions governed by the different random fields, the packets are overwhelmingly likely to have different amplitudes. The interference pattern resulting from all these different evolutions will be slightly "washed out" compared to the classic two–slit interference pattern, with lowered contrast (the ratio of peak to trough in the pattern).

One might consider this as a paradigmatic experiment to test dynamical collapse. Unlike peripheral phenomena, such as random walk or heating or spontaneous radiation, this is a direct test of the collapse of the superposition $\frac{1}{\sqrt{2}}[|L\rangle + |R\rangle]$ analyzed in Section 4.3.

An experiment designed to observe the two-slit interference pattern of a neutron[48] was used to put a limit on the collapse rate of a dynamical collapse theory which predated CSL,[49] but it applies to CSL as well. In this experiment, the two slits were made by putting a 104 μm wire in the center of a 148 μm single slit. The screen was 5 m from the slit plane. The neutron speed was $\approx$ 200 m/s, so the time of flight from slits to screen was $\approx$ 0.025 s.

The observed interference pattern was deemed to agree with the quantum theory prediction to $\approx 0.5\%$ accuracy. Attributing the deviation to CSL collapse taking place

---

[48]Zeilinger (1986).
[49]Pearle (1984).

over the full flight time, we obtain the limit $e^{-\lambda 0.025} \approx 1 - \lambda 0.025 > 1 - 0.005$ or $\lambda < .2$ s^{-1}.

Since that earliest experiment on interference of a fairly massive object, remarkable progress has been made on interfering larger and larger objects. In one experiment,[50] diffraction was observed of complex organic molecules, of diameter $\approx 5 \times 10^{-7}$cm, $C_{284}H_{190}F_{320}N_4S_{12}$ made up of 810 atoms containing 10,123 nucleons. Rather than have the molecules pass through slits, a reflection grating was used. The "grating" was a green–light laser standing wave (produced by laser light incident on, and reflected from, a mirror). The varied intensity along the grating instilled a phase variation in each molecule's wave function due to electric dipole polarization. The molecules were detected at various locations following their departure from the grating, with varied intensities depending upon location. This observed interference pattern was a near–field (Fresnel) "Talbot" pattern which consists of regular repetitions of the grating image. The effect of collapse would be to alter the pattern's intensity. Within experimental accuracy, such an alteration was not seen. Analysis yielded the limit $\lambda \lesssim 5 \times 10^{-6}$ s^{-1}.[51]

An experiment[52] has been proposed to put a tiny mirror into a superposition of spatially distinguishable states. A photon passes through a beam splitter. On one path it takes, it collides with the mirror, mounted on a cantilever, which is then set into oscillation. On the other path, the photon does not hit the mirror, which then stays in place. The photon packets recombine and the photon's time–dependent interference pattern is observed: it would be affected were collapse dynamics between the moving and stationary mirror to take place. This is quite a technically difficult experiment, but it has been pursued.

6.8.1 Concluding Comment

There are quite a variety of different kinds of experiments that can test CSL. There is no experiment yet that contradicts standard quantum theory. (Such an occurrence would be tremendously exciting!) But, also, it has been over three decades since the inception of CSL and there has been no experiment yet that rules out CSL. That is rather remarkable, considering experimental activity and variety.

These experiments place limits on the CSL parameters. Perhaps some day it will be shown that there are no viable parameters, and the theory will have been experimentally falsified. But, perhaps some day CSL effects will actually be seen!

Notes

[A] As an aside, if perception is the criterion, it could be argued that one should put the perceiver into the state vector and then consider the combined collapse rate of a superposition of two states, each with a different location for something perceived and a different physiological response of the perceiver to the two perceptions. Such an analysis has been done[53] with regard to perception of light flashes arising from different locations.[54] We will not pursue such considerations here. We incline toward

[50] Eibenberger et al. (2013).

[51] Toros and Bassi (2018).

[52] Marshall (2002)

[53] Ghirardi et al (1991).

[54] Albert and Vaidman (1989).

presenting a picture of reality that is observer–independent, but consistent with our observations, hence the theoretical constraint. However, regarding the human observer as a part of an experimental apparatus, and thus subject to CSL collapse dynamics, is a legitimate point of view.

[B]This is Eq. (3.14) in Fu's paper, multiplied by CSL's mass–proportionality factor $(m_e/M)^2$, since Fu calculated the non–mass-proportional rate. It is also his Eq. (3.14) multiplied by 4π, since Fu used non–rationalized units, as pointed out by Adler (2007).

7
Interpretational Remarks

If I say electrons behave like particles I give the wrong impression; also if I say they behave like waves. They behave in their own inimitable way, which technically could be called a quantum mechanical way. They behave in a way that is like nothing that you have seen before.[1]

But if you say ψ is not real, I will ask: what is real in your theory? ... If your wave function is not real, you must tell me what is.[2]

The point of the CSL modification of standard quantum theory is to allow the state vector of the theory to correspond to reality, to the physical state of nature. But, how do we extract from the state vector what we see around us, observable reality? This section discusses one approach, in greater depth than the overview given in Section 1.3.3.

7.1 Stuff

We begin with the idea of "stuff"[3].

Consider any operator $\hat{O}$ with eigenbasis $|o_i\rangle$, labeled by eigenvalues o_i. Usually there is degeneracy. That is, this labeling is not sufficient to specify a complete basis, so we include a degeneracy index. So the complete set is labeled $|o_i, j\rangle$, and the complete set of basis states satisfies $\sum_{i,j} |o_i, j\rangle\langle o_i, j| = \hat{1}$.

In homage to Schrödinger and Bell[4] we call the complete set of $|\langle o_i, j|\psi, t\rangle|^2$ the "O-stuff" or the "O-stuff distribution". An individual term labeled by i, j is an "element" of the stuff distribution.

Although one may work with any stuff distribution at all, some may prove more useful than others. We have a special interest in knowing what exists in our three-dimensional space at any time t. In what follows, we concentrate on the stuff corresponding to a special set of operators that is helpful in this regard.

7.2 Particle Number Stuff

In Chapter 4, Section 4.1, we discussed the particle number density operators $\hat{D}_\alpha(\mathbf{x})$, and defined the operator representing the number of particles of type α in a volume

---

[1] Feynman (1985), in his last Messenger lecture.

[2] Bell (1989).

[3] Pearle (2009), Myrvold (2018).

[4] Bell (1990), "In the beginning, Schrödinger tried to interpret his wave function as giving somehow the density of stuff of which the world is made."

V, namely $\hat{N}_V \equiv \int_V d\mathbf{x}\hat{D}_\alpha(\mathbf{x})$. For simplicity, here we will consider just one type of particle, "the nucleon", (the proton and neutron thought of as one kind of particle), and drop the subscript α.[A]

The eigenstates of $\hat{N}_V$ are integers representing the number of nucleons in V, $\hat{N}_V|n, j\rangle_V = n|n, j\rangle_V$ with $n = 0, 1, 2, ...$

Given a state vector $|\psi, t\rangle$, we focus on the particle number in V–stuff,

$$|_V\langle n, j|\psi, t\rangle|^2. \tag{7.1}$$

In classical physics, we can say that any volume V contains a definite number of (point) particles. We are making a *related* statement for CSL–modified quantum theory. V contains something, the particle number in V-stuff. This stuff description is part of our ontology, our *picture of reality*.[B] What we want to do is find the circumstances under which it is possible to make the *same* statement as classical physics. That is, when can we say that V contains a definite number of particles?

7.3   Observability

Not every particle number in V-stuff distribution allows us to say that there is a definite number of particles in V. But, that is effectively what we see. The theory has to tell us what we see.[C] We will use the particle number-stuff to provide what I call *observability*.

In certain circumstances, the particle number in V-stuff is to be promoted to observability status, that is, it is such that we can say there is a definite number of particles *in* V. To achieve this, a criterion will be given. If we know the particle numbers in a wide range of volumes V, this gives us effectively what we observe.

Let us stand back for a moment and review the set of ideas presented so far. The problem we are addressing is how to use the state vector, a rather abstract mathematical entity that represents reality, to obtain very concrete statements about the world we see.

We began with the primary entity of the theory, the state vector, the thing that corresponds to reality.

We next introduced a secondary thing, the particle number in V-stuff distribution. It is constructed from the state vector, and contains less information than the state vector contains.

Finally, we introduce a tertiary thing, a restriction to special particle number in V-stuff distributions, those that satisfy a criterion for observability status.

In summary, as for any theory, we are selecting quantities of interest to us. The theory is objective, the questions we choose to ask of the theory are, of course, subjective.

The criterion operates as follows. Any $\hat{O}$-stuff distribution has the mathematical properties of a probability distribution, that is, positive elements that sum to 1. Therefore, one may use it to calculate a mean $\bar{n}_V$ and a standard deviation σ_V:

$$\bar{n}_V \equiv \langle \psi, t | \hat{N}_V | \psi, t \rangle = \langle \psi, t | \hat{N}_V \sum_{nj} |n,j\rangle_{VV} \langle n,j| |\psi, t \rangle$$

$$= \sum_{n,j} n \, |_V \langle n,j | \psi, t \rangle|^2,$$

$$\sigma_V^2 \equiv \langle \psi, t | [\hat{N}_V - \overline{N}_V]^2 | \psi, t \rangle = \sum_{n,j} n^2 \, |_V \langle n,j | \psi, t \rangle|^2 - \bar{n}_V^2 \qquad (7.2)$$

So, here is our criterion. If $\sigma_V/\bar{n}_V$ is "sufficiently small" (to be elucidated below), we say that the number of particles in V has achieved observability status and *is observable*: we say the number of particles in V is, $\bar{n}_V$.

If this criterion is not met, we cannot say anything about the number of particles in V. In classical physics, one can specify the number of particles within every volume. Here we are saying that sometimes such a specification cannot be made. Standard quantum physics has taught us that some questions, framed within a classical world view, do not have answers, such as "what is the position and momentum of that particle?". The same thing is true about CSL-modified quantum mechanics, where the question that (sometimes) has no answer is "how many particles are in V?".

To give some intuitive understanding for this criterion, recall that, in standard quantum theory, the $\hat{O}$ distribution is interpreted as the probability distribution of outcomes of as yet unperformed future repeated measurements of the eigenvalues of $\hat{O}$. Therefore, according to standard quantum theory, our criterion says that, were such measurements to be performed, the *observation* would always be the same value (the mean), with very small variation. Thus one could reasonably say that this (mean) value *is* the value of the physical variable represented by $\hat{O}$. The justification for this criterion is that it agrees with the experimental observation.

So, how do we determine if $\sigma_V/\bar{n}_V$ is sufficiently small?

In standard quantum theory, the quantity $|\langle a_n | \psi, t \rangle|^2$ ($\hat{A}|a_n\rangle = a_n|a_n\rangle$) provides what Schrödinger[5] called the "catalog of expectations." That is just the Born rule: if in the future, you were to perform an experiment to measure the physical quantity corresponding to $\hat{A}$, you could *expect* to obtain the outcome a_n with probability $|\langle a_n | \psi, t \rangle|^2$. (Of course, in CSL, the outcomes are actually obtained, due to the w-field engendered collapse dynamics.) We utilize this truth in the following way.

Suppose the number of particles in V were to be measured by the most accurate experimental technique available, with result $\overline{N}_V \pm \Delta N_V = \overline{N}_V[1 \pm \Delta N_V/\overline{N}_V]$. If the fractional error in the measurement is larger than the fractional standard deviation provided by the stuff, that is, if $\Delta N_V/\overline{N}_V > \sigma_V/\bar{n}_V$, we say that observability status has been reached and we say the number of particles in V is $\bar{n}_V$. To put it simply, you can say that the number of nucleons is $\bar{n}_V$ if no experiment could prove you wrong!

Typically, due to collapse dynamics, $\sigma_V/\bar{n}_V$ decreases with increasing time. If it is above the threshold $\Delta N_V/\overline{N}_V$, one cannot say what is the number of nucleons in V. The instant it crosses the threshold, the number $\bar{n}_V$ suddenly becomes observable. It is an indication of the tertiary nature of the status of observability that we have to frame it as a sudden transition. There is no sudden transition in nature. We just want

---

[5]Schrödinger (1935).

to ensure that the statement of observability does not conflict with experience. One could just arbitrarily select an upper limit on $\sigma_V / \bar{n}_V$ and, if proven wrong, say "oops" and choose a smaller one. The point is, there *is* an upper limit on $\sigma_V / \bar{n}_V$ below which the statement "there are $\bar{n}_V$ particles in V" conforms with experience. So, one might as well make an informed choice of the best upper limit, based upon experimental accuracy.

We conclude this chapter by giving three examples of how these ideas work.[D]

The first example considers the state vector describing an ordinary piece of matter. We show, for the overwhelming majority of volumes, that one obtains the same number of nucleons inside V as one would expect from a classical point of view.

The second is essentially the von Neumann model of a measurement, where the apparatus is a "pointer" that indicates the result of a two-outcome measurement. As the Hamiltonian evolves the state vector into a superposition describing the two outcomes, the collapse dynamics takes over, acting on the different mass density distributions of the pointer states and so undoing the superposition, generating collapse to one of the outcomes. The upshot is that the initially observable pointer can become unobservable for a brief time interval and then it returns to observability status. This time interval is shorter than human perception time, so the lack of observability is experimentally unobservable.

The third example illustrates the "tails" issue. That is, collapse never goes to completion in a finite time. Thus, for example, the CSL description of an experiment, evolves into a superposition of a state vector describing a macroscopic state with numerical coefficient very, very close to 1, added to additional terms with very, very small numerical coefficients. We argue that one may treat this as if there were not the extra "tail" terms, because they contribute zero particle number to any volume V. That is, the tail terms are unobservable: the space they occupy is observably empty space.

7.4   Stuff for a Macroscopic Object

We start with a conceptually simple class of volumes V, and afterwards consider the general case.

In normal matter, atoms are separated by distances of the order of a few Bohr radii. Therefore, one can construct many different volumes V of arbitrary shape that lie completely within the object, such that each V's surface lies at a distance from every nucleus greater than, say, 10 nuclear diameters (for example, let V's surface roughly split the distance between the atoms). This allows one to consider that each nucleon is either completely inside or outside V, the wave function involving each nucleon being utterly negligible on one side of the surface. If $|\psi, t\rangle$ is the state vector describing the complete object, it can be written as the direct product of $|\psi, t\rangle'$, the normalized state vector describing just the nucleons inside V and $|\psi, t\rangle''$, the normalized state vector describing just the nucleons outside V.

Take the number of nucleons inside V to be N, which means to an excellent approximation that

$$\sum_j |_V \langle n, j | \psi, t \rangle'|^2 = \delta_{nN}. \tag{7.3}$$

Since $N_V|\psi, t\rangle = N|\psi, t\rangle$, we easily find the mean and standard deviation:

$$\bar{n}_V \equiv \langle\psi, t|\hat{N}_V|\psi, t\rangle = N,$$
$$\sigma_V^2 \equiv \langle\psi, t|[\hat{N}_V - \overline{N}_V]^2|\psi, t\rangle = 0. \qquad (7.4)$$

Eqs. (7.4) describe as narrow a distribution as can be! Since $\sigma_V/\bar{n} = 0$, the observability criterion is met, we say that there *are* N nucleons in V.

Turn now to the general case, of arbitrary V. If the surface of V cuts through one or more nucleon wave functions, then $\bar{n}_V < N$. However, if the volume is sufficiently large, if N is large enough so that the number of nucleons whose wave functions are cut through by the surface of V is small compared to the number of the rest of the nucleons well within V, then essentially $\bar{n}_V \approx N$ and $\sigma_V/\bar{n}$, while not 0, is negligibly small. Thus, in this situation there is also observability of the number of nucleons within V: no difference from N can be experimentally detected.

7.5 Observability of a Pointer during a Measurement

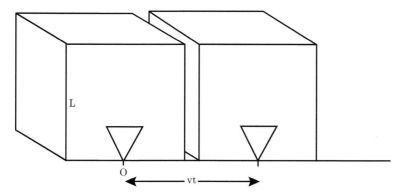

Fig. 7.1 The two superposed states of a macroscopic pointer cube with side length L, displaced by distance vt due to a measurement.

A new consideration arises if the macroscopic object is an apparatus described by the state vector, put into a superposition by a quantum experiment.

To illustrate, we consider a "pointer" that, for simplicity, we take to be a homogeneous cube of side length L containing N nucleons, described by the state $|d\rangle$, with d the location of the pointer center along a dial. The experiment, beginning at time $t = 0$, is to have one of two possible results, 0 or 1, with equal likelihood.

If the result is 0, the pointer does not budge, and is described by the state vector $|0\rangle$. If the result is 1, the pointer state is $|vt\rangle$: the pointer moves with speed v to the right toward the number 1 on the dial, that is far enough away so that the two pointer volumes soon do not overlap. It is imagined that the microscopic system that is being measured, and any mechanism involved, displace much less mass than the pointer, so the pointer motion is responsible for the collapse (see Fig. 7.1).

We consider V to occupy precisely the volume of the pointer before the measurement begins at $t = 0$. Thus, $\hat{N}_V|0\rangle = N|0\rangle$, The state vector describing the pointer at location vt is an eigenstate of $\hat{N}_V$ with eigenvalue equal to the number of nucleons in V:

$$\hat{N}_V|vt\rangle = N\left[\frac{L - vt}{L}\right]|vt\rangle \text{ for } 0 \le t < \frac{L}{v}; \ \hat{N}_V|vt\rangle = 0 \text{ for } t > \frac{L}{v}. \tag{7.5}$$

Since the states $|0\rangle$ and $|vt\rangle$ are eigenstates of $\hat{N}_V$, with different eigenvalues for $t > 0$, they are othogonal, $\langle 0|vt\rangle = 0$, for $t > 0$.

Here is what we will show..

When the situation is described by standard quantum theory (when there is no collapse), as the experiment gets under way and the superposition develops, very rapidly the criterio for observability is not met and forever after one cannot say how many nucleons lie inside V. forever after.

With collapse added, with some choices of L and v, the number of nucleons inside V remains *always* observable. But, for other choices of L and v, the number of nucleons inside V becomes unobservable for a very brief interval, and then becomes observable again, and forever after.

7.5.1   Evolution with No Collapse

Without collapse, for $t > 0$, the state vector representing the situation is

$$|\psi, t\rangle = \frac{1}{\sqrt{2}}[|0\rangle + |vt\rangle]. \tag{7.6}$$

We therefore calculate from (7.5), (7.6), for $0 \le t \le \frac{L}{v}$, while the stationary pointer and the moving pointer overlap,

$$\bar{n}_V \equiv \langle\psi, t|\hat{N}_V|\psi, t\rangle = \frac{1}{2}N + \frac{1}{2}N\left[1 - \frac{vt}{L}\right] = N\left[1 - \frac{vt}{2L}\right],$$

$$\sigma_V^2 \equiv \langle\psi, t|\hat{N}_V^2|\psi, t\rangle - \overline{N}_V^2 = \frac{1}{2}N^2 + \frac{1}{2}N^2\left[1 - \frac{vt}{L}\right]^2 - N^2\left[1 - \frac{vt}{2L}\right]^2 = \left[\frac{Nvt}{2L}\right]^2$$

$$\frac{\sigma_V}{\bar{n}_V} = \frac{vt/2L}{1 - (vt/2L)}. \tag{7.7}$$

and for $t > \frac{L}{v}$,

$$\bar{n}_V = \frac{N}{2}; \ \sigma_V^2 = \frac{N^2}{4}; \ \frac{\sigma_V}{\bar{n}_V} = 1. \tag{7.8}$$

Let us see what these equations tell us in the context of specific choices for the makeup of the pointer. We will choose the pointer to be as small as possible so as to "maximally challenge" the theory.

Like the barely visible sphere of Section 6.1, we choose the pointer cube to be made of carbon, with $\approx 10^{24}$ nucleons/cc, of barely visible (in a microscope) side length $L = 4 \times 10^{-5}$ cm. Then, the number of nucleons in the cube is $N \approx 10^{11}$. Also, we suppose the speed v of the moving pointer is such that the two-cube overlap is just ended at the human perception time $T = .1$ s, so $v = L/T = 4 \times 10^{-4}$ cm/s.

Also, we will take the fractional experimental accuracy of observing the pointer to be 1%.

Now, as the experiment commences, for a brief interval, according to the observability criterion, one can still say that the number of nucleons in V is $\bar{n} \approx 10^{11}$. But, for how long? Since $\frac{v}{2L} = 5$, then according to (7.7),

$$\frac{\sigma_V}{\bar{n}_V} = 5t/(1 - 5t) \approx 5t.$$

This grows larger than experimental accuracy 0.01 at $t \geq 0.002$ s (when the pointer displacement is only 8×10^{-7} cm). After this time, forever after, there is no observability about the number of nucleons in V. Of course, this is what one expects from standard quantum theory: its state vector does not describe what is observed. One sees the pointer at one location or the other, but quantum theory keeps it in an equal-amplitude superposition forever.

7.5.2 Evolution with CSL Collapse

Now we apply this identical example to the situation when there is dynamical collapse. For each sample random field driving the collapse, the state vector at time $t > 0$ has the form

$$|\psi, t\rangle_N = \alpha(t)|0\rangle + \beta(t)|vt\rangle, \tag{7.9}$$

$(\alpha(t)^2 + \beta(t)^2 = 1, \alpha(0) = \beta(0) = \frac{1}{\sqrt{2}})$ where one can think of the (real) amplitudes $\alpha(t), \beta(t)$ playing the gambler's ruin game. Then, using $\frac{v}{L} = 10$, we can calculate the necessary quantities as in (7.7), (7.8), but using the state vector (7.9). We express everything in terms of $\beta^2(t)$:

$$\bar{n}_V = N\alpha(t)^2 + N(1 - 10t)\beta^2(t) = N\left[1 - \beta^2(t)10t\right] \text{ for } 0 < t \leq .1\text{s},$$

$$= N\alpha^2 = N\left[1 - \beta^2(t)\right] \text{ for } t \geq .1\text{s}, \tag{7.10a}$$

$$\sigma_V^2 = (1 - \beta^2(t))\beta^2(t)[N10t]^2 \text{ for } 0 < t \leq .1\text{s},$$

$$= N^2(1 - \beta^2(t))\beta^2(t) \text{ for } t \geq .1\text{s and so,}$$

$$\frac{\sigma_V}{\bar{n}} = \beta(t)10t\frac{\sqrt{1 - \beta^2(t)}}{1 - \beta^2(t)10t} \text{ for } 0 < t \leq .1\text{s},$$

$$= \frac{\beta(t)}{\sqrt{1 - \beta^2(t)}} \text{ for } t \geq .1\text{s}. \tag{7.10b}$$

We now estimate $\beta(t)$ according to CSL. In Section 4.4.2 et. seq., we showed that the off-diagonal element of the density matrix describing the collapse of an object in a superposition of two places $|L\rangle, |R\rangle$ obeys the evolution equation

$$\frac{d}{dt}\langle L|\rho(t)|R\rangle = -(4\pi)^{3/2}\lambda n_a N_{nol}\langle L|\rho(t)|R\rangle. \tag{7.11}$$

N_{nol} is the number of nucleons in one image of the pointer that is not overlapped by the other image of the pointer, and n_a is the number of nucleons in a volume a^3,

$$n_a = N\frac{a^3}{L^3} = 10^{11}\left[\frac{10^{-5}}{4\times 10^{-5}}\right]^3 \approx 10^9$$

We use this result, supposing it is a good approximation at every instant. Thus, in (7.11), we set

$$N_{nol}(t) = N\frac{vt}{L} = N10t \text{ for } 0 \le t \le 0.1 \text{ s}; \quad N_{nol}(t) = N \text{ for } t > .01 \text{ s}. \tag{7.12}$$

Moreover, we can write $\rho(t) = \overline{|\psi,t\rangle\langle\psi,t|}$, that is, as the ensemble average (over all random fields) of state vectors, half of which collapse to $|0\rangle$ and half to $|1\rangle$ (since the initial states in the superposition had equal amplitudes). Therefore,

$$\langle L|\rho(t)|R\rangle = \langle 0|\rho(t)|vt\rangle = \overline{\langle 0|[\alpha(t)|0\rangle + \beta(t)|vt\rangle][\alpha(t)\langle 0| + \beta(t)\langle vt|]|vt\rangle}$$

$$= \overline{\alpha(t)\beta(t)} = \frac{1}{2}\overline{\alpha(t)\beta(t)}_{\alpha\to 1,\beta\to 0} + \frac{1}{2}\overline{\alpha(t)\beta(t)}_{\alpha\to 0,\beta\to 1}.$$

$$= \overline{\alpha(t)\beta(t)}_{\alpha\to 1,\beta\to 0}$$

where the last step can be taken since both terms in this sum are numerically the same, because of the symmetry. The matrix element of $\rho(t)$ has been written in terms of those state vectors in the ensemble for which the collapse is to the state $|0\rangle$ because we are focusing on the question of the evolution of observational reality for the unmoved pointer.

Next, although this is the ensemble average, we take it to represent a "typical" state vector, so we make the approximation $\overline{\alpha(t)\beta(t)} \approx \alpha(t)\beta(t)$.

Now, since the coefficient $\beta(t)$ rapidly gets small and $\alpha(t)$ rapidly gets close to 1, we can make the approximation that $\alpha(t)\beta(t) \approx \beta(t)$. We are interested in how long it takes for $\beta(t)$ to become so small that the number of nucleons in V becomes observable again.

Thus, we set $\langle L|\rho(t)|R\rangle \approx \beta(t)$ in (7.11). The solution is:

$$\beta(t) = \frac{1}{\sqrt{2}}e^{-(4\pi)^{3/2}\lambda n_a \int_0^t dt' N_{nol}(t')}.$$

Therefore,

$$\beta(t) = \frac{1}{\sqrt{2}}e^{-(4\pi)^{3/2}\lambda n_a \int_0^t dt' N10t'} = \frac{1}{\sqrt{2}}e^{-(4\pi)^{3/2}\lambda n_a N5t^2}$$

$$\approx \frac{1}{\sqrt{2}}e^{-2\times 10^0 t^2} \text{ for } t \le 0.1\text{s},$$

$$= \frac{1}{\sqrt{2}}e^{-(4\pi)^{3/2}\lambda n_a N[5\times.1^2 + (t-0.1)]} = \frac{1}{\sqrt{2}}e^{-(4\pi)^{3/2}\lambda n_a N[t-0.05]}$$

$$\approx \frac{1}{\sqrt{2}}e^{-4\times 10^5[t-0.05]} \text{ for } t \ge 0.1\text{s}. \tag{7.13}$$

When $\beta(t)$ is small, (7.10b) combined with (7.13) gives, for $t \lesssim 0.1$s,

$$\frac{\sigma_n}{n} \approx \beta(t)10t \approx 7te^{-2\times 10^6 t^2} \tag{7.14}$$

This rises from 0 at $t = 0$ to a maximum $\approx 2\times 10^{-3}$ at $t = 5\times 10^{-4}$s and exponentially fades thereafter.

We now can apply the observability criterion, which depends upon the percentage accuracy of the best possible measurement. If the fractional measurement error is larger than 2×10^{-3}, that is, a 0.2% error experiment or worse is the best that can be done, the number of nucleons within V *always* has observability, with value $\bar{n} = N \approx 10^{11}$.

However, if the fractional measurement error is smaller than 2×10^{-3}, then there is a time interval during which there is not observability, *when we cannot say what the number of nucleons is within V*. How long is this time interval?

Even if, to make this time interval as long as possible, we choose the fractional experimental error to equal the smallest ever achieved for any experiment, 10^{-15} (fractional error in the hyperfine frequency of Rb), the interval is less than human perception time.

Here is the scenario. Put this observability condition threshold into (7.14):

$$10^{-15} \geq 7te^{-2 \times 10^6 t^2}.$$

At $t = 0$ of course the pointer is observable but, at $t \approx \frac{1}{7} \times 10^{-15}$ s, the threshold is crossed and the pointer is no longer observable. This status continues as t passes 5×10^{-4} s where, as mentioned, the right side of this equation has its maximum, until t is large enough for the right side to once again equal the threshold, which is $t \approx 3 \times 10^{-3}$ s. Then observability kicks in once again, and exists forever after.

So, to summarize, even with numerical choices made that are worst possible for the theory, once the pointer starts to register an experimental response, there is a 0.003 s interval during which the pointer is not observable, much less than the 0.1 s human response time.

We emphasize: the possibility of such a lapse of observability during an experiment is a *necessary* feature of the ideas presented here.

7.6   Tail

Another necessary feature entailed by CSL has to do with collapse never going quite to completion. For example, in the situation discussed above, using (7.13) for $t \geq 0.1$ s,

$$|\psi, t\rangle \approx |0\rangle + \beta(t)|vt\rangle$$
$$\approx |0\rangle + \frac{1}{\sqrt{2}}e^{-4 \times 10^5 [t-0.05]}|vt\rangle. \tag{7.15}$$

The amplitude of the collapsed ("tail") state is small, and getting smaller as time progresses, but it never reaches exactly 0. The concern expressed by some is that, according to the theory, there is something more that exists in nature than just the pointer pointing to 0.

Moreover, they are further concerned that the state vector could be extremely complicated, for example, it could contain a human observer (although then the amplitude of the tail state would be much smaller than that in the illustrative example (7.15) because many more nucleons are involved).

Abner Shimony, in personal communication a decade before I found CSL, criticized this feature of the collapse theories I was working on at the time. He voiced this criticism post–CSL, in a paper[6] giving 8 desiderata for such theories, among which was:

> *If a stochastic dynamical theory is used to account for the outcome of a measurement, it should not permit excessive indefiniteness of the outcome, where "excessive" is defined by considerations of sensory discrimination.* This desideratum tolerates outcomes in which the the apparatus variable does not have a sharp point value, but it does not tolerate "tails" which are so broad that different parts of the range of the variable can be discriminated by the senses, even if very low probability amplitude is assigned to the tail.

My response to this is that CSL fulfills the italicised part of this desideratum. But, I disagree with the unitalicized addition, which implies that it is legitimate in CSL to view the tail by itself. In CSL, *one cannot view the tail state by itself.*

This implicitly takes the tail to be like a world in many-worlds,[7] for which one cares about the structure of the world regardless of its amplitude. It is implied that, out of the state vector, one rips the tail vector corresponding to a world and normalizes that to 1. Then, one thinks of this as somehow a structure given by the theory. But it is *not* a structure given by the theory. One cannot rip apart a state vector in CSL. It is the whole state vector that corresponds to our single world. A piece of the state vector by itself corresponds to nothing physical.

Properly using the whole state vector, there can be no sensory discrimination of a structure in the tail if no particles are there. In the example above, where the end result of collapse is $|0\rangle$, let V' be the "tail region", a volume surrounding the moving pointer. The expectation value of the number of nucleons seen in V' at $t = 0.1$ s is, from (7.15),

$$\langle \psi, t | \hat{N}_{V'} | \psi, t \rangle = N\beta^2(0.1)$$

$$= 10^{11} \left[\frac{1}{\sqrt{2}} e^{-2\times 10^4} \right]^2 \approx 10^{-4\times 10^4}. \tag{7.16}$$

This is a very small number, to say the least, certainly consistent with the statement that there are 0 nucleons in the tail region of space.

7.7 Concluding Remarks

We have to find a new view of the world that has to agree with everything that is known, but disagree in its predictions somewhere, otherwise it is not interesting. And in that disagreement it must agree with nature. If you can find any other view of the world which agrees over the entire range where things have already been observed, but disagrees somewhere else, you have made a great discovery. It is very nearly impossible, but not quite, to find any theory which agrees with experiments over the entire range in which all theories have been checked, and yet gives different consequences in some other range, even a theory whose different consequences do not turn out to agree with nature.[8]

[6]Shimony (1991).
[7]McQueen (2015).
[8]Feynman (1985) in his 1964 Messenger Lectures.

Experimental tests of CSL and theoretical developments are proceeding fairly actively nowadays. The signal achievement of dynamical collapse is to describe the occurrence of events. That is something lacking in standard quantum theory. And, it may be argued, CSL does this with a certain amount of grace.

While there has been no positive sighting of a dynamical collapse effect, neither has there yet been an experimental refutation of the theory's predictions. This is rather remarkable for an alternative to quantum theory that has been around for over thirty years.

Einstein once was quoted as saying he would be sorry for the "dear Lord" if he did not use General Relativity in designing the world. I certainly would not go this far concerning CSL. But, events do occur, and if the dear Lord doesn't use dynamical collapse to explain events, what does He use?

Notes

[A]Chapter 14, the Supplement to this chapter gives two examples of its ideas in the context of a single particle. The first treats an electron in the ground state of Hydrogen, with V a sphere centered on the nucleus, and discusses for what radii one can say the observable reality is that the electron is *in* V. The second discusses the stuff description of a particle undergoing two–slit interference,. The observable reality is that a particle did get past the slits, but there is no observable reality associated with the question, which slit.

[B]Defined succinctly by Rod Tumulka (Tumulka (2018)): "The "ontology'"of a theory means what exists, according to that theory." I believe this second phrase, the theory-dependent qualification, has an appropriate modesty, in that discussion of what exists is inevitably circumscribed by our conceptual limitations, by the theory we have been able to create. Thus, I believe a distinction should be made between "what's out there," and our description of it. The elements of our ontology necessarily, possesses a human imprimatur. Abner Shimony liked to quote this poem by Robert Frost: "We dance round in a ring and suppose, But the Secret sits in the middle and knows."

[C]Of course, photons are literally responsible for most of our observations. We note that an operator representing the number of photons in a volume V is not what we actually discern. Accordingly, the Supplement to this chapter, Section 14.3, shows how to construct an operator we may think of as corresponding to what we observe, the stuff associated to the number of photons in a narrow wavelength range traveling along the axis of a "tube" (which one may imagine held to one's eye). This could be used to define observability, but we do not do so here because the use of particle number stuff has a more direct connection to the operators that appear in non-relativistic CSL.

[D]Two other pictures of observable reality associated to collapse theories, called "primitive ontologies," have been presented. In one, applicable to the SL theory, due to Rod Tumulka,[9] enlarging on a suggestion by Bell, the centers of the hits are called "flashes." This evokes a darkened space-time where only the flashes appear, and that represents observable reality. For example for a table, the flashes continue at a great rate, effectively limning the shape of the table. My view is that this is not a rich enough picture of reality, since there is a lot more to the table than its shape (materials? colors?), which one cannot reconstruct from the flashes. (Moreover, small enough objects, especially moving ones, can have flashes too infrequent to present anything useful.) In the other picture, applicable to CSL, due to GianCarlo Ghirardi and co-workers,[10] when the mean value of the smeared mass density is much larger than its standard deviation, this quantity is said to be "accessible", that is, to represent observable reality. Again, my objection is that I believe observable reality is richer than what is described by smeared mass density.

[9]Tumulka (2018).
[10]Benatti, Ghirardi, and Grassi (1995), Ghirardi and Weber (1997).

8
Supplement to Chapter 1

8.1 Gambler's Ruin Criteria Expressed as Ensemble Averages

The Gambler's Ruin game, discussed in Section 1.5, has a strong analogy to the mechanism of dynamical collapse. If, in collapse dynamics, $|\psi, t\rangle = c_1(t)|1\rangle + c_2(t)|2\rangle$ and $x_i(t) \equiv |c_i(t)|^2$, the analog to $x_i(t)$ is the fraction of the total money in the game possessed by the ith gambler.

It was proved that the game satisfies three conditions (the first is obvious, the rest not so):

1) The normalization condition (1.4):

$$1 = x_1(t) + x_2(t).$$

2) The collapse condition (1.5):

$$either\ x_1(\infty) = 1, x_2(\infty) = 0,\ or\ x_1(\infty) = 0, x_2(\infty) = 1. \tag{8.1}$$

3) The Born rule condition (1.6):

$$x_1(\infty) = 1, x_2(\infty) = 0,\ \text{occurs with probability } x_1(0),$$
$$x_1(\infty) = 0, x_2(\infty) = 1,\ \text{occurs with probability } x_2(0). \tag{8.2}$$

Endnote C of Chapter 1 asserted that the two statements (1.1), (1.2) about *individual* games can be arrived at starting from two postulates involving *averages* over the complete ensemble of games. This assertion is proved here.

Actually, we will prove it for a more general case. It is easy to generalize the Gambler's Ruin game from two to N gamblers. One can imagine the gamblers pairing off in a random way, each pair tossing a coin and exchanging a dollar according to the result. When a gambler loses all of his/her money, he/she leaves the game, which goes on until there is just one gambler left, who possesses all the money. $x_n(t)$ is the fraction of the total money possessed by gambler n at time t.

We do this as an analogy to dynamical collapse of N states,, where $|\psi, t\rangle = \sum_{n=1}^{N} c_n(t)|a_n\rangle$ and $x_n(t) = |c_n(t)|^2$). Eq. (1.4), the statement that the state is normalized to 1,

$$\sum_{n=1}^{N} x_n(t) = 1, \tag{8.3}$$

obviously applies to the game, as the statement that the money in the game is constant. This is not a statement about averages, as it holds for all t for each individual realization of the game.

8.1.1 Average Statements Lead to Individual Consequences

The two equations involving averages, mentioned above, that are equivalent to (8.1) and (8.2) (generalized to N states) are:

$$\overline{x_n(t)x_m(t)} \xrightarrow{t \to \infty} 0, \text{ where } n \neq m \tag{8.4}$$

$$\frac{d}{dt}\overline{x_n(t)} = 0. \tag{8.5}$$

First, let's see the implications of Eq. (8.4) for individual games. The average in Eq. (8.4) is calculated by multiplying the probability of each possible game (characterized by its unique set of toss outcomes) by the resulting $x_n(\infty)x_m(\infty)$ for that game and adding up each game's contribution. Thus, every term in the sum is the product of three non-negative numbers. The only way such a sum can vanish is if each term in the sum, representing each possible game's contribution, vanishes. If there is even one game for which there is just one n and one m such that $x_n(\infty)x_m(\infty) \neq 0$, then (8.4) is violated. So, for *each* game there can be *no more than one* $x_n(\infty)$ that does not vanish.

There cannot be a game in which *all* the $x_n(\infty)$ vanish, since that violates (8.3): the money cannot disappear! Therefore, for each individual game, there is precisely one $x_n(\infty)$ that does not vanish and, by (8.3), that non–vanishing $x_n(\infty)$ must equal 1.

Thus, we obtain from (8.4) (the generalization to N states of) what we call the collapse postulate, Eq. (1.1):

$$\text{for each game, for } \textit{some } n, \quad x_n(\infty) = 1 \text{ and}$$
$$x_1(\infty) = 0, ...x_{n-1}(\infty) = 0, x_{n+1}(\infty) = 0, ...x_N(\infty) = 0. \tag{8.6}$$

Next, we can see how (8.5) leads to the Born rule condition (8.2). (8.5) connects the initial and final average values of $x_n(t)$:

$$\overline{x_n(0)} = \overline{x_n(\infty)}. \tag{8.7}$$

We are certain about the initial condition, since all the repeated games are to start the same, with each gambler having a specified sum: $\overline{x_n(0)} = x_n(0)$. Now, denote by P_n the probability that a game will end up with $x_n(\infty) = 1$. So, $1 - P_n$ is the probability that a game will end up with $x_n(\infty) = 0$. These are the only possibilities. Then, from (8.6), (8.7),

$$x_n(0) = \overline{x_n(\infty)} = 1 \times P_n + 0 \times [1 - P_n] = P_n, \tag{8.8}$$

which is Born's rule, Eq. (1.2), generalized to the case of N gamblers.

8.2 Fair Game Considerations

Eq. (8.5), the constancy of $\overline{x_n(t)}$ can be called the "fair game" postulate. Why?

In the context of a two gambler game with total money involved = $100 (as in the game between Willie Oneka and Tillie Tuesday described in Section 1.5) if the coin is a fair coin (equal probability 1/2 of obtaining heads or tails), the mean amount of money (averaged over all games) possessed by a gambler is constant over time. We'll concentrate upon Tillie (the argument is identical for Willie).

Consider the probability $q(x, t_k)$ of Tillie possessing the fraction x of the total sum on the kth toss. This is the sum of probabilities of all games played—of all coin toss sequences—that lead to this result. Of course, $\sum_x q(x, t_k) = 1$, since it is certain that Tillie has *some* value of x, even if it may be 0.

Here, x takes on the values 0.00, 0.01, 0.02, ... 0.99, 1.00. There are only two routes for Tillie to have the fraction x at time t_{k+1} from what she had at time t_k. At time t_k, she could have had one less dollar, i.e., have the fraction $x - 0.01$ and win the toss, which occurs with probability 1/2. Or, she could have had one more dollar, have had the fraction $x + 0.01$ and lose the toss, which occurs with probability 1/2. This implies

$$q(x, t_{k+1}) = \frac{1}{2}q(x - .01, t_k) + \frac{1}{2}q(x + .01, t_k). \tag{8.9}$$

It is useful to include results that cannot be achieved, $x = -0.01, 1.01$ when calculating $\bar{x}(t)$ (setting $q(-0.01, t) = q(1.01, t) = 0$) so we need not worry about end points in the sum we are about to calculate.

We now can calculate $\overline{x(t_{k+1})}$ using (8.9) (first line of Eq. (8.10)), and then perform some manipulations:

$$\overline{x(t_{k+1})} \equiv \sum_x xq(x, t_{k+1}) = \sum_x x\left[\frac{1}{2}q(x - .01, t_k) + \frac{1}{2}q(x + .01, t_k)\right]$$

$$= \sum_x [(x - .01) + .01]\frac{1}{2}q(x - .01, t_k) + \sum_x [(x + .01) - .01]\frac{1}{2}q(x + .01, t_k)$$

$$= \frac{1}{2}\sum_{x'} x'q(x', t_k) + .01\sum_x \frac{1}{2}q(x - .01, t_k)$$

$$+ \frac{1}{2}\sum_{x'} x'_n q(x', t_k) - .01\sum_x \frac{1}{2}q(x - .01, t_k)$$

$$= \frac{1}{2}\overline{x(t_k)} + \frac{.01}{2} \cdot 1 + \frac{1}{2}\overline{x(t_k)} - \frac{.01}{2} \cdot 1 = \overline{x(t_k)}. \tag{8.10}$$

In the second line, we replaced x by $(x \mp .01) \pm .01$. In the third line, in two of the sums, the replacements $x' = x \pm .01$ have been made.

This concludes the proof that shows $\bar{x}(t_k)$ does not change with time if it is a fair game. We note that, were it not a fair game, if the first 1/2 in the square bracket in the first line of (8.10) was replaced by $1/2 + \epsilon$, and the second 1/2 in the square bracket was replaced by $1/2 - \epsilon$, the last term in (8.10) would have $.02\epsilon$ added to it, and (8.10) would not hold.

9

Supplement to Chapter 2

9.1 How CSL Works: Collapse to Position Eigenstates

Section 2.1.1 shows how the CSL evolution equation (2.2),

$$\psi(x,t) = e^{-\frac{1}{4\lambda t}[B(t)-2\lambda\hat{A}t]^2}|\psi,0\rangle \qquad (9.1)$$

and the probability rule (2,3),

$$P(B(t))dB(t) = \frac{1}{\sqrt{2\pi\lambda t}}\langle\psi,t|\psi,t\rangle dB(t), \qquad (9.2)$$

together lead to collapse obeying the Born probability rule when the collapse-generating operator $\hat{A}$ has a discrete eigenvalue spectrum. The case of just two eigenstates was discussed in great detail.

As mentioned in the footnote at the end of Section 2.1.1, we treat here the case of a continuous spectrum. So, we consider that the collapse-generating operator is $\hat{A} = \hat{X}$, the position operator for a particle moving in one dimension.

For simplicity, choose the initial wave function to be a Gaussian centered on x_0:

$$\psi(x,0) = \frac{1}{[2\pi\sigma^2]^{1/4}}e^{-\frac{(x-x_0)^2}{4\sigma^2}}. \qquad (9.3)$$

The points made are that the wave function asymptotically approaches an eigenfunction of the position operator and the probability of that outcome is dx multiplied by the squared initial wave function at that position, the Born probability.

The evolution equation (9.1), in the position basis, using $\langle x|f(\hat{X}) = f(x)\langle x|$, becomes

$$\psi(x,t) = \langle x|\psi,t\rangle = \langle x|e^{-\frac{1}{4\lambda t}[B(t)-2\lambda\hat{X}t]^2}|\psi,0\rangle$$

$$= e^{-\frac{1}{4\lambda t}[B(t)-2\lambda xt]^2}\frac{1}{[2\pi\sigma^2]^{1/4}}e^{-\frac{(x-x_0)^2}{4\sigma^2}}. \qquad (9.4)$$

The probability rule Eq. (9.2) is

$$P(B(t))dB(t) = \frac{dB(t)}{\sqrt{2\pi\lambda t}}\int_{-\infty}^{\infty}dx|\psi(x,t)|^2$$

$$= \frac{dB(t)}{\sqrt{2\pi\lambda t}}\int_{-\infty}^{\infty}dx e^{-2\lambda t[x-\frac{B(t)}{2\lambda t}]^2}\frac{1}{\sqrt{2\pi\sigma^2}}e^{-\frac{(x-x_0)^2}{2\sigma^2}}$$

$$= \frac{dB(t)}{2\lambda t}\frac{1}{\sqrt{2\pi(\sigma^2+(4\lambda t)^{-1})}}e^{-\frac{\left[x_0-\frac{B(t)}{2\lambda t}\right]^2}{2[\sigma^2+(4\lambda t)^{-1}]}}, \qquad (9.5)$$

where the integral Eq. (A.5), of the product of two Gaussians, has been employed in going from the second line to the third.

Changing variable $B(t)$ to $b(t) \equiv \frac{B(t)}{2\lambda t}$, so $db(t) \equiv \frac{dB(t)}{2\lambda t}$ (note, t is a constant), we rewrite Eqs. (9.4), (9.5) as

$$\psi(x,t) = e^{-\frac{[x-b(t)]^2}{(\lambda t)^{-1}}} \frac{1}{[2\pi\sigma^2]^{1/4}} e^{-\frac{(x-x_0)^2}{4\sigma^2}} \tag{9.6a}$$

$$P(b(t))db(t) = db(t)\frac{1}{\sqrt{2\pi(\sigma^2 + (4\lambda t)^{-1})}} e^{-\frac{[x_0-h(t)]^2}{2[\sigma^2+(4\lambda t)^{-1}]}}. \tag{9.6b}$$

Let's look at the asymptotic behavior. For large enough t so that $(\lambda t)^{-1} << \sigma^2$, in (9.6a) the first Gaussian is so sharply peaked around $x = b(t)$ that we may make that replacement in the second Gaussian. Then, (9.6a) and (9.6b) (neglecting $(4\lambda t)^{-1}$ compared to σ^2) become

$$\psi(x,t) \approx e^{-\lambda t[x-b(t)]^2} \frac{1}{[2\pi\sigma^2]^{1/4}} e^{-\frac{(b(t)-x_0)^2}{4\sigma^2}} \tag{9.7a}$$

$$P(b(t))db(t) \approx db(t)\frac{1}{\sqrt{2\pi\sigma^2}} e^{-\frac{[b(t)-x_0]^2}{2\sigma^2}} = db(t)|\psi(b(t),0)|^2 \tag{9.7b}$$

From (9.7a), the normalized wave function is

$$\psi_N(x,t) = \psi(x,t)/\left[\int_{-\infty}^{\infty} dx |\psi(x,t)|^2\right]^{1/2} = \left[\frac{2\lambda t}{\pi}\right]^{1/4} e^{-\lambda t[x-b(t)]^2}. \tag{9.8}$$

Its square

$$|\psi_N(x,t)|^2 = \frac{1}{\sqrt{2\pi(4\lambda t)^{-1}}} e^{-\frac{[x-b(t)]^2}{2(4\lambda t)^{-1}}}. \tag{9.9}$$

approaches $\delta(x - b(t))$ as $t \to \infty$, that is, it is in the form of a normalized Gaussian since it has area 1, and its width goes to zero in the infinite time limit.

Therefore, $\psi_N(x,t)$ may be thought of as asymptotically approaching the square root of a delta function! But, it may also be thought of as approaching a delta function, with asymptotically vanishingly small amplitude, for (9.8) may also be written as

$$\psi_N(x,t) = \left[\frac{2\pi}{\lambda t}\right]^{1/4}\left\{\frac{1}{\sqrt{2\pi(2\lambda t)^{-1}}} e^{-\frac{[x-b(t)]^2}{2(2\lambda t)^{-1}}}\right\}. \tag{9.10}$$

where the expression in the curly brackets is also an asymptotic delta function. This asymptotic delta function $\delta(x - b(t))$ is an eigenstate of $\hat{X}$.

Moreover, from (9.7b), we see that the probability density of occurrence of the asymptotic eigenstate with eigenvalue $b(t)$ is $|\psi(b(t),0)|^2$, the absolute magnitude squared of the initial wave function at $x = b(t)$.

To summarize, we have proved for this example of the collapse–generating operator having a continuous spectrum what was proved in Section 2.1.1 for an example of a discrete spectrum. The asymptotic (as t approaches infinity) states, the possible collapse outcomes, are eigenstates of the collapse–generating operator, the position operator. And the Born rule is obeyed.

10

Supplement to Chapter 3

10.1   Particle in One Dimension, Position Collapse

Section 3.5 summarized the behavior of the wave function for a free particle undergoing collapse in position. The complete solution, upon which that summary was based, is given here.

We are considering a free particle initially in a Gaussian state centered on $x = 0$:

$$\langle x|\psi',0\rangle = \frac{1}{[2\pi\sigma^2]^{1/4}} e^{-\frac{x^2}{4\sigma^2}} \tag{10.1}$$

which satisfies the CSL-modified Schrödinger equation (3.16):

$$\frac{\partial}{\partial t}\langle x|\psi',t\rangle = \left[\frac{i}{2m}\frac{\partial^2}{\partial x^2} + w(t)x - \lambda x^2\right]\langle x|\psi',t\rangle. \tag{10.2}$$

As explained in Section 3.4, Eq. (10.2) is written in terms of $\langle x|\psi',t\rangle$ rather than in terms of $\langle x|\psi,t\rangle$, where $\langle x|\psi,t\rangle = e^{-\frac{1}{4\lambda}\int_0^t dt' w^2(t')}\langle x|\psi',t\rangle$, because (10.2) would then contain a term $\sim w^2(t)$ which is anathema to mathematicians. Actually, it doesn't matter which we use! The probability rule can be written in terms of either, $P_t(w) = \langle\psi,t|\psi,t\rangle$ or $P_t(w) = \langle\psi',t|\psi',t\rangle e^{-\frac{1}{2\lambda}\int_0^t dt' w^2(t')}$.

We guess (correctly!) that the solution has a Gaussian form

$$\langle x|\psi',t\rangle = e^{-A(t)x^2+B(t)x+C(t)}. \tag{10.3}$$

Putting (10.3) into (10.2), gives three equations, the coefficients of x^2, x and 1:

$$\frac{d}{dt}A(t) = -\frac{2i}{m}A^2(t) + \lambda, \tag{10.4a}$$

$$\frac{d}{dt}B(t) = -\frac{2i}{m}A(t)B(t) + w(t), \tag{10.4b}$$

$$\frac{d}{dt}C(t) = \frac{i}{2m}[-2A(t) + B^2(t)], \tag{10.4c}$$

which we proceed to solve, and then interpret. The answers are expressed in terms of a characteristic time $\tau \equiv \sqrt{\frac{m}{\lambda\hbar}}$ and characteristic squared length $s^2 \equiv \sqrt{\frac{\hbar}{m\lambda}}$ (note, $\frac{\hbar}{m} = \frac{s^2}{\tau}$).

10.1.1   Solution for A(t)

Equation Eq. (10.4a) is a *nonlinear* first order differential equation for the single variable $A(t)$. It is an example of what is called a Ricatti equation. By a change of variable, it can be converted to a *linear* second order equation as follows. Write $A(t) = \alpha \frac{\dot z(t)}{z(t)}$, and substitute this into Eq.(10.4a):

$$\alpha\left[\frac{\ddot z(t)}{z(t)} - \frac{\dot z^2(t)}{z^2(t)}\right] = -\frac{2i}{m}\alpha^2\frac{\dot z^2(t)}{z^2(t)} + \lambda. \tag{10.5}$$

When the choice

$$\alpha = -i\frac{m}{2} \tag{10.6}$$

is made, the nonlinear terms cancel, and we are left with the linear equation

$$\ddot z(t) = i\frac{2\lambda}{m}z(t) \equiv i\frac{2}{\tau^2}z(t), \tag{10.7}$$

with solution

$$z(t) = c_1 e^{(1+i)\frac{t}{\tau}} + c_2 e^{-(1+i)\frac{t}{\tau}}. \tag{10.8}$$

Therefore,

$$\begin{aligned} A(t) &= -i\frac{m}{2}(1+i)\frac{1}{\tau}\frac{c_1 e^{(1+i)\frac{t}{\tau}} - c_2 e^{-(1+i)\frac{t}{\tau}}}{c_1 e^{(1+i)\frac{t}{\tau}} + c_2 e^{-(1+i)\frac{t}{\tau}}} \\ &= \frac{(1-i)}{2s^2}\frac{e^{(1+i)\frac{t}{\tau}} - \frac{c_2}{c_1}e^{-(1+i)\frac{t}{\tau}}}{e^{(1+i)\frac{t}{\tau}} + \frac{c_2}{c_1}e^{-(1+i)\frac{t}{\tau}}}. \end{aligned} \tag{10.9}$$

The constant $\frac{c_2}{c_1}$ is a function of the initial width σ, which may be found by setting $A(0)$ in (10.9) equal to $1/4\sigma^2$ (from (10.1)). We won't need $\frac{c_2}{c_1}$ because, no matter what the initial width, the growing exponential dominates, and the particle's wave packet width approaches an equilibrium value

$$A(t) \xrightarrow[t\to\infty]{} \frac{1-i}{2s^2}. \tag{10.10}$$

   We could more simply have obtained this equilibrium value by just assuming A is independent of t, setting $dA(t)/dt = 0$ in Eq. (10.4a), obtaining (10.10) as the value of A. We gave this detailed solution to show how the packet width grows or shrinks until the wave packet expansion due to the Hamiltonian evolution is precisely cancelled by the wave packet shrinking due to the collapse evolution.[1]

---

[1]When $\hat H = 0$ (achieved for example by setting the mass $m = \infty$), so there is pure collapse dynamics, no mechanism to make the wave function spread, the initial Gaussian wave function evolves under the ensemble of $w(t)$'s into an ensemble of approximate position eigenfunctions (precise position eigenstates in the $t \to \infty$ limit). And, the centers of the resulting wave packets have a probability distribution that obeys the Born rule. This is shown in Chapter 9 above.

10.1.2 Solution for $B(t)$

We hereafter take $A(t)$ to have its equilibrium value, that is, we will suppose that $\sigma^2 = s^2(1+i)/4$ so $c_2 = 0$, and the initial wave function is centered at $x = 0$. Thereafter, the width of the particle wave packet is constant. What then is interesting is how the mean position and momentum evolve according to the dynamics. Thus, we next solve for $B(t)$.

Eq. (10.4b), with (10.10) inserted is

$$\frac{d}{dt}B(t) = -\frac{2i}{m}\frac{1-i}{2s^2}B(t) + w(t),$$

with solution

$$B(t) = \int_0^t dt' w(t')e^{-(1+i)\frac{(t-t')}{\tau}} \equiv B_R(t) + iB_I(t), \text{ where}$$

$$B_R(t) \equiv \int_0^t dt' w(t')e^{-\frac{(t-t')}{\tau}}\cos\frac{(t-t')}{\tau}, B_I(t) \equiv -\int_0^t dt' w(t')e^{-\frac{(t-t')}{\tau}}\sin\frac{(t-t')}{\tau}.$$

$$(10.11)$$

So, we see that both $B_R(t), B_I(t)$ have a random nature, with ensemble mean zero.

Since we know $A(t), B(t)$, using $\psi(x,t) = e^{-\frac{1}{4\lambda}\int_0^t dt' w^2(t')}\psi'(x,t)$ we can write $|\psi(x,t)|^2$ as

$$|\psi(x,t)|^2 = e^{-\frac{1}{2\lambda}\int_0^t dt' w^2(t')-\frac{1}{s^2}x^2+2xB_R(t)+2C_R(t)}$$

$$= e^{-\frac{1}{s^2}[x-s^2B_R(t)]^2-\frac{1}{2\lambda}\int_0^t dt' w^2(t')+s^2B_R^2(t)+2C_R(t)}. \quad (10.12)$$

Even if we don't yet know $C(t)$, we can read out from the first term on the right hand side of (10.12) the particle's mean position and its variance:

$$\langle\hat{X}\rangle(t) = s^2B_R(t), \quad (10.13a)$$
$$\langle\hat{X}^2\rangle(t) - \langle\hat{X}\rangle^2(t) = s^2/2. \quad (10.13b)$$

We see from (10.13a) that the mean position has a random nature, and the mean value of the ensemble position is zero, the same as the mean position of the initial Gaussian wave packet. Eq. (10.13b) confirms our choice of equilibrium wave packet width.

10.1.3 Probability Rule

The probability rule says that

$$P(w) = \int_{-\infty}^{\infty} dx |\psi(x,t)|^2 \sim e^{-\frac{1}{2\lambda}\int_0^t dt' w^2(t')+s^2B_R^2(t)+2C_R(t)}, \quad (10.14)$$

which follows from integrating (10.12). (We don't need to keep track of the normalization factor, since that is determined by $\int Dw P(w) = 1$). We now proceed to express $P(w)$ in terms of $B_R(t)$.

We need $C_R(t)$, as (10.14) shows, but we don't need $C_I(t)$ as that is a pure time-dependent phase factor, which plays no role in any calculation. We find from (10.4c),

$$\frac{d}{dt}C_R(t) = \frac{s^2}{\tau}A_I(t) - \frac{s^2}{\tau}B_R(t)B_I(t)$$

so, upon integrating,

$$2C_R(t) = -\frac{t}{\tau} - 2\frac{s^2}{\tau}\int_0^t dt' B_R(t')\bar{B}_I(t'). \tag{10.15}$$

So, we now have $C_R(t)$ expressed in terms of $B_R(t)$, $B_I(t)$.

$B_I(t)$ can be expressed in terms of $B_R(t)$ and its time-derivative. By taking the time derivative of (10.11), we find that

$$\frac{d}{dt}B_R(t) = \frac{d}{dt}\int_0^t dt' w(t')e^{-\frac{(t-t')}{\tau}}\cos\frac{(t-t')}{\tau} = w(t) - \frac{1}{\tau}B_R(t) + \frac{1}{\tau}B_I(t). \tag{10.16}$$

We use this result to find an expression for $s^2 B_R^2(t)$ which appears in the exponent of (10.14):

$$s^2 B_R^2(t) \equiv s^2 \int_0^t dt' \frac{d}{dt'}B_R^2(t') = 2s^2 \int_0^t dt' B_R(t')[w(t') - \frac{1}{\tau}B_R(t') + \frac{1}{\tau}B_I(t')]. \tag{10.17}$$

Putting (10.15) and (10.17) into (10.14), using $s^2 = \frac{1}{\lambda\tau}$, we see the integrals $\int_0^t dt' B_R(t')B_I(t')$ from $2C_R(t)$ and $s^2 B_R^2(t)$ cancel, and so

$$P(w) \sim e^{-\frac{1}{2\lambda}\int_0^t dt' w^2(t') + 2\frac{1}{\lambda\tau}\int_0^t dt' B_R(t')[w(t') - \frac{1}{\tau}B_R(t')]}e^{-\frac{t}{\tau}}$$

$$= e^{-\frac{1}{2\lambda}\int_0^t dt'[w(t') - \frac{2}{\tau}B_R(t')]^2}e^{-\frac{t}{\tau}}, \tag{10.18}$$

where we have replaced the proportionality sign by an equality. That is, $\int Dw P(w) = 1$ which we now pause to show.

10.1.4 Proof that $\int Dw P(w) = 1$

We need to show when we integrate (10.18) over Dw, we get

$$e^{\frac{t}{\tau}} = \int Dw e^{-\frac{1}{2\lambda}\int_0^t dt'[w(t') - \frac{2}{\tau}B_R(t')]^2}$$

$$= \int_{-\infty}^{\infty} \frac{1}{\sqrt{2\pi\lambda/dt}}dw(t)e^{-\frac{1}{2\lambda}dt[w(t) - \frac{2}{\tau}B_R(t)^2]}$$

$$\cdot \int_{-\infty}^{\infty} \frac{1}{\sqrt{2\pi\lambda/dt}}dw(t - dt)e^{-\frac{1}{2\lambda}dt[w(t-dt) - \frac{2}{\tau}B_R(t-dt)^2]}\dots$$

$$\tag{10.19}$$

Let's look at the first integral in (10.19), the integral over $dw(t)$. The rest of the integrals, over $dw(t - dt)$, ... go the same way. From Eq.(10.11), we display $B_R(t)$'s dependence on $w(t)$ by writing it

$$B_R(t) = \int_{t-dt}^{t} dt' w(t') e^{-\frac{(t-t')}{\tau}} \cos \frac{(t-t')}{\tau} + \int_{0}^{t-dt} dt' w(t') e^{-\frac{(t-t')}{\tau}} \cos \frac{(t-t')}{\tau})$$

$$= dt w(t) + B_R(t - dt).$$

Only the first term depends upon $w(t)$, and we can then perform the Gaussian integral in (10.19). That is correct—except that this resulting expression is wrong: it should be $B_R(t) = \frac{1}{2} dt w(t) + B_R(t - dt)$.

This can be rigorously justified because the integral over $w(t)$ is not a Riemann integral, which we have been assuming. It is a Stratonovich integral (see Chapter 15 on SDEs). However, the $1/2$ factor can be made plausible by comparing the integral over $w(t)$ and the delta function, both singular functions. Consider $\int_{-\infty}^{t} dt' \delta(t - t')$. The only contribution is from the neighborhood of $t' = t$, where one might think the delta function could be written as $\frac{1}{dt}$ and the integral as $dt \frac{1}{dt} = 1$. However, if one regards the delta function $\delta(t - t')$ as the limit of a symmetric function such as $\lim_{\sigma \to 0} \frac{1}{\sqrt{2\pi\sigma^2}} e^{-(t-t')^2/2\sigma^2}$, and does the above integral before taking the limit, one is integrating the area over only half the Gaussian, and the result is $1/2$.

Using this, the first integral in (10.19) is

$$I(t) = \int_{-\infty}^{\infty} \frac{1}{\sqrt{2\pi\lambda/dt}} dw(t) e^{-\frac{1}{2\lambda} dt [w(t) - \frac{dt}{\tau} w(t) - 2\frac{dt}{\tau} B_R(t-dt)]^2}$$

$$= \int_{-\infty}^{\infty} \frac{1}{\sqrt{2\pi\lambda/dt}} dw(t) e^{-\frac{1}{2\lambda} dt [1 - \frac{dt}{\tau}]^2 [w(t) - 2\frac{dt}{\tau[1-\frac{dt}{\tau}]} B_R(t-dt)]^2}$$

$$= \frac{1}{1 - \frac{dt}{\tau}} \approx e^{\frac{dt}{\tau}}. \tag{10.20}$$

Then, the product of all the integrals in (10.19) is

$$I(t) I(t - dt)... = \left[e^{\frac{dt}{\tau}} \right]^{\frac{t}{dt}} = e^{\frac{t}{\tau}}, \tag{10.21}$$

which proves (10.19).

10.1.5 Introducing a New White Noise $v(t)$

The probability expression (10.18) for $w(t)$ is simpler if we define a new white noise function $v(t)$:

$$v(t) = w(t) - \frac{2}{\tau} B_R(t) = w(t) - \frac{2}{\tau} \int_{0}^{t} dt' w(t') e^{-\frac{t-t'}{\tau}} \cos \frac{t-t'}{\tau}, \tag{10.22}$$

because

$$Dw P(w) = Dw e^{-\frac{1}{2\lambda} \int_{0}^{t} dt' [w(t') - \frac{2}{\tau} B_R(t')]^2} e^{-\frac{t}{\tau}},$$

$$\equiv Dv P(v) = Dv e^{-\frac{1}{2\lambda} \int_{0}^{t} dt' v^2(t')}, \tag{10.23}$$

that is, $v(t)$ is standard white noise!

This does require showing that $Dwe^{-\frac{t}{\tau}} = Dv$. That is, in going from one set of integration variables to another requires calculating the Jacobian J, where $Dv = JDw$, so we must show $J = e^{-\frac{t}{\tau}}$.

We have just seen that

$$v(t') \equiv w(t') - \frac{2}{\tau}B_R(t') \approx w(t')\left[1 - \frac{dt}{\tau}\right] + f(w(t-dt), w(t-2dt), ...w(0)). \quad (10.24)$$

Therefore, the row of the Jacobian determinant $\frac{\partial}{\partial w(t'')}v(t')$ has zeros for $t'' > t'$ and has the value $\left[1 - \frac{dt}{\tau}\right]$ on the diagonal $t'' = t'$ (and $\frac{\partial}{\partial w(t'')}f$ for $t'' < t'$ on the rest of the row). The determinant with zeros below the diagonal is just the product of the diagonal elements, so

$$J = \left[1 - \frac{dt}{\tau}\right]^{\frac{t}{dt}} = e^{-\frac{t}{\tau}}, \quad \text{QED.} \quad (10.25)$$

Next, we invert Eq. (10.22) and obtain $w(t)$ in terms of $v(t)$. To do this, take the derivative of (10.22), and then the derivative of that, using (3.26) for $\frac{d}{d}B_R(t)$ and a similar expression for $\frac{d}{d}B_I(t)$:

$$\dot{v}(t) = \dot{w}(t) - \frac{2}{\tau}\left[w(t) + \frac{1}{\tau}\left[-B_R(t) + B_I(t)\right]\right],$$

$$\ddot{v}(t) = \ddot{w}(t) - \frac{2}{\tau}\left[\dot{w}(t) + \frac{1}{\tau}\left[-\left[w(t) + \frac{1}{\tau}[-B_R(t) + B_I(t)]\right] - \frac{1}{\tau}[B_I(t) + B_R(t)]\right]\right]$$

$$= \ddot{w}(t) - \frac{2}{\tau}\dot{w}(t) + \frac{2}{\tau^2}w(t) + \frac{4}{\tau^3}B_I(t)$$

$$= \ddot{w}(t) - \frac{2}{\tau}\dot{w}(t) + \frac{2}{\tau^2}w(t) + \frac{2}{\tau}\left[-\dot{v} + \dot{w}(t) - \frac{2}{\tau}w(t) + \frac{2}{\tau^2}B_R(t)\right]$$

$$= \ddot{w}(t) - \frac{2}{\tau}\dot{v} - \frac{2}{\tau^2}\left[w(t) - \frac{2}{\tau}B_R(t)\right]$$

$$= \ddot{w}(t) - \frac{2}{\tau}\dot{v} - \frac{2}{\tau^2}v(t). \quad (10.26)$$

In going from the third line to the fourth line, we have replaced $B_I(t)$ by its equivalent, found from the first line, by solving it for $B_I(t)$. Defining the Brownian motion $b(t)$ associated to the white noise, $v(t) \equiv \frac{d}{dt}b(t)$, we Integrate (10.26) twice, with the result

$$w(t) = v(t) + \frac{2}{\tau}b(t) + \frac{2}{\tau^2}\int_0^t dt'b(t'). \quad (10.27)$$

Equating the expressions for $w(t) - v(t)$ in Eqs. (10.22) and (10.27), we find

$$B_R(t) = b(t) + \frac{1}{\tau}\int_0^t dt'b(t'). \quad (10.28)$$

To find $B_I(t)$, take its derivative in Eq. (10.11):

$$\frac{d}{dt}B_I(t) = \frac{d}{dt}\left[-\int_0^t dt' w(t')e^{-(t-t')/\tau}\sin(t-t')/\tau\right]$$

$$= -\frac{1}{\tau}[B_I(t) + B_R(t)]. \tag{10.29}$$

The solution of this first order differential equation for $B_I(t)$ is

$$B_I(t) = -\frac{1}{\tau}\int_0^t dt' B_R(t')e^{-(t-t')/\tau}$$

$$= -\frac{1}{\tau}\int_0^t dt'\left[b(t') + \frac{1}{\tau}\int_0^{t'} dt'' b(t'')\right]e^{-(t-t')/\tau}, \tag{10.30}$$

using (10.28) to express $B_R(t)$ in terms of $b(t)$, By exchanging the order of integration in the last integral in (10.30), one obtains

$$\frac{1}{\tau}\int_0^t dt'\int_0^{t'} dt'' b(t'')e^{-(t-t')/\tau} = \frac{1}{\tau}\int_0^t dt'' b(t'')\int_{t''}^t dt' e^{-(t-t')/\tau}$$

$$= \int_0^t dt'' b(t'')[1 - e^{-(t-t'')/\tau}]. \tag{10.31}$$

Putting (10.31) into (10.30) results in

$$B_I(t) = -\frac{1}{\tau}\int_0^t dt' b(t')e^{-(t-t')/\tau} - \frac{1}{\tau}\int_0^t dt'' b(t'')[1 - e^{-(t-t'')/\tau}]$$

$$= -\frac{1}{\tau}\int_0^t dt' b(t'). \tag{10.32}$$

The problem is completely solved! In the solution (10.3), we know $A(t)$ (given by (10.10)), $B(t)$ (given by (10.28), (10.32)), $C_R(t)$ (given by (10.15)), and we don't care about $C_I(t)$, all expressed in terms of standard white noise $v(t)$ (or, rather, in terms of its time-integral $b(t)$). Now we can see how the particle behaves.

10.1.6 Free Particle: Mean Position Behavior

The expectation value of the position is given in terms of $B_R(t)$ in (10.13a) so, by (10.28), we have

$$\langle \hat{X}\rangle(t) = s^2\left[b(t) + \frac{1}{\tau}\int_0^t dt' b(t')\right]. \tag{10.33}$$

So, the mean position undergoes Brownian motion plus! This result, the significance of the added integral over $b(t)$ in (10.28), is discussed in Section 3.5.1.

10.1.7 Free Particle: Mean Momentum Behavior

Next, we calculate the expectation value of the momentum for the wave function (10.3):

$$\langle \hat{P} \rangle(t) \equiv \frac{\int dx \langle \psi, t|x \rangle \frac{1}{i}\frac{d}{dx}\langle x|\psi, t\rangle}{\int dx \langle \psi, t|x \rangle \langle x|\psi, t\rangle}$$

$$= \frac{1}{i}\frac{\int dx|\langle x|\psi, t\rangle|^2[-2Ax + B]}{\int dx|\langle x|\psi, t\rangle|^2}$$

$$= \frac{1}{i}[-2A\langle \hat{X}\rangle + B] = \frac{1}{i}[-2\frac{1-i}{2s^2}s^2 B_R + B_R + iB_I]$$

$$= B_R + B_I. \tag{10.34}$$

Therefore, from (10.28), (10.32):

$$\langle \hat{P} \rangle(t) = \hbar b(t) \tag{10.35}$$

(in the last line, the condition $\hbar = 1$ has been removed).

Thus, the mean momentum simply undergoes Brownian motion. Like the mean position, the ensemble average of this mean momentum vanishes, since the particle is as likely to be moving to the left as to the right. The momentum distribution spreads with time since the probability of $b(t)$ is

$$P(b(t))db(t) = db(t)\frac{1}{\sqrt{2\pi\lambda t}}e^{-\frac{1}{2\lambda t}b^2(t)}.$$

This says that large $b(t)$, and therefore large momenta, become increasingly more probable. Physically, this is because the collapse continuously acts to narrow the position spread, and so, by the uncertainty principle, it increases the momentum spread.

10.1.8   Free Particle: Mean Energy Behavior

Therefore, the particle's energy will increase with time. So, let's calculate the expectation value of the energy:

$$\frac{1}{2m}\langle \hat{P}^2 \rangle(t) \equiv -\frac{1}{2m}\frac{\int dx \langle \psi, t|x \rangle \frac{d^2}{dx^2}\langle x|\psi, t\rangle}{\int dx \langle \psi, t|x \rangle \langle x|\psi, t\rangle}$$

$$= -\frac{1}{2m}\frac{\int dx|\langle x|\psi, t\rangle|^2[(-2Ax + B(t))^2 - 2A]}{\int dx|\langle x|\psi, t\rangle|^2}$$

$$= \frac{1}{2m}[2A - B^2(t) + 4AB(t)\langle \hat{X}\rangle(t) - 4A^2\langle \hat{X}^2\rangle]$$

$$= \frac{1}{2m}[2A - (B_R(t) + iB_I(t))^2 + 4A(B_R(t) + iB_I)(t)s^2 B_R(t)$$

$$-4A^2(s^4 B_R(t)^2 + \frac{1}{2}s^2)]$$

$$= \frac{1}{2m}\left[(B_R(t) + B_I(t))^2 + \frac{1}{s^2}\right] = \frac{\hbar^2}{2m}\left[b^2(t) + \frac{1}{s^2}\right]. \tag{10.36}$$

In going from the third step to the fourth step, Eqs. (10.13a), (10.13b) were used. In the last step, we have used the expressions (10.28) and (10.32) for $B_R(t), B_I(t)$.

From (10.35), (10.36), we find the standard deviation of the momentum is, like that of the position, constant:

$$\langle \hat{P}^2 \rangle(t) - \langle \hat{P} \rangle^2(t) = \frac{\hbar^2}{s^2}.$$

The ensemble average of the energy (10.36),

$$\frac{1}{2m}\overline{\langle \hat{P}^2 \rangle}(t) = \frac{\hbar^2}{2m}\left[\overline{b^2(t)} + \frac{1}{s^2}\right] = \frac{\hbar^2}{2m}\left[\lambda t + \frac{1}{s^2}\right].\tag{10.37}$$

increases linearly with time, which is the universal behavior, with or without a potential, as shown in the Supplement to Chapter 4 and discussed in Section 4.4.

We have completely solved this problem.This concludes our presentation of the free particle wave function calculation.

10.2 Free Particle Density Matrix

There is nothing that can be learned from the density matrix that cannot be found by taking the stochastic average of something related to the wave function which we now know.

Nonetheless, it is interesting to see what can be learned from the density matrix evolution equation alone. This is a useful exercise, for there are problems where the state vector is difficult to find, but for which the density matrix is tractable. Or, problems for which one really is only interested in the ensemble behavior anyway.

For example, we will use the density matrix to calculate $\langle \hat{X}^2 \rangle(t)$ (this was done using the wave function at the end of Section 3.5.1).

10.2.1 Getting Moments from the Lindblad Equation

The density matrix evolution equation (Lindblad equation) is, according to Eq. (3.12),

$$\frac{d}{dt}\hat{\rho}(t) = -i\frac{1}{2m}[\hat{P}^2, \hat{\rho}(t)] - \frac{\lambda}{2}[\hat{X}, [\hat{X}, \hat{\rho}(t)]] \text{ or }\tag{10.38a}$$

$$\frac{\partial}{\partial t}\langle x|\hat{\rho}(t)|x'\rangle = i\frac{1}{2m}\left[\frac{\partial^2}{\partial x^2} - \frac{\partial^2}{\partial x'^2}\right]\langle x|\hat{\rho}(t)|x'\rangle - \frac{\lambda}{2}(x - x')^2\langle x|\hat{\rho}(t)|x'\rangle.\tag{10.38b}$$

A useful thing about the density matrix evolution equation is that sometimes one does not need to actually solve it to calculate something. That is the case here. Manipulations using the evolution equation alone suffice.

Thus, from (10.38a), and the cyclic property of the trace operation (see Eq. (E4)), which is used in the second and fourth lines below:

$$\frac{d}{dt}\overline{\langle \hat{P}\rangle}(t) \equiv \frac{d}{dt}\text{Tr}\left(\hat{P}\hat{\rho}(t)\right) = -i\text{Tr}\left(\hat{P}\frac{1}{2m}[\hat{P}^2, \hat{\rho}(t)]\right) - \text{Tr}\left(\hat{P}\frac{\lambda}{2}[\hat{X}, [\hat{X}, \hat{\rho}(t)]]\right)$$

$$= -\frac{i}{2m}\text{Tr}\left(\hat{\rho}(t)[\hat{P}, \hat{P}^2]\right) - \frac{\lambda}{2}\text{Tr}\left(\hat{\rho}(t)[\hat{X}, [\hat{X}, \hat{P}]]\right) = 0,$$

so $\overline{\langle \hat{P}\rangle}(t) = \langle \hat{P}\rangle(0)$. $\hspace{3cm}$ (10.39a)

$$\frac{d}{dt}\overline{\langle \hat{X}\rangle}(t) = \frac{d}{dt}\text{Tr}\left(\hat{X}\hat{\rho}(t)\right) = -\frac{i}{2m}\text{Tr}\left(\hat{\rho}(t)[\hat{X}, \hat{P}^2]\right) - \frac{\lambda}{2}\text{Tr}\left(\hat{\rho}(t)[\hat{X}, [\hat{X}, \hat{X}]]\right)$$

$$= \frac{1}{m}\text{Tr}\left(\hat{\rho}(t)\hat{P}\right) \equiv \langle \hat{P}\rangle(t),$$

so $\overline{\langle \hat{X}\rangle}(t) = \langle \hat{P}\rangle(0)t + \langle \hat{X}\rangle(0)$. $\hspace{2.5cm}$ (10.39b)

Since initially $\psi(x,0) = \frac{1}{[\pi s^2]^{1/4}}e^{-x^2/2s^2}$ (the wave function's initial width is its equi-librium width), then $\langle \hat{P}\rangle(0) = 0$ and $\langle \hat{X}\rangle(0) = 0$, so $\overline{\langle \hat{X}\rangle}(t) = 0, \overline{\langle \hat{P}\rangle}(t) = 0$. So, without knowing the density matrix, we have found that the ensemble average of the expectation value of position and momentum vanishes for all time. Physically, this is because, for any possible wave function motion to the left, there is an identical motion of another wave function to the right. Of course, this agrees with the ensemble average of the expectation values of position and momentum found from the wave function, (10.33) and (10.35), since $\overline{b(t)} = 0$.

More complicated, but still not as complicated as solving for the state vector or density matrix, is

$$\frac{d}{dt}\overline{\langle \hat{X}^2\rangle}(t) = -\frac{i}{2m}\text{Tr}\left(\hat{\rho}(t)[\hat{X}^2, \hat{P}^2]\right) - \frac{\lambda}{2}\text{Tr}\left(\hat{\rho}(t)[\hat{X}, [\hat{X}, \hat{X}^2]]\right)$$

$$= \frac{1}{m}\langle \hat{X}\hat{P} + \hat{P}\hat{X}\rangle(t), \hspace{2cm} (10.40a)$$

$$\frac{d}{dt}\overline{\langle \hat{X}\hat{P} + \hat{P}\hat{X}\rangle}(t) = \frac{2}{m}\langle \hat{P}^2\rangle(t), \hspace{2cm} (10.40b)$$

$$\frac{d}{dt}\overline{\langle \hat{P}^2\rangle}(t) = \lambda \hspace{4cm} (10.40c)$$

So, we have three coupled first order differential equations, which means that we can solve for these quantities, given the initial conditions.

Using the initial wave function we find the initial values $\langle \hat{P}^2\rangle(0) = \frac{1}{s^2}$, $\langle \hat{X}\hat{P} + \hat{P}\hat{X}\rangle(0) = 1$, $\langle \hat{X}^2\rangle(0) = \frac{s^2}{2}$. Then, upon integrating these equations (in reverse order, (10.40c) first, (10.40a) last) and utilizing $\frac{1}{\tau} = \sqrt{\frac{\lambda}{m}}$, $s^2 = \frac{1}{\sqrt{m\lambda}}$, we have:

$$\overline{\langle \hat{X}^2\rangle}(t) = s^2\left[\frac{1}{3}\left(\frac{t}{\tau}\right)^3 + \left(\frac{t}{\tau}\right)^2 + \frac{t}{\tau} + \frac{1}{2}\right], \hspace{1.5cm} (10.41a)$$

$$\overline{\langle \hat{X}\hat{P} + \hat{P}\hat{X}\rangle}(t) = \left(\frac{t}{\tau}\right)^2 + 2\frac{t}{\tau} + 1, \hspace{2cm} (10.41b)$$

$$\overline{\langle \hat{P}^2\rangle}(t) = \lambda t + \frac{1}{s^2}. \hspace{3cm} (10.41c)$$

Some of these results were obtained previoously.

10.2.2   Free Particle Density Matrix

The three moments (10.41a), (10.41b), (10.41c) *are* sufficient to completely determine
the density matrix itself, to find the solution of (10.38b). Because the individual wave
functions are Gaussians, the density matrix in the position basis has to have the
Hermitian Gaussian form,

$$\langle x|\hat{\rho}(t)|x'\rangle = \sqrt{\frac{2\alpha_R(t) - \beta(t)}{\pi}}e^{-\alpha(t)x^2 - \alpha^*(t)x'^2 + \beta(t)xx'}, \tag{10.42}$$

depending upon three real functions of t, $\alpha_R(t), \alpha_I(t), \beta(t)$ (the coefficient in front of
the exponential follows from demanding that $\text{Tr}\hat{\rho}(t) = \int_{-\infty}^{\infty} dx\langle x|\hat{\rho}(t)|x\rangle = 1$).
    We use (10.42) to calculate the three quantities (10.41a), (10.41b), (10.41c), which
determines $\alpha_R(t), \alpha_I(t), \beta(t)$:

$$\overline{\langle \hat{X}^2\rangle}(t) = \sqrt{\frac{2\alpha_R(t) - \beta(t)}{\pi}} \int_{-\infty}^{\infty} dx x^2 e^{-[2\alpha_R(t) - \beta(t)]x^2} = \frac{1}{2[2\alpha_R(t) - \beta(t)]}, \tag{10.43a}$$

$$\overline{\langle \hat{X}\hat{P} + \hat{P}\hat{X}\rangle}(t) = -i + 2\overline{\langle \hat{X}\hat{P}\rangle}(t)$$

$$= -i + 2\sqrt{\frac{2\alpha_R(t) - \beta(t)}{\pi}} \int_{-\infty}^{\infty} dx x \frac{1}{i}[-2\alpha x + \beta x]e^{-[2\alpha_R(t) - \beta(t)]x^2}$$

$$= -i - 2i\frac{[-2\alpha + \beta]}{2[2\alpha_R(t) - \beta(t)]} = -\frac{2\alpha_I}{[2\alpha_R(t) - \beta(t)]}, \tag{10.43b}$$

$$\overline{\langle \hat{P}^2\rangle}(t) = 2\alpha - \frac{[-2\alpha + \beta]^2}{2[2\alpha_R(t) - \beta(t)]}$$

$$= \frac{2\alpha_R(t) + \beta(t)}{2} + \frac{2\alpha_I^2(t)}{2\alpha_R(t) - \beta(t)}. \tag{10.43c}$$

The solutions of these equations for $\alpha_R(t), \alpha_I(t), \beta(t)$ are

$$\alpha_R(t) = \frac{\overline{\langle \hat{P}^2\rangle}(t)}{2} + \frac{1}{8\overline{\langle \hat{X}^2\rangle}(t)} - \frac{\overline{\langle \hat{X}\hat{P} + \hat{P}\hat{X}\rangle}(t)^2}{8\overline{\langle \hat{X}^2\rangle}(t)},$$

$$\alpha_I(t) = -\frac{\overline{\langle \hat{X}\hat{P} + \hat{P}\hat{X}\rangle}(t)}{4\overline{\langle \hat{X}^2\rangle}(t)},$$

$$\beta(t) = 2\alpha_R - \frac{1}{2\overline{\langle \hat{X}^2\rangle}(t)}, \tag{10.44}$$

Putting (10.41a), (10.41b), (10.41c) into Eqs. (10.44) gives $\alpha_R(t), \alpha_I(t), \beta(t)$, which
we insert into the expression for the density matrix (10.42), and thus the problem is
completely solved.

10.2.3 Density Matrix at Small and Large Times

These expressions are fairly complicated, so we will only consider them in limiting cases. For small t/τ, we get:

$$\alpha_R(t) \xrightarrow[t\to0]{} \frac{1}{2s^2}[1 + o(t/\tau)],$$

$$\alpha_I(t) \xrightarrow[t\to0]{} -\frac{1}{2s^2}[1 + o(t/\tau)^2],$$

$$\beta(t) \xrightarrow[t\to0]{} \frac{1}{2s^2}[o(t/\tau)^2] \tag{10.45}$$

so, using these values in (10.42), the initial density matrix is, correctly, $\langle x|\rho(0)|x'\rangle = \psi(x,0)\psi^*(x',0) = e^{-Ax^2}e^{-A^*x'^2}$.

For large t/τ,

$$\alpha_R(t) \xrightarrow[t\to\infty]{} \frac{1}{8s^2}[t/\tau + o(1)],$$

$$\alpha_I(t) = \xrightarrow[t\to\infty]{} \frac{1}{8s^2}[0 + (t/\tau)^{-1}]$$

$$\beta(t) \xrightarrow[t\to\infty]{} \frac{1}{4s^2}[t/\tau + o(1)]. \tag{10.46}$$

The expression for (10.42) may be written as

$$\langle x|\rho(t)|x'\rangle = \frac{1}{\sqrt{2\pi\langle\hat{X}^2\rangle(t)}}e^{-i\alpha_I(t)[x^2-x'^2]}e^{-\alpha_R(t)(x-x')^2}e^{-\frac{xx'}{2\langle\hat{X}^2\rangle(t)}}. \tag{10.47}$$

From (10.47), (10.43a), (10.46), we see that, on the diagonal, for large t,

$$\langle x|\rho(t)|x\rangle = \frac{1}{\sqrt{2\pi\langle\hat{X}^2\rangle(t)}}e^{-\frac{1}{2\langle\hat{X}^2\rangle(t)}x^2} \xrightarrow[t\to\infty]{} \frac{1}{\sqrt{(2/3)\pi s^2(t/\tau)^3}}e^{-\frac{1}{(2/3)s^2(t/\tau)^3}x^2}, \tag{10.48}$$

there is a very large spread of the ensemble, which occurs because the centers of the wave packets have been undergoing random walk.

Also for large t, the off-diagonal elements become

$$\langle x|\rho(t)|x'\rangle \to \frac{1}{\sqrt{(2/3)\pi s^2(t/\tau)^3}}e^{-\frac{t/\tau}{8s^2}(x-x')^2}. \tag{10.49}$$

This displays the asymptotic vanishing of the off-diagonal elements of the density matrix, characteristic of collapse.

This concludes our discussion of the free particle density matrix.

10.3 Harmonic Oscillator Problem

In a footnote at the beginning of Section 3.5, it was stated that the density matrix treatment of the one-dimensional harmonic oscillator subjected to collapse in position would be discussed here.

There is no obvious collapse behavior here, because the Hamiltonian part of the evolution dramatically interferes. What is most notable here is that the collapse, in continually narrowing the wave packet, gives a continual energy increase to the particle.

For this problem without collapse, it is well known that the one-dimensional harmonic oscillator is most simply and transparently discussed in terms of the annihilation and creation operators $\hat{a}^\dagger, \hat{a}$, (obeying $[\hat{a}, \hat{a}^\dagger] = 1$) rather than in terms of the position and momentum operators $\hat{X} = \sqrt{\frac{\hbar}{2m\omega}}(\hat{a}^\dagger + \hat{a})$, $\hat{P} = i\sqrt{\frac{m\omega}{2\hbar}}(\hat{a}^\dagger - \hat{a})$, and this is true of the collapse problem as well. This is because the energy eigenstates are written simply as $|n\rangle = \frac{1}{\sqrt{n!}}\hat{a}^{\dagger n}|0\rangle$, where $|0\rangle$ is the lowest energy state vector (ground state).

We consider the harmonic oscillator to be initially in the ground state, $|\psi, 0\rangle = |0\rangle$ (whose wave function is $\langle x|\psi, 0\rangle \sim e^{-m\omega x^2/2}$). Each state vector in the ensemble of state vectors has the form $|\psi, t\rangle = \sum_{n=0}^{\infty} c_n(t)|n\rangle$. since the $|n\rangle$ are a complete set of (orthonormal) vectors Here we will discuss the density matrix in the energy eigenstate basis $|n\rangle$, to see how collapse of the whole ensemble excites the initial ground state.

Eq. (3.12) is the density matrix evolution equation to be solved:

$$\frac{d}{dt}\hat{\rho}(t) = -i[\frac{\hat{P}^2}{2m} + \frac{m\omega^2\hat{X}^2}{2}, \hat{\rho}(t)] - \frac{\lambda}{2}[\hat{X}, [\hat{X}, \hat{\rho}(t)]] \text{ or}$$

$$= -i[\omega\hat{a}^\dagger\hat{a}, \hat{\rho}(t)] - \frac{\lambda'}{2}[\hat{a}^\dagger + \hat{a}, [\hat{a}^\dagger + \hat{a}, \hat{\rho}(t)]], \qquad (10.50)$$

(defining $\lambda' \equiv \lambda\frac{\hbar}{2m\omega}$). The initial density matrix is $\hat{\rho}(0) = |0\rangle\langle 0|$ satisfying $\hat{a}\hat{\rho}(0) = \hat{\rho}(0)\hat{a}^\dagger = 0$.

10.3.1 Expectation Values of the Linear Terms

With use of (10.50), the expectation values of $\hat{a}^\dagger, \hat{a}$ for the ensemble of state vectors are seen to obey

$$\frac{d}{dt}\overline{\langle\hat{a}\rangle(t)} \equiv \frac{d}{dt}\text{Tr}\hat{a}\hat{\rho}(t) = -i\omega\text{Tr}\hat{a}[\hat{a}^\dagger\hat{a}, \hat{\rho}(t)] - \frac{\lambda'}{2}\text{Tr}\hat{a}[\hat{a}^\dagger + \hat{a}, [\hat{a}^\dagger + \hat{a}, \hat{\rho}(t)]]$$

$$= -i\omega\text{Tr}\hat{\rho}(t)[\hat{a}, \hat{a}^\dagger\hat{a}] - \frac{\lambda'}{2}\text{Tr}\hat{\rho}(t)][\hat{a}^\dagger + \hat{a}, [\hat{a}^\dagger + \hat{a}, \hat{a}] = -i\omega\overline{\langle\hat{a}\rangle(t)},$$

$$\frac{d}{dt}\overline{\langle\hat{a}^\dagger\rangle(t)} = i\omega\overline{\langle\hat{a}^\dagger\rangle(t)}. \qquad (10.51)$$

(in the second line we have used the cyclic property of the trace to pull $\hat{\rho}(t)$ out of the commutator).

The solutions are $\overline{\langle\hat{a}\rangle(t)} = \langle\hat{a}\rangle(0)e^{-i\omega t} = 0$, $\overline{\langle\hat{a}^\dagger\rangle(t)} = \langle\hat{a}^\dagger\rangle(0)e^{i\omega t} = 0$: these vanish because the initial state is the ground state, so $\langle\hat{a}^\dagger\rangle(0) \equiv \langle 0|\hat{a}^\dagger|0\rangle = 0$, $\langle\hat{a}\rangle(0) \equiv \langle 0|\hat{a}|0\rangle = 0$.

10.3.2 Expectation Values of the Quadratic Terms

The quadratic expectation values, as in the case of the free particle, satisfy a closed set of first order differential equations, which can be solved subject to the initial conditions, and from which the whole density matrix may be constructed.

Multiplying (10.50) by respectively $\hat{a}^2$ and $\hat{a}^\dagger \hat{a}$ and taking the trace, followed by cyclic permutation within the trace to get $\hat{\rho}(t)$ outside the commutators, as was done in (10.51), results in

$$\frac{d}{dt}\overline{\langle \hat{a}^2 \rangle}(t) = -2i\omega\overline{\langle \hat{a}^2 \rangle}(t) - \lambda',$$

$$\frac{d}{dt}\overline{\langle \hat{a}^\dagger a \rangle}(t) = \lambda', \tag{10.52}$$

with solutions that vanish at $t = 0$ (since $\overline{\langle \hat{a}^2 \rangle}(0) \equiv \langle 0|\hat{a}^2|0\rangle = 0$, etc.):

$$\overline{\langle \hat{a}^2 \rangle}(t) = -\lambda' e^{-i\omega t}\frac{\sin \omega t}{\omega}, \ \overline{\langle \hat{a}^{\dagger 2} \rangle}(t) = \overline{\langle \hat{a}^2 \rangle}(t)^*, \overline{\langle \hat{a}^\dagger \hat{a} \rangle}(t) = \lambda' t. \tag{10.53}$$

In particular, we note again that the energy increases linearly with time, $\hbar\omega\overline{\langle \hat{a}^\dagger \hat{a} \rangle}(t) = \hbar\omega\lambda' t = \lambda t\frac{\hbar^2}{2m}$, and is independent of the potential (evidenced by no dependence on ω).

10.3.3   Constructing the Density Matrix

Now we look for a density matrix giving these expectation values. The formal solution of (10.50) looks like the exponential of a quadratic form. We make the Ansatz

$$\hat{\rho}(t) = C(t)e^{R(t)\hat{a}^{\dagger 2}}e^{S(t)\hat{a}^\dagger_L \hat{a}_R}\hat{\rho}(0)e^{R^*(t)\hat{a}^2}, \tag{10.54}$$

where, as discussed in Section 2.4, the subscripts L and R mean

$$e^{S(t)a^\dagger_L \hat{a}_R}\hat{\rho}(0) \equiv \sum_{n=0}^{\infty}\frac{S^n(t)}{n!}\hat{a}^{\dagger n}\rho(0)\hat{a}^n. \tag{10.55}$$

The three expectation values (10.53) will enable us to find the three parameters $R(t)$ (real and imaginary parts are two parameters) and the real $S(t)$. The normalization factor $C(t)$ is to be found from $\mathrm{Tr}\hat{\rho}(t) = 1$.

To find the expectation value of the quadratic operators with the density matrix (10.51), we will need to know $\hat{a}\hat{\rho}(t)$. This requires knowing how $\hat{a}$ acts on the two factors to the left of $\hat{\rho}(0)$ in the expression (10.54) for $\hat{\rho}(t)$:

$$\hat{a}\hat{F}(t) \equiv \hat{a}e^{R(t)\hat{a}^{\dagger 2}}e^{S(t)\hat{a}^\dagger_L \hat{a}_R}\hat{\rho}(0) = e^{R(t)\hat{a}^{\dagger 2}}e^{-R(t)\hat{a}^{\dagger 2}}\hat{a}e^{R(t)\hat{a}^{\dagger 2}}e^{S(t)\hat{a}^\dagger_L \hat{a}_R}\hat{\rho}(0)$$

$$= e^{R(t)\hat{a}^{\dagger 2}}[\hat{a} + 2R(t)\hat{a}^\dagger]\sum_{n=0}^{\infty}\frac{S^n(t)}{n!}\hat{a}^{\dagger n}\rho(0)\hat{a}^n$$

$$= e^{R(t)\hat{a}^{\dagger 2}}S(t)\sum_{n=1}^{\infty}\frac{S^{n-1}(t)}{(n-1)!}\hat{a}^{\dagger(n-1)}\rho(0)\hat{a}^{n-1}\hat{a} + 2R\hat{a}^\dagger \hat{F}(t)$$

$$= 2R(t)\hat{a}^\dagger \hat{F}(t) + S(t)\hat{F}(t)\hat{a}, \quad \text{so}$$

$$\hat{a}\hat{\rho}(t) = 2R(t)\hat{a}^\dagger \hat{\rho}(t) + S(t)\hat{\rho}(t)\hat{a}. \tag{10.56}$$

In the first line of (10.56), we have simply put $\hat{1} = e^{R(t)\hat{a}^{\dagger 2}}e^{-R(t)\hat{a}^{\dagger 2}}$ in front of $\hat{a}$. In the second line, we have used the identity

$$e^{-\hat{Z}}\hat{a}e^{\hat{Z}} = \hat{a} + [\hat{Z},\hat{a}] + \frac{1}{2!}[\hat{Z},[\hat{Z},\hat{a}]] + ...$$

In the third line we have used

$$\hat{a}\hat{a}^{\dagger n}\hat{\rho}(0) = \{\hat{a}\hat{a}^{\dagger n} - \hat{a}^{\dagger n}\hat{a} + \hat{a}^{\dagger n}\hat{a}\}\hat{\rho}(0) = [\hat{a},\hat{a}^{\dagger n}]\hat{\rho}(0) + 0 = n\hat{a}^{\dagger(n-1)}\hat{\rho}(0).$$

Now we are in a position to evaluate the traces of the quadratic forms, expressed in terms of R, S. Then, we equate the results to the time-dependent expressions (10.53), and so obtain R, S and thus the density matrix, in terms of t.

Using (10.56)^2 we obtain

$$\begin{aligned}
\overline{\langle\hat{a}^\dagger\hat{a}\rangle}(t) &\equiv \mathrm{Tr}\hat{a}^\dagger\hat{a}\rho(t) = \mathrm{Tr}\hat{a}^\dagger[2R(t)\hat{a}^\dagger\hat{\rho}(t) + S(t)\hat{\rho}(t)\hat{a}] \\
&= 2R(t)\overline{\langle\hat{a}^{\dagger 2}\rangle}(t) + S(t)\overline{\langle\hat{a}\hat{a}^\dagger\rangle}(t) \\
&= 2R(t)\overline{\langle\hat{a}^{\dagger 2}\rangle}(t) + S(t)[1 + \overline{\langle\hat{a}^\dagger\hat{a}\rangle}(t)], \\
\overline{\langle\hat{a}^2\rangle}(t) &= \mathrm{Tr}\hat{a}[2R(t)\hat{a}^\dagger\hat{\rho}(t) + S(t)\hat{\rho}(t)\hat{a}] \\
&= 2R(t)[1 + \overline{\langle\hat{a}^\dagger\hat{a}\rangle}(t)] + S(t)\overline{\langle\hat{a}^2\rangle}(t).
\end{aligned} \tag{10.57}$$

These are two linear inhomogeneous equations for $R(t), S(t)$. The solution is

$$\begin{aligned}
R(t) &= \frac{1}{2}\frac{\overline{\langle\hat{a}^2\rangle}(t)}{[1 + \overline{\langle\hat{a}^\dagger\hat{a}\rangle}(t)]^2 - \overline{\langle\hat{a}^{\dagger 2}\rangle}(t)\overline{\langle\hat{a}^2\rangle}(t)]} \\
&= -\frac{1}{2}\frac{\lambda'e^{-i\omega t}}{[1 + \lambda't]^2 - [\frac{\lambda'}{\omega}\sin\omega t]^2}\frac{\sin\omega t}{\omega},
\end{aligned}$$

$$\begin{aligned}
S(t) &= 1 - 2R(t)\frac{1 + \overline{\langle\hat{a}^\dagger\hat{a}\rangle}(t)}{\overline{\langle\hat{a}^2\rangle}(t)} \\
&= 1 - \frac{1 + \lambda't}{[1 + \lambda't]^2 - [\frac{\lambda'}{\omega}\sin\omega t]^2}
\end{aligned} \tag{10.58}$$

where we have inserted the expressions from Eqs. (10.53). We note that $R(0) = S(0) = 0$, so the density matrix is correct at $t = 0$, provided $C(0) = 1$.

To find $C(t)$, we use $\mathrm{Tr}\rho(t) = 1$. Taking the time derivative of the trace of (10.54), using $R(t) = \frac{1}{2}\frac{\overline{\langle\hat{a}^2\rangle}(t)}{D(t)}$, $S = 1 - \frac{1+\lambda't}{D(t)}$, where $D(t) \equiv [1 + \lambda't]^2 - \overline{\langle\hat{a}^{\dagger 2}\rangle}(t)\overline{\langle\hat{a}^2\rangle}(t)$, we get

---

[2]By taking the trace of (10.56) we obtain $(1 - S)\hat{a}(t) = 2R\overline{\hat{a}^\dagger}(t)$. The solution of this and its complex conjugate $(1 - S^*)\overline{\hat{a}^\dagger}(t) = 2R^*\overline{\hat{a}}(t)$ is $\hat{a}(t) = \hat{a}^\dagger(t) = 0$, as required by (10.51) et seq.

$$\frac{d}{dt}\text{Tr}\rho(t) = 0 = \frac{\dot{C}(t)}{C(t)} + \dot{R}(t)\overline{\langle \hat{a}^{\dagger 2}\rangle}(t) + \dot{S}(t)\overline{\langle \hat{a}\hat{a}^{\dagger}\rangle}(t) + \dot{R}^*(t)\overline{\langle \hat{a}^2\rangle}(t) \text{ so}$$

$$-\frac{\dot{C}(t)}{C(t)} = \frac{1}{2}\left[\overline{\langle \hat{a}^{\dagger 2}\rangle}(t)\frac{d}{dt}\frac{\overline{\langle \hat{a}^2\rangle}(t)}{D(t)} + \overline{\langle \hat{a}^2\rangle}(t)\frac{d}{dt}\frac{\overline{\langle \hat{a}^{\dagger 2}\rangle}(t)}{D(t)}\right] - (1+\lambda't)\frac{d}{dt}\frac{1+\lambda't}{D(t)}$$

$$= -\frac{D(t)}{2}\frac{d}{dt}\frac{D(t)}{D(t)^2} = \frac{1}{2D(t)}\frac{d}{dt}D(t). \tag{10.59}$$

Therefore,

$$C(t) = \frac{1}{\sqrt{D(t)}} = \frac{1}{\sqrt{[1+\lambda't]^2 - [\frac{\lambda'}{\omega}\sin\omega t]^2}}. \tag{10.60}$$

where $C(0) = 1$

Thus, knowing $R(t)$, $S(t)$ (Eqs.(10.58)) and $C(t)$ (Eq.(10.60)), we have obtained the complete density matrix (10.54). One can directly show, with a great deal of computation, that it satisfies the density matrix evolution equation (10.50).

Now we can draw some consequences.

First, we can calculate the ensemble average occupation number $\langle n|\rho(t)|n\rangle$ of any state $|n\rangle = \frac{1}{\sqrt{n!}}\hat{a}^{\dagger n}|0\rangle$, by expanding the exponentials and keeping only terms $\sim \langle n|\hat{a}^{\dagger n}\rho(0)\hat{a}^n|n\rangle$. Thus,

$$\langle 0|\hat{\rho}(t)|0\rangle = C(t), \tag{10.61a}$$

$$\langle 1|\hat{\rho}(t)|1\rangle = C(t)S(t), \tag{10.61b}$$

$$\langle 2|\hat{\rho}(t)|2\rangle = C(t)\langle 0|\frac{\hat{a}^2}{\sqrt{2}}\left[R(t)\hat{a}^{\dagger 2}\rho(0)R^*(t)\hat{a}^2 + \frac{S^2(t)}{2}\hat{a}^{\dagger 2}\rho(0)\hat{a}^2\right]\frac{\hat{a}^{\dagger 2}}{\sqrt{2}}|0\rangle$$

$$= C(t)[2|R(t)|^2 + S^2(t)], \tag{10.61c}$$

$$\langle 3|\hat{\rho}(t)|3\rangle = C(t)[3!|R(t)|^2 S(t) + S^3(t)], \tag{10.61d}$$

etc.

We see from (Eq.(10.61a)) that the initial no-particle state is excited, since $C(t)$ falls from its initial value of 1. When, say, $t \geq 10/\omega$, by Eq.(10.60), $C(t) \approx \frac{1}{1+\lambda't}$. The other states, initially unoccupied ($R(0) = S(0) = 0$) grow and then diminish. Since, for large t, $R(t) \to 0, S(t) \to \frac{\lambda't}{1+\lambda't}$, it follows from (10.54), (10.55) that, asymptotically, all states become equally populated.

$$\langle n|\hat{\rho}(t)|n\rangle \xrightarrow[t\to\infty]{} \frac{1}{\lambda't}\langle n|e^{S\hat{a}_L^{\dagger}\hat{a}_R}\hat{\rho}(0)|n\rangle$$

$$= \frac{1}{\lambda't}\sum_{m=0}^{\infty}\frac{S^m}{m}\langle n|\hat{a}^{\dagger m}|0\rangle\langle 0|\hat{a}^m|n\rangle = \frac{1}{\lambda't}\sum_{m=0}^{\infty}S^m\langle n|m\rangle\langle m|n\rangle$$

$$= \frac{1}{\lambda't}\left[\frac{\lambda't}{1+\lambda't}\right]^n = \frac{1}{\lambda't}\left[1+\frac{1}{\lambda't}\right]^{-n} = \frac{1}{\lambda't}\left[1-\frac{n}{\lambda't}+\frac{n(n+1)}{2!(\lambda't)^2}-\cdots\right]$$

$$\to \frac{1}{\lambda't} \text{ for } n \ll \lambda't \tag{10.62}$$

(using $|m\rangle = \frac{\hat{a}^{m\dagger}}{\sqrt{m!}}|0\rangle$ and $\langle n|m\rangle = \delta_{nm}$ in the second line).

Calculations are greatly simplified if we only calculate at times that are an integer multiple of the oscillator period $T = \frac{2\pi}{\omega}$, for then $R(t) = 0$ exactly, not just for large t. Alternatively, for $t \gg T$, one can set $R(t) \approx 0$ as then $R(t) \ll S(t)$). Then we have the simpler expression for the density matrix:

$$\hat{\rho}(t) = \frac{1}{1+\lambda't}e^{\frac{\lambda't}{1+\lambda't}\hat{a}_L^\dagger \hat{a}_R}|0\rangle\langle 0| \equiv \frac{1}{1+\lambda't}\sum_{n=0}^{\infty}\left[\frac{\lambda't}{1+\lambda't}\right]^n \frac{1}{n!}\hat{a}^{n\dagger}|0\rangle\langle 0|\hat{a}^n$$

$$= \frac{1}{1+\lambda't}\sum_{n=0}^{\infty}\left[\frac{\lambda't}{1+\lambda't}\right]^n |n\rangle\langle n|, \tag{10.63}$$

Although this density matrix is diagonal, that does *not* mean it describes an ensemble of energy eigenstates $|n\rangle$. The states that make up the ensemble are *superpositions* of energy eigenstates and, as in the example given in Section 2.4, the off-diagonal terms contrive to cancel. However, if at time t suddenly a measurement is performed to determine which state $|n\rangle$ is occupied, the coefficient of $|n\rangle\langle n|$ is its probability of occurrence $p_n(t)$:

$$p_n(t) = \frac{(\lambda't)^n}{(1+\lambda't)^{n+1}} = \frac{1}{1+\lambda't}\frac{1}{[\frac{1}{\lambda't}+1]^n} \xrightarrow{t\to\infty} \frac{1}{\lambda't}e^{-\frac{n}{\lambda't}} \tag{10.64}$$

(using the approximation $[1+\epsilon]^n \approx e^{\epsilon n}$). Setting $n = 0$, we see that the ground state probability diminishes, inversely proportional to t, from its initial value 1 as the oscillator gets excited. The maximum value of the probability occurs at $\lambda't = n$, when the probability has the value $p_n(n/\lambda') = \frac{1}{n}e^{-1}$. The lower energy states are more likely to get populated early. Roughly speaking, the ensemble of possible solutions has all states with $n < \lambda't$ tending toward equal probability, and states with $n > \lambda't$ waiting to be excited,

Interestingly, the density matrix (10.63) has the same form as the *thermal density matrix*. This is the density matrix describing an oscillator in a thermal bath, that is, the oscillator in thermal equilibrium at temperature T.

According to statistical physics, a system in thermal equilibrium that is described by the Hamiltonian $\hat{H}$ has probability $p(E) \sim e^{-\frac{E}{kT}}$ (k is Boltzmann's constant) of being in the energy eigenstate $|E\rangle$ (eigenvalues E). Therefore, the system's density matrix is

$$\rho = C\sum_E e^{-\frac{E}{kT}}|E\rangle\langle E|$$

where C is the inverse of the trace of the sum. For the harmonic oscillator then, this is

$$\hat{\rho}_{th} = [1-e^{-\frac{\hbar\omega}{kT}}]\sum_{n=0}^{\infty} e^{-\frac{\hbar\omega}{kT}n}|n\rangle\langle n|. \tag{10.65}$$

This is the same as (10.63) provided we identify

$$kT = \frac{\hbar\omega}{\ln[1+\frac{1}{\lambda't}]}. \tag{10.66}$$

Therefore, the CSL excitation process may be described as the same as if the oscillator was in a thermal bath, always in thermal equilibrium, as the temperature is raised from $T = 0$ to $T = \infty$ as time increases: for large $\lambda' t$, from (10.66), we have $kT \approx \hbar\omega\lambda' t$.

The mean energy in the thermal state is, using $\hat{H}|n\rangle = \hbar\omega|n\rangle$ (having subtracted the zero point energy $\hbar\omega/2$), from (10.65),

$$\text{tr} H\rho_{th} = [1 - e^{-\frac{\hbar\omega}{kT}}]\hbar\omega \sum_{n=0}^{\infty} n e^{-\frac{\hbar\omega n}{kT}}.$$

Replacing the factor n by a derivative, and using $\sum_{n=0}^{\infty} z^n = \frac{1}{1-z}$, we obtain

$$\text{tr}\hat{H}\rho_{th} = -\hbar\omega[1 - e^{-\frac{\hbar\omega}{kT}}]\frac{d}{d\frac{\hbar\omega}{kT}}\left[\frac{1}{1 - e^{-\frac{\hbar\omega}{kT}}}\right] = \hbar\omega\frac{1}{e^{\frac{\hbar\omega}{kT}} - 1}. \tag{10.67}$$

In terms of the collapse parameters, this is

$$\text{tr}\hat{H}\rho_{th} = \hbar\omega\frac{1}{1 + \frac{1}{\lambda' t} - 1} = \lambda\frac{\hbar^2}{2m}t. \tag{10.68}$$

Once again, we see the linear increase with time of energy. As we have shown must be the case, the result (10.68) is independent of the frequency of the harmonic oscillator, and the same as for the free particle.

This concludes our discussion of the harmonic oscillator undergoing collapse in position. For a treatment of the individual wave functions for this problem, see Chapter 18.

11
Supplement to Chapter 4

11.1 Mean Energy Increase of an Object

The equation describing the mean energy rate of increase of a collection of identical particles of mass m, following from the CSL Lindblad equation for the density matrix, was given in Eq.(4.15) as

$$\frac{d}{dt}\mathrm{Tr}\hat{H}\hat{\rho}(t) = -\mathrm{Tr}\hat{\rho}(t)\frac{\lambda}{2}\left[\frac{m}{M}\right]^2 \int d\mathbf{z}' \int d\mathbf{z}\, e^{-\frac{(\mathbf{z}-\mathbf{z}')^2}{4a^2}}[\hat{D}(\mathbf{z}'),[\hat{D}(\mathbf{z}),\hat{K}]]. \tag{11.1}$$

$\hat{D}(\mathbf{z})$ is the number density operator for these particles and $\hat{K}$ is their kinetic energy operator. The reason that only the kinetic energy operator appears in the commutator on the right side of this equation is that the potential energy terms all are functions of $\hat{D}(\mathbf{x})$, and so their commutator with $\hat{D}(\mathbf{z})$ vanishes.

We are going to do the evaluation of the right hand side of Eq. (11.1) here: the result obtained was discussed in Section 4.5. It turns out to be a constant.

This calculation is most easily done if one expresses $\hat{D}(\mathbf{z})$ and $\hat{K}$ in terms of two operators: $\hat{\xi}^{\dagger}(\mathbf{x})$, which creates a particle at a point $\mathbf{x}$ of space, and $\hat{\xi}(\mathbf{x})$, which annihilates the particle there. Accordingly, we first provide an introduction to these operators (good stuff to know!) and then proceed to evaluate (11.1).

11.1.1 Some Useful Quantum Fields

For a spinless particle moving in one dimension, along the x-axis, the most general quantum operator is Hermitian, of the form $F(\hat{X},\hat{P})$. We need operators that additionally depend upon $\mathbf{x}$, of the form $F(\hat{X},\hat{P},\mathbf{x})$. In classical physics, a quantity that depends upon position is called a classical field. Here we have an operator in quantum theory that depends upon position, and so it is called a quantum field.

The number density operator $\hat{D}(\mathbf{x})$, discussed in Section 4.1 and appearing in the density matrix evolution equation, is a quantum field. If there are many bosonic particles, with position operators $\hat{\mathbf{X}}_1, \hat{\mathbf{X}}_2, ...$, the density matrix is $\hat{D}(\mathbf{x}) \equiv \delta(\hat{\mathbf{X}}_1-\mathbf{x})+\delta(\hat{\mathbf{X}}_2-\mathbf{x})+...$ It turns out that it helps simplify calculations to replace $D(\mathbf{x})$ iby its equivalent expression $D(\mathbf{x}) = \hat{\xi}^{\dagger}(\mathbf{x})\hat{\xi}(\mathbf{x})$, as shown below.

The operators $\hat{\xi}^{\dagger}(\mathbf{x}), \hat{\xi}(\mathbf{x})$ behave pretty much like the creation and annihilation operators $\hat{a}^{\dagger}, \hat{a}$, of the one-dimensional harmonic oscillator problem, but they do so at each point of space.

For the harmonic oscillator, the ground state is $|0\rangle$, the commutator is $[\hat{a},\hat{a}^{\dagger}] = 1$ (all other commutators vanish), a annihilates the ground state, $a|0\rangle = 0$, the nth excited state is $\frac{1}{\sqrt{n!}}\hat{a}^{\dagger n}|0\rangle = |n\rangle$ and the state number operator is $\hat{a}^{\dagger}\hat{a}$: $\hat{a}^{\dagger}\hat{a}|n\rangle = n|n\rangle$.

Analogously, the no–particle state (vacuum state) is denoted $|0\rangle$, the commutator is[1] $[\hat{\xi}(\mathbf{x}), \hat{\xi}^\dagger(\mathbf{x}')] = \delta(\mathbf{x} - \mathbf{x}')$ (all other commutators vanish), $\hat{\xi}(\mathbf{x})$ annihilates the no-particle state, $\hat{\xi}(\mathbf{x})|0\rangle = 0$, the state representing n particles at locations $x_1...x_n$ is

$$|\mathbf{x}_1, \mathbf{x}_2, ...\mathbf{x}_n\rangle = \frac{1}{\sqrt{n!}} \hat{\xi}^\dagger(\mathbf{x}_1)\hat{\xi}^\dagger(\mathbf{x}_2)...\hat{\xi}^\dagger(\mathbf{x}_n)|0\rangle. \tag{11.2}$$

and the particle number density at $\mathbf{x}$ operator is

$$\hat{D}(\mathbf{x}) \equiv \hat{\xi}^\dagger(\mathbf{x})\hat{\xi}(\mathbf{x}). \tag{11.3}$$

Let's check some of the details, to see that these operators behave as claimed. The number density operator commutator vanishes,

$$\begin{aligned}
[\hat{D}(\mathbf{x}), \hat{D}(\mathbf{x}')] &= [\hat{\xi}^\dagger(\mathbf{x})\hat{\xi}(\mathbf{x}), \hat{\xi}^\dagger(\mathbf{x}')\hat{\xi}(\mathbf{x}')] \\
&= \hat{\xi}^\dagger(\mathbf{x})[\hat{\xi}(\mathbf{x}), \hat{\xi}^\dagger(\mathbf{x}')]\hat{\xi}(\mathbf{x}') + \hat{\xi}^\dagger(\mathbf{x}')[\hat{\xi}^\dagger(\mathbf{x}), \hat{\xi}(\mathbf{x}')]\hat{\xi}(\mathbf{x}) \\
&= \hat{\xi}^\dagger(\mathbf{x})\delta(\mathbf{x} - \mathbf{x}')\hat{\xi}(\mathbf{x}') - \hat{\xi}^\dagger(\mathbf{x}')\delta(\mathbf{x} - \mathbf{x}')\hat{\xi}(\mathbf{x}) = 0. \tag{11.4}
\end{aligned}$$

where, in going from the first to the second line, the identity $[AB, C] = A[B, C] + [A, C]B$ leads to the identity we have used, $[\hat{A}\hat{B}, \hat{C}\hat{D}] = \hat{A}[\hat{B}, \hat{C}]\hat{D} + \hat{A}\hat{C}[\hat{B}, \hat{D}] + [\hat{A}, \hat{C}]\hat{D}\hat{B} + \hat{C}[\hat{A}, \hat{D}]\hat{B}$, with the two middle commutators vanishing.

That they are properly normalized can be verified:

$$\begin{aligned}
\langle \mathbf{x}_1'|\mathbf{x}_1\rangle &= \langle 0|\hat{\xi}(\mathbf{x}_1')\hat{\xi}^\dagger(\mathbf{x}_1)|0\rangle = \langle 0|\left[[\hat{\xi}(\mathbf{x}_1'), \hat{\xi}^\dagger(\mathbf{x}_1)] + \hat{\xi}^\dagger(\mathbf{x}_1)\hat{\xi}(\mathbf{x}_1')\right]|0\rangle \\
&= \delta(\mathbf{x}_1' - \mathbf{x}_1)\langle 0|0\rangle = \delta(\mathbf{x}_1' - \mathbf{x}_1), \\
\langle \mathbf{x}_1', \mathbf{x}_2'|\mathbf{x}_1, \mathbf{x}_2\rangle &= \frac{1}{2}\langle 0|\hat{\xi}(\mathbf{x}_2')\hat{\xi}(\mathbf{x}_1')\hat{\xi}^\dagger(\mathbf{x}_1)\hat{\xi}^\dagger(\mathbf{x}_2)|0\rangle \\
&= \frac{1}{2}\langle 0|\hat{\xi}(\mathbf{x}_2')\left[[\hat{\xi}(\mathbf{x}_1'), \hat{\xi}^\dagger(\mathbf{x}_1)] + \hat{\xi}^\dagger(\mathbf{x}_1)\hat{\xi}(\mathbf{x}_1')\right]\hat{\xi}^\dagger(\mathbf{x}_2)|0\rangle \\
&= \frac{1}{2}\langle 0|\hat{\xi}(\mathbf{x}_2')\left[\delta(\mathbf{x}_1' - \mathbf{x}_1)\hat{\xi}^\dagger(\mathbf{x}_2)|0\rangle + \hat{\xi}^\dagger(\mathbf{x}_1)\delta(\mathbf{x}_1' - \mathbf{x}_2)|0\rangle\right] \\
&= \frac{1}{2}[\delta(\mathbf{x}_1' - \mathbf{x}_1)\delta(\mathbf{x}_2' - \mathbf{x}_2) + \delta(\mathbf{x}_1' - \mathbf{x}_2)\delta(\mathbf{x}_2' - \mathbf{x}_1)], \text{ etc.} \tag{11.5}
\end{aligned}$$

where one uses the commutation relation to move the annihilation operator to the right, generating delta functions along the way, until it annihilates the vacuum state.

That these are the number density eigenstates with the correct eigenvalues can be seen from:

$$\begin{aligned}
\hat{D}(\mathbf{x})|\mathbf{x}_1, \mathbf{x}_2, ...\mathbf{x}_n\rangle &= \hat{\xi}^\dagger(\mathbf{x})\hat{\xi}(\mathbf{x}) \cdot \frac{1}{\sqrt{n!}}\hat{\xi}^\dagger(\mathbf{x}_1)\hat{\xi}^\dagger(\mathbf{x}_2)...\hat{\xi}^\dagger(\mathbf{x}_n)|0\rangle \\
&= \left[\sum_{i=1}^{n}\delta(\mathbf{x} - \mathbf{x}_i)\right]\frac{1}{\sqrt{n!}}\hat{\xi}^\dagger(\mathbf{x}_1)\hat{\xi}^\dagger(\mathbf{x}_2)...\hat{\xi}^\dagger(\mathbf{x}_n)|0\rangle \tag{11.6}
\end{aligned}$$

where, again, the commutation relations have been used to move $\hat{\xi}(\mathbf{x})$ to the right, annihilating a creation operator and generating a delta function as it goes along until

---

[1]For simplicity, we will only do calculations for bosons. For fermions, $\{\hat{\xi}(\mathbf{x}), \hat{\xi}^\dagger(\mathbf{x}')\} = \delta(\mathbf{x} - \mathbf{x}')$ ($\{,\}$ is the anti-commutator), and all other anti-commutators vanish.

it annihilates at the vacuum. With use of $\hat{\xi}^\dagger(\mathbf{x})\delta(\mathbf{x} - \mathbf{x}_i) = \hat{\xi}^\dagger(\mathbf{x}_i)\delta(\mathbf{x} - \mathbf{x}_i)$, we see that the state vector on the right is unchanged.

This shows the particle position eigenstates and the particle number density operator are properly expressed by (11.2) and (11.6) respectively.

If one has a properly symmetrized boson wave function $\psi(\mathbf{x}_1, ...\mathbf{x}_n)$ for n particles (that is, the wave function is unchanged if any pair of position arguments are interchanged), the n particle state vector is

$$|\psi\rangle = \int d\mathbf{x}_1...d\mathbf{x}_n |\mathbf{x}_1, \mathbf{x}_2, ...\mathbf{x}_n\rangle\langle\mathbf{x}_n, ...\mathbf{x}_2, ...\mathbf{x}_1|\psi\rangle$$

$$= \frac{1}{\sqrt{n!}} \int d\mathbf{x}_1...d\mathbf{x}_n \hat{\xi}^\dagger(\mathbf{x}_1)...\hat{\xi}^\dagger(\mathbf{x}_n)|0\rangle\psi(\mathbf{x}_1, ...\mathbf{x}_n). \tag{11.7}$$

A special case is where the wave function $\psi(\mathbf{x}_1, ...\mathbf{x}_n)$ is made up of n orthonormal one-particle wave functions. It is the sum of $n!$ terms (multiplied by the factor $\frac{1}{\sqrt{n!}}$), each term is $\phi_1...\phi_n$ with a different ordering (permutation) of the n particle location arguments. Then, the state vector expression is simply

$$|\psi\rangle = \int d\mathbf{x}_1...d\mathbf{x}_n \hat{\xi}^\dagger(\mathbf{x}_1)...\hat{\xi}^\dagger(\mathbf{x}_n)|0\rangle\phi_1(\mathbf{x}_1)...\phi_n(\mathbf{x}_n). \tag{11.8}$$

Parallel to the particle number density operator (11.3) is the momentum density operator:

$$\hat{\mathbf{D}}_{mom}(\mathbf{x}) \equiv \frac{\hbar}{2i}\left[\hat{\xi}^\dagger(\mathbf{x})\nabla\hat{\xi}(\mathbf{x}) - (\nabla\hat{\xi}^\dagger(\mathbf{x}))\hat{\xi}(\mathbf{x})\right]. \tag{11.9}$$

The expectation value of the momentum density for the one-particle state with wave function $\psi(\mathbf{x})$ is

$$\langle\psi|\hat{\mathbf{D}}_{mom}(\mathbf{x})|\psi\rangle = \langle 0| \int d\mathbf{z}\hat{\xi}(\mathbf{z})\psi^*(\mathbf{z})\frac{\hbar}{2i}\left[\hat{\xi}^\dagger(\mathbf{x})\nabla\hat{\xi}(\mathbf{x}) - (\nabla\hat{\xi}^\dagger(\mathbf{x}))\hat{\xi}(\mathbf{x})\right] \int d\mathbf{z}'\psi(\mathbf{z}')\hat{\xi}^\dagger(\mathbf{z}')|0\rangle$$

$$= \frac{\hbar}{2i} \int d\mathbf{z} \int d\mathbf{z}'\psi^*(\mathbf{z})\left[\delta(\mathbf{z} - \mathbf{x})\nabla\delta(\mathbf{x} - \mathbf{z}') - \nabla\delta(\mathbf{z} - \mathbf{x}))\delta(\mathbf{x} - \mathbf{z}')\right]\psi(\mathbf{z}')$$

$$= \frac{\hbar}{2i}\left[\psi^*(\mathbf{x})\nabla\psi(\mathbf{x}) - \psi(\mathbf{x})\nabla\psi^*(\mathbf{x})\right]. \tag{11.10}$$

To see that this makes sense, if, in the neighborhood of $\mathbf{x}$ the wave function is approximately $\psi(\mathbf{x}) = e^{i\mathbf{k}\cdot\mathbf{x}}|\psi(\mathbf{x})|$, then $\langle\psi|\hat{\mathbf{D}}_{mom}(\mathbf{x})|\psi\rangle \approx \hbar\mathbf{k}|\psi(\mathbf{x})|^2$, essentially the particle's momentum at the point multiplied by its probability density of being at the point.

Of course, the integral of the momentum density (11.9) over all space is the momentum operator:

$$\hat{\mathbf{P}} = \int d\mathbf{x}\hat{\mathbf{D}}_{mom}(\mathbf{x}) = \int d\mathbf{x}\frac{\hbar}{2i}\left[\hat{\xi}^\dagger(\mathbf{x})\nabla\hat{\xi}(\mathbf{x}) - (\nabla\hat{\xi}^\dagger(\mathbf{x}))\hat{\xi}(\mathbf{x})\right]$$

$$= \int d\mathbf{x}\frac{\hbar}{i}\hat{\xi}^\dagger(\mathbf{x})\nabla\hat{\xi}(\mathbf{x}) \tag{11.11}$$

following an integration by parts (assuming that any wave function this operates on vanishes at infinity). For example, the one-particle matrix element is

$$\langle\phi|\hat{\mathbf{P}}|\psi\rangle = \langle 0|\int d\mathbf{z}\hat{\xi}(\mathbf{z})\phi^*(\mathbf{z})\int d\mathbf{x}\frac{\hbar}{i}\left[\hat{\xi}^\dagger(\mathbf{x})\nabla\hat{\xi}(\mathbf{x})\right]\int d\mathbf{z}'\hat{\xi}^\dagger(\mathbf{z}')\psi(\mathbf{z}')|0\rangle$$

$$= \int d\mathbf{x}\frac{\hbar}{i}\phi^*(\mathbf{x})\nabla\psi(\mathbf{x}), \tag{11.12}$$

which is precisely the usual expression.

The operator we need for our calculation is the kinetic energy operator. Just as in the case of the momentum, we start by defining the kinetic energy density operator, show that it is reasonable, and then integrate it over all space to see that we get the kinetic energy operator. The kinetic energy density operator is

$$\hat{D}_{KE}(\mathbf{x}) \equiv \frac{-\hbar^2}{4m}[\hat{\xi}^\dagger(\mathbf{x})\nabla_{\mathbf{x}}^2\hat{\xi}(\mathbf{x}) + \nabla_{\mathbf{x}}^2\hat{\xi}^\dagger(\mathbf{x})\hat{\xi}(\mathbf{x}) \tag{11.13}$$

We calculate the expectation value of the kinetic energy density, for one particle, as we did for the momentum density:

$$\langle\psi|\hat{D}_{KE}(\mathbf{x})|\psi\rangle = \frac{-\hbar^2}{4m}[\psi^*(\mathbf{x})\nabla_{\mathbf{x}}^2\psi(\mathbf{x}) + (\nabla_{\mathbf{x}}^2\psi^*(\mathbf{x}))\psi(\mathbf{x})]. \tag{11.14}$$

To see how this works, if in the neighborhood of $\mathbf{x}$ the wave function is approximately $\psi(\mathbf{x}) = e^{i\mathbf{k}\cdot\mathbf{x}}|\psi(\mathbf{x})|$, and moreover $|\nabla_{\mathbf{x}}|\psi(\mathbf{x})|| << |\mathbf{k}||\psi(\mathbf{x})|$, $\nabla_{\mathbf{x}}^2|\psi(\mathbf{x})| << k^2|\psi(\mathbf{x})|$ then $\langle\psi|\hat{D}_{KE}(\mathbf{x})|\psi\rangle| \approx \frac{\hbar^2 k^2}{2m}|\psi(\mathbf{x})|^2$, essentially the particle's kinetic energy at the point multiplied by its probability density of being at the point.

The integral of the kinetic energy density (11.13) over all space is the kinetic energy operator:

$$\hat{K} = \int d\mathbf{x}\hat{D}_{KE}(\mathbf{x}) = \int d\mathbf{x}\frac{-\hbar^2}{4m}[\hat{\xi}^\dagger(\mathbf{x})\nabla_{\mathbf{x}}^2\hat{\xi}(\mathbf{x}) + \nabla_{\mathbf{x}}^2\hat{\xi}^\dagger(\mathbf{x})\hat{\xi}(\mathbf{x})]$$

$$= \frac{-\hbar^2}{2m}\int d\mathbf{x}\hat{\xi}^\dagger(\mathbf{x})\nabla_{\mathbf{x}}^2\hat{\xi}(\mathbf{x}). \tag{11.15}$$

The matrix element for one particle is

$$\langle\phi|\hat{K}|\psi\rangle = \langle 0|\int d\mathbf{z}\hat{\xi}(\mathbf{z})\phi^*(\mathbf{z})\frac{-\hbar^2}{2m}\int d\mathbf{x}\hat{\xi}^\dagger(\mathbf{x})\nabla_{\mathbf{x}}^2\hat{\xi}(\mathbf{x})\int d\mathbf{z}'\hat{\xi}^\dagger(\mathbf{z}')\psi(\mathbf{z}')|0\rangle$$

$$= \frac{\hbar^2}{2m}\int d\mathbf{x}\phi^*(\mathbf{x})\nabla_{\mathbf{x}}^2\psi(\mathbf{x}), \tag{11.16}$$

which is precisely the usual expression.

We conclude this discussion by displaying the non–relativistic Schrödinger equation in this field operator language, for particles in an external potential $W(\mathbf{x})$ and mutually interacting with potential $V(\mathbf{x} - \mathbf{x}')$:

$$i\hbar\frac{d}{dt}|\psi,t\rangle = \left\{\frac{-\hbar^2}{2m}\int d\mathbf{x}\hat{\xi}^\dagger(\mathbf{x})\nabla_{\mathbf{x}}^2\hat{\xi}(\mathbf{x}) + \int d\mathbf{x}W(\mathbf{x})\hat{\xi}^\dagger(\mathbf{x})\hat{\xi}(\mathbf{x})\right.$$

$$\left. + \frac{1}{2}\int d\mathbf{x}d\mathbf{x}'V(\mathbf{x} - \mathbf{x}')\hat{\xi}^\dagger(\mathbf{x})\hat{\xi}^\dagger(\mathbf{x}')\hat{\xi}(\mathbf{x})\hat{\xi}(\mathbf{x}')\right\}|\psi,t\rangle. \tag{11.17}$$

To see this gives the usual nonrelativistic Schrödinger equation,[2] take the scalar product of (11.17) with $\langle \mathbf{x}_n, ... \mathbf{x}_2, \mathbf{x}_1 |$:

$$
i\hbar \frac{d}{dt} \langle \mathbf{x}_n, ... \mathbf{x}_2, \mathbf{x}_1 | \psi, t \rangle = \frac{-\hbar^2}{2m} \int d\mathbf{x} \langle \mathbf{x}_n, ... \mathbf{x}_2, \mathbf{x}_1 | \hat{\xi}^\dagger(\mathbf{x}) \nabla_\mathbf{x}^2 \hat{\xi}(\mathbf{x}) | \psi, t \rangle
$$

$$
+ \int d\mathbf{x} W(\mathbf{x}) \langle \mathbf{x}_n, ... \mathbf{x}_2, \mathbf{x}_1 | \hat{\xi}^\dagger(\mathbf{x}) \hat{\xi}(\mathbf{x}) | \psi, t \rangle
$$

$$
+ \frac{1}{2} \int d\mathbf{x} d\mathbf{x}' V(\mathbf{x} - \mathbf{x}') \langle \mathbf{x}_n, ... \mathbf{x}_2, \mathbf{x}_1 | \hat{\xi}^\dagger(\mathbf{x}) \hat{\xi}^\dagger(\mathbf{x}') \hat{\xi}(\mathbf{x}) \hat{\xi}(\mathbf{x}') | \psi, t \rangle
$$

$$
= \frac{-\hbar^2}{2m} \int d\mathbf{x} \sum_{i=1}^{n} \delta(\mathbf{x} - \mathbf{x}_i) \nabla_\mathbf{x}^2 \langle \mathbf{x}_n, ... \mathbf{x}, ... \mathbf{x}_1 | \psi, t \rangle
$$

$$
+ \int d\mathbf{x} W(\mathbf{x}) \sum_{i=1}^{n} \delta(\mathbf{x} - \mathbf{x}_i) \langle \mathbf{x}_n, ... \mathbf{x}, ... \mathbf{x}_1 | \psi, t \rangle
$$

$$
+ \frac{1}{2} \int d\mathbf{x} d\mathbf{x}' V(\mathbf{x} - \mathbf{x}') \sum_{i=1, j=1}^{n} \delta(\mathbf{x} - \mathbf{x}_i) \delta(\mathbf{x}' - \mathbf{x}_j) \langle \mathbf{x}_n, ... \mathbf{x}, ... \mathbf{x}'... \mathbf{x}_1 | \psi, t \rangle
$$

$$
= \left\{ \frac{-\hbar^2}{2m} \sum_{i=1}^{n} \nabla_{\mathbf{x}_i}^2 + \sum_{i=1}^{n} W(\mathbf{x}_i) + \frac{1}{2} \sum_{i=1, j=1}^{n} V(\mathbf{x}_i - \mathbf{x}_j) \right\} \langle \mathbf{x}_n, ... \mathbf{x}_2, \mathbf{x}_1 | \psi, t \rangle \quad (11.18)
$$

11.1.2 Mean Energy Increase

We are to evaluate Eq. (11.1), with the number density operator and the kinetic energy operator expressed in terms of the particle creation and annihilation operators:

$$
\frac{d}{dt} \text{Tr} \hat{H} \hat{\rho}(t) = -\text{Tr} \hat{\rho}(t) \frac{\lambda}{2} \left[\frac{m}{M} \right]^2 \int d\mathbf{z}' \int d\mathbf{z} e^{-\frac{(\mathbf{z} - \mathbf{z}')^2}{4a^2}}
$$

$$
\cdot [\hat{\xi}^\dagger(\mathbf{z}') \hat{\xi}(\mathbf{z}'), [\hat{\xi}^\dagger(\mathbf{z}) \hat{\xi}\mathbf{z}), -\frac{\hbar^2}{2m} \int d\mathbf{x} \hat{\xi}^\dagger(\mathbf{x}) \nabla_x^2 \hat{\xi}(\mathbf{x})]]. \quad (11.19)
$$

The first commutator (using $[\hat{A}\hat{B}, \hat{C}] = \hat{A}[\hat{B}, \hat{C}] + [\hat{A}, \hat{C}]\hat{B}$ and noting that an operator that depends upon $\mathbf{z}$ can be slid past ∇_x^2) is:

---

[2]We could have written a more general Hamiltonian, if the particles are charged and are subject to an external vector potential $\mathbf{A}(\mathbf{x}, t)$. Then, the kinetic energy term should be replaced by $\frac{1}{2m} \int d\mathbf{x} \hat{\xi}^\dagger(\mathbf{x}) \left[\frac{\hbar}{i} \nabla - \frac{q}{c} \mathbf{A}(\mathbf{x}, t) \right]^2 \hat{\xi}(\mathbf{x})$.

$$\int d\mathbf{z} \int d\mathbf{x} e^{-\frac{(\mathbf{z}-\mathbf{z}')^2}{4a^2}} [\hat{\xi}^\dagger(\mathbf{z})\hat{\xi}(\mathbf{z}), \hat{\xi}^\dagger(\mathbf{x})\nabla_\mathbf{x}^2\hat{\xi}(\mathbf{x})]$$

$$= \int d\mathbf{z} \int d\mathbf{x} e^{-\frac{(\mathbf{z}-\mathbf{z}')^2}{4a^2}} \left\{ \xi^\dagger(\mathbf{z})\delta(\mathbf{z}-\mathbf{x})\nabla_\mathbf{x}^2\hat{\xi}(\mathbf{x}) - \xi^\dagger(\mathbf{x})\hat{\xi}(\mathbf{z})\nabla_\mathbf{x}^2\delta(\mathbf{z}-\mathbf{x}) \right\}$$

$$= \int d\mathbf{z} e^{-\frac{(\mathbf{z}-\mathbf{z}')^2}{4a^2}} \xi^\dagger(\mathbf{z})\nabla_\mathbf{z}^2\hat{\xi}(\mathbf{z}) - \int d\mathbf{x}\xi^\dagger(\mathbf{x})\nabla_\mathbf{x}^2 \int d\mathbf{z} e^{-\frac{(\mathbf{z}-\mathbf{z}')^2}{4a^2}} \hat{\xi}(\mathbf{z})\delta(\mathbf{z}-\mathbf{x})$$

$$= \int d\mathbf{z} e^{-\frac{(\mathbf{z}-\mathbf{z}')^2}{4a^2}} \xi^\dagger(\mathbf{z})\nabla_\mathbf{z}^2\hat{\xi}(\mathbf{z}) - \int d\mathbf{x}\zeta^\dagger(\mathbf{x})\nabla_\mathbf{x}^2 [e^{-\frac{(\mathbf{x}-\mathbf{z}')^2}{4u^7}}\hat{\zeta}(\mathbf{x})]$$

$$= -\int d\mathbf{x}\xi^\dagger(\mathbf{x})\hat{\xi}(\mathbf{x})\nabla_\mathbf{x}^2 e^{-\frac{(\mathbf{x}-\mathbf{z}')^2}{4a^2}} - 2\int d\mathbf{x}\xi^\dagger(\mathbf{x})\nabla_\mathbf{x} e^{-\frac{(\mathbf{x}-\mathbf{z}')^2}{4a^2}} \cdot \nabla_\mathbf{x}\hat{\xi}(\mathbf{x}). \qquad (11.20)$$

In going to the last equation, we have used

$$\nabla_\mathbf{x}^2[f(x)g(x)] = g(x)\nabla_\mathbf{x}^2 f(x) + f(x)\nabla_\mathbf{x}^2 g(x) + 2\nabla_\mathbf{x}f(x)\cdot\nabla_\mathbf{x}g(x).$$

Applying now the second commutator, we see that the first term in Eq. (11.20) makes no contribution since it commutes with $\hat{\xi}^\dagger(\mathbf{z}')\hat{\xi}(\mathbf{z}')$, so we need only calculate the commutator of the second term:

$$\int d\mathbf{z}' [\hat{\xi}^\dagger(\mathbf{z}')\hat{\xi}(\mathbf{z}'), -2\int d\mathbf{x}\xi^\dagger(\mathbf{x})\nabla_\mathbf{x} e^{-\frac{(\mathbf{x}-\mathbf{z}')^2}{4a^2}} \cdot \nabla\hat{\xi}(\mathbf{x})]$$

$$= -2\int d\mathbf{z}' \int d\mathbf{x} \left\{ \hat{\xi}^\dagger(\mathbf{z}')\delta(\mathbf{z}'-\mathbf{x})\nabla_\mathbf{x} e^{-\frac{(\mathbf{x}-\mathbf{z}')^2}{4a^2}} \cdot \nabla_\mathbf{x}\hat{\xi}(\mathbf{x}) \right.$$

$$\left. -\hat{\xi}^\dagger(\mathbf{x})\nabla_\mathbf{x} e^{-\frac{(\mathbf{x}-\mathbf{z}')^2}{4a^2}} \cdot \nabla_\mathbf{x}\delta(\mathbf{z}'-\mathbf{x})\hat{\xi}(\mathbf{z}') \right\}$$

$$= -2\int d\mathbf{x}\hat{\xi}^\dagger(\mathbf{x}) \left\{ \nabla_\mathbf{x} e^{-\frac{(\mathbf{x}-\mathbf{z}')^2}{4a^2}} \big|_{\mathbf{z}'=\mathbf{x}} \cdot \nabla_\mathbf{x}\hat{\xi}(\mathbf{x}) \right.$$

$$\left. -\int d\mathbf{z}'\nabla_\mathbf{x} e^{-\frac{(\mathbf{x}-\mathbf{z}')^2}{4a^2}} \cdot \nabla_\mathbf{x}\delta(\mathbf{z}'-\mathbf{x})\hat{\xi}(\mathbf{z}') \right\}. \qquad (11.21)$$

In the second integral of (11.21), we write $\nabla_\mathbf{x}\delta(\mathbf{z}'-\mathbf{x}) = -\nabla_{\mathbf{z}'}\delta(\mathbf{z}'-\mathbf{x})$, and then integrate by parts with respect to $\mathbf{z}'$:

$$= -2\int d\mathbf{x}\hat{\xi}^\dagger(\mathbf{x}) \left\{ \nabla_\mathbf{x} e^{-\frac{(\mathbf{x}-\mathbf{z}')^2}{4a^2}} \big|_{\mathbf{z}'=\mathbf{x}} \cdot \nabla_\mathbf{x}\hat{\xi}(\mathbf{x}) \right.$$

$$\left. -\int d\mathbf{z}'\delta(\mathbf{z}'-\mathbf{x})\nabla_{\mathbf{z}'}\cdot[\nabla_\mathbf{x} e^{-\frac{(\mathbf{x}-\mathbf{z}')^2}{4a^2}}\hat{\xi}(\mathbf{z}')] \right\}$$

$$= 2\int d\mathbf{x}\hat{\xi}^\dagger(\mathbf{x})\hat{\xi}(\mathbf{x})\nabla_{\mathbf{z}'}\cdot\nabla_\mathbf{x} e^{-\frac{(\mathbf{x}-\mathbf{z}')^2}{4a^2}} \big|_{\mathbf{z}'=\mathbf{x}} = -2\int d\mathbf{x}\hat{\xi}^\dagger(\mathbf{x})\hat{\xi}(\mathbf{x})\frac{-3}{2a^2}$$

$$= \hat{N}\frac{3}{a^2}. \qquad (11.22)$$

where $\hat{N} = \int d\mathbf{x}\hat{\xi}^\dagger(\mathbf{x})\hat{\xi}(\mathbf{x})$ is the particle number operator. All that calculation has ended up with this simple result!

Putting this result into Eq. (11.19), we obtain

$$\frac{d}{dt}\langle H \rangle = \mathrm{Tr}\hat{\rho}(t)\hat{N}\frac{3}{a^2}\frac{\lambda}{2}\left[\frac{m}{M}\right]^2\frac{\hbar^2}{2m} = N\lambda\left[\frac{m}{M}\right]^2\frac{3\hbar^2}{4ma^2}, \qquad (11.23)$$

where we have written $\mathrm{Tr}\hat{\rho}(t)\hat{N} = N$, the (fixed) number of particles of mass m in the system.

This result is discussed in Section 4.5.[3]

11.2 Galilean Invariance of Non–Relativistic CSL

The evolution equation for the state vector in non–relativistic CSL is given in Eq. (4.4). That equation is reproduced here in a slightly more general form,

$$|\psi, t_1\rangle = Te^{-i\int_{t_0}^{t_1} dt\hat{H}}$$

$$\cdot e^{-\int_{t_0}^{t_1} dt \int_{-\infty}^{\infty} d\mathbf{x}\frac{1}{4\lambda}\left[w(\mathbf{x},t) - 2\lambda\sum_{\alpha}\frac{m_\alpha}{M}\frac{1}{[\pi a^2]^{3/4}}\int_{-\infty}^{\infty} d\mathbf{z}e^{-\frac{(\mathbf{x}-\mathbf{z})^2}{2a^2}}\hat{D}_\alpha(\mathbf{z})\right]^2}|\psi, t_0\rangle, \quad (11.24)$$

allowing general initial and final times. The (infinite) limits of the spatial integrals are also explicitly shown, and the integration variables $(\mathbf{x}', t')$ have been changed to $(\mathbf{x}, t)$.

We will prove what might be called stochastic Galilean invariance.

Two reference frames are considered: an initial reference frame (coordinates $\mathbf{x}, t$), and a Galilean transformed reference frame (coordinates $\mathbf{x}', t'$). The Galilean transformations are time-translation, space translation and "boosting" (the primed coordinate frame moves with constant velocity with respect to the initial frame).[4]

For all three cases, it is shown that the vector evolution equation Eq. (11.24) is invariant (has the same form in both frames), except that the scalar field $w(\mathbf{x}, t)$ in the two equations is different. However, since w can be any function whatsoever, the *ensemble* of all evolution equations, with all possible w's, is Galilean invariant, which is what is meant by stochastic Galilean invariance.

11.2.1 Time–Translation Invariance

First, consider time–translation invariance.

The coordinate transformation is $t' = t + \tau$. where τ is a constant. In Eq. (11.24) ireplace t by $t' - \tau$. The integration variable is $dt = dt'$. The limits on the time integral become $t_0' \equiv t_0 + \tau$ and $t_1' \equiv t_1 + \tau$. The time–translated state vectors are related to the original frame's state vectors by $|\psi, t_0' - \tau\rangle \equiv |\psi, t_0'\rangle'$ and $|\psi, t_1' - \tau\rangle \equiv |\psi, t_1\rangle'$. And we write $w'(\mathbf{x}, t') \equiv w(\mathbf{x}, t' - \tau)$.

When all these changes have been made, we see that the form of (11.24) is unchanged: unprimed variables have been replaced by primed variables.

The only way the two equations differ is that w' is not the same function of its time coordinate as w is of its time coordinate. But, as mentioned earlier, what is time–translation invariant is the set of evolution equations with all possible w functions. In

---

[3]It can be shown that this result is also obtained for fermions, as well as for the more general Hamiltonian mentioned in footnote 3, containing an external vector potential

[4]A complete discussion would include rotational invariance as well, but that will be left to the reader!

this set, there is a function w' that has identical values at t' that w has at the same numerical value of its time argument.

Therefore, we have proven stochastic time-t-ranslation invariance of (11.24).

11.2.2 Space–Translation Invariance

Next, consider space–translation invariance.

The coordinate transformation is $\mathbf{x}' = \mathbf{x} + \mathbf{a}$, where $\mathbf{a}$ is a constant vector. The integration variables are $d\mathbf{x}' = d\mathbf{x}$. In all expressions, replace $\mathbf{x}$ by $\mathbf{x}' - \mathbf{a}$. The limits on the spatial integral, $\pm\infty$, are unchanged.

What is changed is the argument of the Gaussian, which becomes $(\mathbf{x}' - \mathbf{a} - \mathbf{z})^2$. Making a change of variable to $\mathbf{z}' = \mathbf{z} + \mathbf{a}$ changes the particle number density operator term to $\hat{D}_\alpha(\mathbf{z}' - \mathbf{a})$. Also changed is the argument of w, so that $w(\mathbf{x}' - \mathbf{a}, t) \equiv w'(\mathbf{x}', t)$ now appears.

Now, the unitary operator that generates a spatial translation of $\mathbf{a}$ is $U = e^{i\hat{\mathbf{P}}\cdot\mathbf{a}}$, where $\hat{\mathbf{P}}$ is the momentum operator (and we have set $\hbar = 1$). To see this, use Eq. (11.3), $\hat{D}(\mathbf{x}) \equiv \hat{\xi}^\dagger(\mathbf{x})\hat{\xi}(\mathbf{x})$, and Eq. (11.11), $\hat{\mathbf{P}} = -i\int d\mathbf{x}'\hat{\xi}^\dagger(\mathbf{x}')\nabla\hat{\xi}(\mathbf{x}')$:

$$\hat{U}\hat{\xi}(\mathbf{x})\hat{U}^\dagger = \hat{\xi}(\mathbf{x}) + \frac{1}{1!}[\mathbf{a}\cdot\int d\mathbf{x}'\hat{\xi}^\dagger(\mathbf{x}')\nabla\hat{\xi}(\mathbf{x}'), \hat{\xi}(\mathbf{x})] + \frac{1}{2!}\cdots$$

$$= \hat{\xi}(\mathbf{x}) - \frac{1}{1!}[\mathbf{a}\cdot\nabla]\hat{\xi}(\mathbf{x}) + \frac{1}{2!}[\mathbf{a}\cdot\nabla]^2\hat{\xi}(\mathbf{x}) + \dots$$

$$= \hat{\xi}(\mathbf{x} - \mathbf{a}). \tag{11.25}$$

In the first line, we have used the power series expansion of the unitary transformation. In obtaining the second line, the commutator has been evaluated using $[\hat{\xi}^\dagger(\mathbf{x}'), \hat{\xi}(\mathbf{x})] = -\delta(\mathbf{x}' - \mathbf{x})$. Then, we recognize that the second line is a Taylor series expansion, giving the result.

The Hermitian conjugate of (11.25) is $\hat{U}\hat{\xi}^\dagger(\mathbf{x})\hat{U}^\dagger = \hat{\xi}^\dagger(\mathbf{x} - \mathbf{a})$, so (ignoring the subscript α),

$$\hat{D}(\mathbf{z}' - \mathbf{a}) = \hat{\xi}^\dagger(\mathbf{z}' - \mathbf{a})\hat{\xi}(\mathbf{z}' - \mathbf{a})$$

$$- \hat{U}\hat{\xi}^\dagger(\mathbf{z}')\hat{U}^\dagger\hat{U}\hat{\xi}(\mathbf{z}')\hat{U}^\dagger$$

$$= \hat{U}D(\mathbf{z}')\hat{U}^\dagger. \tag{11.26}$$

Thus, Eq. (11.24) expressed in the new coordinate system has replaced D by the expression (11.26).

Expanding the exponential in the evolution equation in a power series results in the product of operators of the form

$$\hat{U}D(\mathbf{z}_1)\hat{U}^\dagger\hat{U}D(\mathbf{z}_2)\hat{U}^\dagger\hat{U}D(\mathbf{z}_3)\hat{U}^\dagger\dots = \hat{U}D(\mathbf{z}_1)D(\mathbf{z}_2)D(\mathbf{z}_3)\hat{U}^\dagger\dots$$

That is, $\hat{U}$ can be taken out of the exponential to the left, and $\hat{U}^\dagger$ can be taken out of the exponential to the right. So far we have

$$|\psi, t_1\rangle = T e^{-i \int_{t_0}^{t_1} dt \hat{H}}$$

$$\cdot \hat{U} e^{- \int_{t_0}^{t_1} dt \int_{-\infty}^{\infty} d\mathbf{x}' \frac{1}{4\lambda} \left[w'(\mathbf{x}',t) - 2\lambda \sum_\alpha \frac{m_\alpha}{M} \frac{1}{[\pi a^2]^{3/4}} \int_{-\infty}^{\infty} d\mathbf{z}' e^{-\frac{(\mathbf{x}' - \mathbf{z}')^2}{2a^2}} \hat{D}_\alpha(\mathbf{z}') \right]^2} \hat{U}^\dagger |\psi, t_0\rangle,$$

$$= \hat{U} T e^{-i \int_{t_0}^{t_1} dt \hat{H}}$$

$$\cdot e^{- \int_{t_0}^{t_1} dt \int_{-\infty}^{\infty} d\mathbf{x}' \frac{1}{4\lambda} \left[w'(\mathbf{x}',t) - 2\lambda \sum_\alpha \frac{m_\alpha}{M} \frac{1}{[\pi a^2]^{3/4}} \int_{-\infty}^{\infty} d\mathbf{z}' e^{-\frac{(\mathbf{x}' - \mathbf{z}')^2}{2a^2}} \hat{D}_\alpha(\mathbf{z}') \right]^2} \hat{U}^\dagger |\psi, t_0\rangle,$$

where, since $[\hat{H}, \hat{P}] = 0$, the unitary operator has been taken completely outside the product of the two exponentials.

Multiplying the equation by $\hat{U}^\dagger$, and defining the state vectors in the translated coordinate system as $\hat{U}^\dagger |\psi, t_0\rangle \equiv |\psi, t_0\rangle'$, $\hat{U}^\dagger |\psi, t_1\rangle \equiv |\psi, t_1\rangle'$, we see that this equation has the same form as the evolution equation in the original coordinate system.

As discussed earlier, $w(\mathbf{x}, t)$ has been replaced by a different function $w'(\mathbf{x}', t)$, so we have proven stochastic space—translation invariance of (11.24).

11.2.3 Boost Invariance

Last, we consider boost invariance.

For simplicity, we consider the case where there is only one kind of particle, of mass m, and drop the subscript α.

The coordinate transformation to a reference frame moving with velocity $\mathbf{v}$ with respect to the original reference frame is $\mathbf{x}' = \mathbf{x} - \mathbf{v}t$. Since $t' = t$, the integration variables satisfy $d\mathbf{x}dt = d\mathbf{x}'dt'$. In all expressions, replace $\mathbf{x}$ by $\mathbf{x}' + \mathbf{v}t'$. The limits on the spatial integral, $\pm\infty$, are unchanged, as are the limits on the time integral.

What is changed is the argument of the Gaussian, which becomes $(\mathbf{x}' + \mathbf{v}t' - \mathbf{z})^2$. Making a change of variable to $\mathbf{z}' = \mathbf{z} - \mathbf{v}t'$ changes the particle number density operator term to $\hat{D}(\mathbf{z}' + \mathbf{v}t')$. (Also changed is the argument of w so that $w(\mathbf{x}' + \mathbf{v}t', t') \equiv w'(\mathbf{x}', t')$ now appears.)

The expression (11.26), with $\mathbf{a}$ replaced by $-\mathbf{v}t'$ is

$$\hat{D}(\mathbf{z}' + \mathbf{v}t') = e^{-i\hat{\mathbf{P}} \cdot \mathbf{v}t'} D(\mathbf{z}') e^{i\hat{\mathbf{P}} \cdot \mathbf{v}t'}. \tag{11.27}$$

Setting

$$\hat{B}(t') \equiv \int_{-\infty}^{\infty} d\mathbf{x}' \frac{1}{4\lambda} \left[w'(\mathbf{x}', t') - 2\lambda \frac{m}{M} \frac{1}{[\pi a^2]^{3/4}} \int_{-\infty}^{\infty} d\mathbf{z}' e^{-\frac{(\mathbf{x}' - \mathbf{z}')^2}{2a^2}} \hat{D}(\mathbf{z}') \right]^2, \tag{11.28}$$

we can write Eq. (11.24) as

$$|\psi, t_1\rangle = T e^{-i \int_{t_0}^{t_1} dt' \hat{H}} e^{- \int_{t_0}^{t_1} dt' e^{-i\hat{\mathbf{P}} \cdot \mathbf{v}t'} \hat{B}(t') e^{i\hat{\mathbf{P}} \cdot \mathbf{v}t'}} |\psi, t_0\rangle. \tag{11.29}$$

We recall the nature of the time-ordering operation:

$$T e^{\int_{t_0}^{t_1} dt' \hat{A}(t')} = e^{dt' \hat{A}(t_1)} \ldots e^{dt' \hat{A}(t_0 + 2dt')} e^{dt' \hat{A}(t_0 + dt')} e^{dt' \hat{A}(t_0)}$$

$$= \prod_{k=0}^{\frac{t_1 - t_0}{dt'}} e^{dt' \hat{A}(t_0 + kdt')}. \tag{11.30}$$

Explicitly writing out the time-ordering operation in (11.29),

$$|\psi, t_1\rangle = \prod_{k=0}^{\frac{t_1-t_0}{dt'}} e^{-idt'\hat{H}} e^{-dt'e^{-i\hat{\mathbf{P}}\cdot\mathbf{v}(t_0+kdt')}\hat{B}(t_0+kdt')e^{i\hat{\mathbf{P}}\cdot\mathbf{v}(t_0+kdt')}} |\psi, t_0\rangle$$

$$= \prod_{k=0}^{\frac{t_1-t_0}{dt'}} e^{-idt'\hat{H}} e^{-i\hat{\mathbf{P}}\cdot\mathbf{v}(t_0+kdt')} e^{-dt'\hat{B}(t_0+kdt')} e^{i\hat{\mathbf{P}}\cdot\mathbf{v}(t_0+kdt')} |\psi, t_0\rangle, \quad (11.31)$$

the last step following from

$$e^{e^{-i\hat{A}}\hat{B}e^{i\hat{A}}} = 1 + \frac{1}{1!} e^{-i\hat{A}}\hat{B}e^{i\hat{A}} + \frac{1}{2!} e^{-i\hat{A}}\hat{B}e^{i\hat{A}} e^{-i\hat{A}}\hat{B}e^{i\hat{A}} + \dots$$

$$= e^{-i\hat{A}} e^{\hat{B}} e^{i\hat{A}}. \quad (11.32)$$

Now, $\hat{H}'$s only dependence on the center–of–mass coordinates $\hat{\mathbf{X}}_{cm}$ and $\hat{\mathbf{P}}$ (commutation relations $[X_{cmi}, P_j] = i\delta_{ij})$ is through the center of mass kinetic energy term $\frac{\hat{\mathbf{P}}^2}{2Nm}$ (N is the number of particles): the potential energy depends on the relative position coordinates, not the center-of-mass coordinate.

Therefore, since $[\hat{H}, \hat{\mathbf{P}}] = 0$, we can slide the momentum-dependent exponential to the left of the Hamiltonian-dependent exponential. Noting that

$$e^{i\hat{\mathbf{P}}\cdot\mathbf{v}(t_0+(k+1)dt')} e^{-i\hat{\mathbf{P}}\cdot\mathbf{v}(t_0+kdt')} = e^{i\hat{\mathbf{P}}\cdot\mathbf{v}dt'},$$

we can write Eq. (11.31) as:

$$|\psi, t_1\rangle = e^{-i\hat{\mathbf{P}}\cdot\mathbf{v}t_1} \left\{ \prod_{k=0}^{\frac{t_1-t_0}{dt'}} e^{i\hat{\mathbf{P}}\cdot\mathbf{v}dt'} e^{-idt'\hat{H}} e^{-dt'\hat{B}(t_0+kdt')} \right\} e^{i\hat{\mathbf{P}}\cdot\mathbf{v}t_0} |\psi, t_0\rangle.$$

$$(11.33)$$

We now show that the generator of boosts is $\mathbf{P}t - m\hat{N}\hat{\mathbf{X}}_{cm} = \mathbf{P}t - m\hat{\mathbf{X}}$, where

$$\hat{N} = \int d\mathbf{x}\,\hat{D}(\mathbf{x})$$

is the particle number operator, and

$$\hat{\mathbf{X}} \equiv \hat{N}\hat{\mathbf{X}}_{cm} = \int d\mathbf{x}\mathbf{x}\hat{D}(\mathbf{x}).$$

That is, we show that the state vectors transform under the boost by

$$e^{i[\mathbf{P}t - m\hat{\mathbf{X}}\cdot\mathbf{v}]}|\psi, t\rangle = |\psi, t\rangle.' \quad (11.34)$$

We will do so by applying this unitary transformation to (11.33) and show that it results in a form-invariant evolution equation. Thus,

$$|\psi, t_1\rangle' = e^{i[\mathbf{P}t_1 - m\hat{\mathbf{X}}\cdot\mathbf{v}]} e^{-i\hat{\mathbf{P}}\cdot\mathbf{v}t_1} \left\{ \prod_{k=0}^{\frac{t_1-t_0}{dt'}} e^{i\hat{\mathbf{P}}\cdot\mathbf{v}dt'} e^{-idt'\hat{H}} e^{-dt'\hat{B}(t_0+kdt')} \right\} e^{i\hat{\mathbf{P}}\cdot\mathbf{v}t_0} |\psi, t_0\rangle$$

$$= e^{-\frac{i}{2}\hat{N}m\mathbf{v}^2 t_1} e^{-im\hat{\mathbf{X}}\cdot\mathbf{v}} \left\{ \prod_{k=0}^{\frac{t_1-t_0}{dt'}} e^{i\hat{\mathbf{P}}\cdot\mathbf{v}dt'} e^{-idt'\hat{H}} e^{-dt'\hat{B}(t_0+kdt')} \right\} e^{i\hat{\mathbf{P}}\cdot\mathbf{v}t_0} |\psi, t_0\rangle$$

$$(11.35)$$

Here we have used the Campbell–Baker–Hausdorff formula;

$$e^{i[\mathbf{P}t - m\hat{\mathbf{X}}\cdot\mathbf{v}]} = e^{-im\hat{\mathbf{X}}\cdot\mathbf{v}} e^{i\mathbf{P}\cdot\mathbf{v}t} e^{-\frac{1}{2}[-im\hat{\mathbf{X}}\cdot\mathbf{v}, i\mathbf{P}\cdot\mathbf{v}t]}$$
$$= e^{-im\hat{\mathbf{X}}\cdot\mathbf{v}} e^{i\mathbf{P}\cdot\mathbf{v}t} e^{-\frac{i}{2}\hat{N}m\mathbf{v}^2 t}$$

(appropriate since $\hat{N}$ has a vanishing commutator with $\hat{\mathbf{X}}, \hat{\mathbf{P}}$).
   Next, we calculate

$$e^{-im\hat{\mathbf{X}}\cdot\mathbf{v}} e^{i\hat{\mathbf{P}}\cdot\mathbf{v}dt'} e^{-idt'\hat{H}} = \left[e^{-im\hat{\mathbf{X}}\cdot\mathbf{v}} e^{i\hat{\mathbf{P}}\cdot\mathbf{v}dt'} e^{im\hat{\mathbf{X}}\cdot\mathbf{v}} \right] \left[e^{-im\hat{\mathbf{X}}\cdot\mathbf{v}} e^{-idt'\hat{H}} e^{im\hat{\mathbf{X}}\cdot\mathbf{v}} \right] e^{-im\hat{\mathbf{X}}\cdot\mathbf{v}}$$

$$= \left[e^{i\hat{\mathbf{P}}\cdot\mathbf{v}dt'} e^{i\hat{N}m\mathbf{v}^2 dt'} \right] \left[e^{-i\hat{\mathbf{P}}\cdot\mathbf{v}dt'} e^{-idt'\hat{H}} e^{-\frac{i}{2}\hat{N}m\mathbf{v}^2 dt} \right] e^{-im\hat{\mathbf{X}}\cdot\mathbf{v}}$$

$$= e^{\frac{i}{2}\hat{N}m\mathbf{v}^2 dt} e^{-idt'\hat{H}} e^{-im\hat{\mathbf{X}}\cdot\mathbf{v}} \qquad (11.36)$$

where we have used (11.32) as

$$e^{-i\hat{A}} \hat{B} e^{i\hat{A}} = e^{e^{-i\hat{A}} B e^{i\hat{A}}} = e^{B + \frac{1}{1!}[-iA,B] + \frac{1}{2!}[-iA,[-iA,B]]}$$

where higher–order commutators vanish.
   Therefore, we can continue running $e^{-im\hat{\mathbf{X}}\cdot\mathbf{v}}$ past all terms in the product (since $\hat{\mathbf{X}}$ is a function of $\hat{D}$, it commutes with B), so (11.35) becomes

$$|\psi, t_1\rangle' = e^{-\frac{i}{2}\hat{N}m\mathbf{v}^2 t_1} \left\{ \prod_{k=0}^{\frac{t_1-t_0}{dt'}} e^{\frac{i}{2}\hat{N}m\mathbf{v}^2 dt'} e^{-idt'\hat{H}} e^{-dt'\hat{B}(t_0+kdt')} \right\} e^{-im\hat{\mathbf{X}}\cdot\mathbf{v}} e^{i\hat{\mathbf{P}}\cdot\mathbf{v}t_0} |\psi, t_0\rangle$$

$$= e^{-\frac{i}{2}\hat{N}m\mathbf{v}^2 t_1} \left\{ \prod_{k=0}^{\frac{t_1-t_0}{dt'}} e^{\frac{i}{2}i\hat{N}m\mathbf{v}^2 dt''} e^{-idt''\hat{H}} e^{-dt''\hat{B}(t_0+kdt'')} \right\} e^{\frac{i}{2}\hat{N}m\mathbf{v}^2 t_0} |\psi, t_0\rangle',$$

$$(11.37)$$

where we have again used the Campbell–Baker–Haussdorff theorem on the product of two exponentials multiplying the initial state vector, the result being the boosted initial state vector (multiplied by a phase factor).
   Converting the bracketed expression in (11.37) back to its time-ordered form, the final result is

$$|\psi, t_1\rangle' = e^{-\frac{i}{2}\hat{N}m\mathbf{v}^2 t_1} e^{\frac{i}{2}\hat{N}m\mathbf{v}^2 (t_1-t_0)} T e^{-i\int_{t_0}^{t_1} dt'\hat{H}} e^{-\int_{t_0}^{t_1} dt'\hat{B}(t')} e^{\frac{i}{2}\hat{N}m\mathbf{v}^2 t_0} |\psi, t_0\rangle'$$

$$= T e^{-i\int_{t_0}^{t_1} dt'\hat{H}} e^{-\int_{t_0}^{t_1} dt'\hat{B}(t')} |\psi, t_0\rangle'. \qquad (11.38)$$

We see that (11.38) has the same form as the evolution equation in the original coordinate system.

As discussed earlier, $w(\mathbf{x},t)$ has been replaced by a different function $w'(\mathbf{x}',t')$, so we have proven stochastic boost invariance of this equation.

This concludes our demonstration of the Galilean invariance of non–relativistic CSL.

12
Supplement to Chapter 5

There is no Supplement to Chapter 5.

13

Supplement to Chapter 6

13.1 Electron Excitation: Spontaneous Radiation in Germanium

The CSL collapse mechanism tends to narrow wave functions. Suppose an electron sits in the ground state of an atom. If its wave function gets slightly narrowed, then that is no longer the ground state. Since all the possible states of the electron are a complete set, the wave function may be expanded in that set. The sum will be the ground state with a squared magnitude coefficient slightly less than 1, and all the other states, bound states and unbound states. These have small but non–zero squared magnitude coefficients. In particular, this means that there is a probability for the electron to have positive energy and escape the atom completely.

Section 6.2.1 discusses an experiment that monitors a chunk of very pure Ge. It looks for one of the two 1s electrons in any of the Ge atoms to be excited by the collapse mechanism and escape the atom.

The notation is as follows. The initial state is $|\phi_0\rangle|\chi_0\rangle$. $|\phi_0\rangle$ is the ground state of all the electrons with respect to the center of mass of the atom. $|\chi_0\rangle$ is the initial cm state. Γ is the probability/sec of a transition from this initial state to the state $|\phi_n\rangle|\chi_k\rangle$ ($n \neq 0$). $|\phi_n\rangle$ denotes an excited state of the electrons of the atom relative to the cm. $|\chi_k\rangle$ is any state of the cm. All states are properly orthonormalized energy eigenstates. The discrete indices are for notational convenience, as the cm states and the free electron states are part of a continuum.

In Section 6.2.1, starting from the CSL evolution equation for the density matrix, we arrived at an equation for Γ (the un-numbered equation before Eq. (6.7)), in the case of a general (not necessarily mass-proportional) coupling:

$$\Gamma = \frac{\lambda}{2a^2}\left[g_e - \frac{m_e}{M}\right]^2 \sum_{i,j=1}^{Z} |\langle\chi_k|\chi_0\rangle|^2 \langle\phi_n|\mathbf{R}_i|\phi_0\rangle\langle\phi_0|\mathbf{R}_j|\phi_n\rangle. \tag{13.1}$$

Here we are restricting $|\phi_n\rangle \equiv |\phi_\mathbf{k}\rangle$ to represent the state describing the atom with all electrons bound except for one free 1s electron of momentum $\mathbf{k}$ and energy $E = \mathbf{k}^2/2m_e$, and $|\chi_k\rangle$ is to represent a freely moving cm state. The $\hat{\mathbf{R}}_i$ are the coordinates (relative to the center of mass) of the Z electrons in the atom.

Since we wish to calculate the excitation rate Γ of the electron regardless of the cm state, we sum Eq. (6.1) over the cm states, using

$$\sum_k |\langle\chi_k|\chi_0\rangle|^2 = \langle\chi_0|\sum_k |\chi_k\rangle\langle\chi_k|]\chi_0\rangle = \langle\chi_0|\hat{1}|\chi_0\rangle = 1.$$

Then, because of the symmetry of the wave function under particle exchange, the excitation rate can be written in terms of just one electron's relative coordinate $\hat{\mathbf{R}}_e$, using $\sum_{i=1}^{Z}\langle\phi_0|\hat{\mathbf{R}}_j|\phi_n\rangle = Z\langle\phi_0|\hat{\mathbf{R}}_e|\phi_n\rangle$:

$$\Gamma_n = \frac{\lambda}{2a^2}\left(g_e - \frac{m_e}{M}\right)^2 Z^2|\langle\phi_n|\hat{\mathbf{R}}_e|\phi_0\rangle|^2. \tag{13.2}$$

One way to evaluate the matrix element is to approximate the initial ground state wave function as of the form $\langle\mathbf{x}|\phi_0\rangle = \frac{1}{\sqrt{Z!}}[\phi_{1s}(\mathbf{x}_1)\phi_2(\mathbf{x}_2)...\phi_Z(\mathbf{x}_Z)\pm...]$, consisting of $Z!$ terms which are permutations (with appropriate sign changes, as the electrons are fermions) of the Z position coordinates among the Z bound states. The 1s state of the electron is labeled, but the other bound states are just numbered. Similarly, the excited state wave function has the form $\langle\mathbf{x}|\phi_n\rangle = \frac{1}{\sqrt{Z!}}[\phi_{\mathbf{k}}(\mathbf{x}_1)\phi_2(\mathbf{x}_2)...\phi_Z(\mathbf{x}_Z)\pm...]$, that is, we assume that the bound electron states other than this 1s electron state are unchanged by the ionization.

Then, the matrix element $\langle\phi_n|\hat{\mathbf{R}}_e|\phi_0\rangle$ for the whole atom is replaced by the matrix element involving just the 1s electron state before the excitation, $|\phi_{1s}\rangle$, and the state $\langle\phi_{\mathbf{k}}|$ of that electron excited out of the atom with momentum $\mathbf{k}$. This is because $\langle\phi_n|\hat{\mathbf{R}}_e|\phi_0\rangle = \frac{1}{Z!}\langle\phi_{\mathbf{k}}|\hat{\mathbf{R}}_e|\phi_{1s}\rangle(Z-1)! = \frac{1}{Z}\langle\phi_{\mathbf{k}}|\hat{\mathbf{R}}_e|\phi_{1s}\rangle$, the factor $\frac{1}{Z!}$ coming from the normalization of the two states, the factor $(Z-1)!$ arising from the permutations of all electrons but the excited 1s electron. Therefore, the Z^2 term in (6.2) is cancelled. Eq. (13.2) becomes:

$$\Gamma_{\mathbf{k}} \approx \frac{\lambda}{2a^2}\left(g_e - \frac{m_e}{M}\right)^2|\langle\phi_{\mathbf{k}}|\hat{\mathbf{R}}_e|\phi_{1s}\rangle|^2. \tag{13.3}$$

The states $|\phi_{1s}\rangle, |\phi_{\mathbf{k}}\rangle$ are a complete orthonormalized set of states and so satisfy $|\phi_{1s}\rangle\langle\phi_{1s}| + \int d\Omega dE|\phi_{\mathbf{k}}\rangle\langle\phi_{\mathbf{k}}| = \hat{1}$, where Ω is the element of solid angle toward which $\mathbf{k}$ points. $\Gamma_{\mathbf{k}}$ then is an infinitesimal probability/sec, of exciting the 1s electron to a momentum state between $E, E+dE$ and $\Omega, \Omega+d\Omega$, so it is more appropriately expressed as an infinitesimal $d\Gamma(\Omega, E)$. And, since one does not measure the angle at which the 1s emerges, but only its energy, one may integrate over Ω to get the excitation rate to any angle, and then it is appropriate to call the resultant rate $d\Gamma(E)$. Thus,

$$\frac{d\Gamma(E)}{dE} \approx \frac{\lambda}{2a^2}\left(g_e - \frac{m_e}{M}\right)^2\int d\Omega|\langle\phi_{\mathbf{k}}|\hat{\mathbf{R}}_e|\phi_{1s}\rangle|^2. \tag{13.4}$$

Actually, in the paper cited in Section 6.2.1, the evaluation of the matrix element was more sophisticated than the approach described above. It was done using the Hartree method where, self-consistently, each electron feels the potential of the nucleus and that of all the other electrons. We will not review the calculation here. However, we can deduce the form of the result using dimensional analysis.

The dimension of the left side of (13.4) is (energy-sec)$^{-1}$. λ provides dimension (sec)$^{-1}$. The right side has the factor a^{-2} (dimension length^{-2}), and the dimension of the squared position operator is length^2). The only length in the problem is the mean radius of the 1s electron, which is $\frac{a_0}{Z}$, where $a_0 \approx .53 \cdot 10^{-8}$cm is the Bohr radius. Therefore, the integrated matrix element in (13.4) must have the form $[\frac{a_0}{Z}]^2\gamma(E)$, where $\gamma(E)$ has dimension energy^{-1}:

Therefore, the equation to be compared with experiment is (13.4) written as

$$\frac{d\Gamma(E)}{dE} = \frac{\lambda}{2} \left(\frac{a_0/Z}{a}\right)^2 \left(g_e - \frac{m_e}{M}\right)^2 \gamma(E). \tag{13.5}$$

Eq. (13.5) appears in Section 6.2.1 as (6.7), and is used there to analyze the experimental data.

13.2   A Relativistic CSL Model

Here we discuss setting the collapse-generating operator equal to the simplest quantum object obeying special relativity, a quantum scalar field operator $\hat{\phi}(\mathbf{x}, t)$, thereby constructing a relativistically invariant CSL model. This illustrates the argument given in Section 6.4 that such a model leads to creation of particles. Here we will see that the rate of energy creation for each momentum particle per unit volume is the same, but since the range of momentum is infinite, this means creating infinite energy/sec per unit volume.

Introductory texts on quantum field theory show that, when there is no interaction, the quantum field (an operator at each point of space and each instant of time) obeys the Klein–Gordon equation and can be written as

$$\hat{\phi}(\mathbf{x}, t) = \frac{1}{(2\pi)^{3/2}} \int \frac{d\mathbf{k}}{\sqrt{2\omega}} \left[\hat{a}(\mathbf{k})e^{i\mathbf{k}\cdot\mathbf{x}-i\omega t} + \hat{a}^\dagger(\mathbf{k})e^{-i\mathbf{k}\cdot\mathbf{x}+i\omega t}\right], \tag{13.6}$$

where $\hat{a}^\dagger(\mathbf{k})$ creates a particle of momentum $\mathbf{k}$ and mass m ($\omega \equiv \sqrt{\mathbf{k}^2 + m^2}$) out of the vacuum state $|0\rangle$ (that is, $\hat{a}^\dagger(\mathbf{k})|0\rangle = |\mathbf{k}\rangle$), and $[\hat{a}(\mathbf{k}), \hat{a}^\dagger(\mathbf{k}')] = \delta(\mathbf{k} - \mathbf{k}')$. The Hamiltonian, $\hat{H} = \int d\mathbf{k}\,\omega\hat{a}^\dagger(\mathbf{k})\hat{a}(\mathbf{k})$, can be thought of as describing a collection of harmonic oscillators. Indeed, writing $\hat{a}(\mathbf{k})\sqrt{d\mathbf{k}} \equiv \hat{a}_\mathbf{k}$, and treating the Hamiltonian as a sum of these new operators, they obey the usual harmonic oscillator commutation relations $[\hat{a}_\mathbf{k}, \hat{a}_{\mathbf{k}'}^\dagger] = \delta_{k,k'}$.

In classical physics, the particle idea and the field idea are two separate concepts. However, in quantum field theory, the two are inextricably connected: in Eq. (13.6), the left side is the quantum field, the right side's operators create and annihilate particles.

Upon replacing $\hat{A}(\mathbf{x})$ by $\hat{\phi}(\mathbf{x}, 0)$ in the state vector evolution Eq. (3.29) and in the density matrix Lindblad equation Eq. (3.31), one obtains[1]

$$|\psi, t\rangle = Te^{-i\int_0^t dt' \hat{H}}e^{\int_0^t dt' \int d\mathbf{x}' \frac{1}{4\lambda}[w(\mathbf{x}',t') - 2\lambda\hat{\phi}(\mathbf{x}',0)]^2}|\psi, 0\rangle, \tag{13.7a}$$

$$\frac{d}{dt}\hat{\rho}(t) = -i[\hat{H}, \hat{\rho}(t)] - \frac{\lambda}{2} \int d\mathbf{x}[\hat{\phi}(\mathbf{x}, 0), [\hat{\phi}(\mathbf{x}, 0), \hat{\rho}(t)]]. \tag{13.7b}$$

The number of particles with momenta between $\mathbf{k}$ and $\mathbf{k} + d\mathbf{k}$ is represented by $\hat{a}^\dagger(\mathbf{k})\hat{a}(\mathbf{k})d\mathbf{k}$. Using (13.7b), we calculate:

---

[1] These equations are Lorentz invariant, but they don't look it because they aren't expressed in the interaction picture. The interaction picture state vector is $|\psi, t\rangle_I \equiv e^{i\hat{H}t}|\psi, t\rangle$. Put that into the Schrödinger equation $\frac{d}{dt}|\psi, t\rangle = -i\hat{H}|\psi, t\rangle - \int d\mathbf{x}' \frac{1}{4\lambda}[w(\mathbf{x}', t) - 2\lambda\hat{\phi}(\mathbf{x}', 0)]^2|\psi, t\rangle$, and use $e^{i\hat{H}t}\hat{\phi}(\mathbf{x}, 0)e^{-i\hat{H}t} = \hat{\phi}(\mathbf{x}, t)$. Then, the solution of the resulting Schrödinger equation is manifestly Lorentz invariant, $|\psi, t\rangle_I = Te^{-\int_0^t dt' d\mathbf{x}' \frac{1}{4\lambda}[w(\mathbf{x}', t') - 2\lambda\hat{\phi}(\mathbf{x}', t')]^2}|\psi, 0\rangle$, with $w(\mathbf{x}, t)$ a Lorentz scalar classical field.

$$\frac{d}{dt}\overline{a^\dagger(\mathbf{k})\hat{a}(\mathbf{k})} \equiv \frac{d}{dt}\mathrm{Tr}a^\dagger(\mathbf{k})\hat{a}(\mathbf{k})\hat{\rho}(t) = -\mathrm{Tr}\hat{\rho}(t)\frac{\lambda}{2}\int d\mathbf{x}[\hat{\phi}(\mathbf{x},0),[\hat{\phi}(\mathbf{x},0),a^\dagger(\mathbf{k})\hat{a}(\mathbf{k})]],$$

(13.8)

where the properties of the trace operation allow the density matrix to be shuffled out of the commutators.

The first commutator is, using $[A, BC] = [A, B]C + B[A, C]$:

$$[\hat{\phi}(\mathbf{x},0), a^\dagger(\mathbf{k})\hat{a}(\mathbf{k})] = \frac{1}{(2\pi)^{3/2}}\frac{1}{\sqrt{2\omega}}\left[\hat{a}(\mathbf{k})e^{i\mathbf{k}\cdot\mathbf{x}} - \hat{a}^\dagger(\mathbf{k})e^{-i\mathbf{k}\cdot\mathbf{x}}\right].$$

The second commutator is:

$$[\hat{\phi}(\mathbf{x},0), \frac{1}{(2\pi)^{3/2}}\frac{1}{\sqrt{2\omega}}\left[\hat{a}(\mathbf{k})e^{i\mathbf{k}\cdot\mathbf{x}} - \hat{a}^\dagger(\mathbf{k})e^{-i\mathbf{k}\cdot\mathbf{x}}\right] = -\frac{1}{(2\pi)^3}\frac{2}{2\omega}.$$

Therefore,

$$\frac{d}{dt}\overline{a^\dagger(\mathbf{k})\hat{a}(\mathbf{k})} = \mathrm{Tr}\{\hat{\rho}(t)\}\frac{\lambda}{2}\int d\mathbf{x}\frac{1}{(2\pi)^3}\frac{2}{2\omega} = \frac{\lambda}{2(2\pi)^3\omega}V,$$

(13.9)

where we have written the volume of space $V \equiv \int d\mathbf{x}$.

T0 find the ensemble mean rate of energy increase per unit volume, we divide Eq. (13.9) by V, and multiply by ω. This expression is a finite constant, independent of $\mathbf{k}$. However, when we multiply by $d\mathbf{k}$ and integrate over all momenta to obtain the total mean energy increase per unit volume,

$$\frac{d}{dt}\overline{H}/V = \frac{\lambda}{2(2\pi)^3}\int d\mathbf{k},$$

(13.10)

the result is infinite.

To summarize, the Hamiltonian is that of a collection of harmonic oscillators, one for each $\mathbf{k}$. Like the oscillator discussed in Section 10.3, each is excited by the collapse. There are an infinite number of these oscillators, each produces the same energy/sec-vol, and so there is an infinite energy/sec-vol produced.

13.3 Random Walk of an Extended Object in CSL

Section 6.5 discusses experimental tests regarding the motion of an isolated extended object. In Section 3.5, and in the associated Supplement, Chapter 10, we treat the motion of a point particle. Here we show that the density matrix evolution equation for an extended object is the same as for the point particle, with appropriate change of mass and collapse rate variables.

We start with the CSL density matrix evolution Eq.(4.13) for N nucleons,

$$\frac{d}{dt}\hat{\rho}(t) = -i[\hat{H}, \hat{\rho}(t)] - \frac{\lambda}{2}\sum_{i,j=1}^{N}\left[e^{-\frac{(\hat{\mathbf{x}}_i - \hat{\mathbf{x}}_j)^2}{4a^2}}\hat{\rho}(t) + \hat{\rho}(t)e^{-\frac{(\hat{\mathbf{x}}_i - \hat{\mathbf{x}}_j)^2}{4a^2}} - 2e^{-\frac{(\hat{\mathbf{x}}_{iL} - \hat{\mathbf{x}}_{jR})^2}{4a^2}}\right]\hat{\rho}(t).$$

(13.11)

Before treating the general case, it is worthwhile to do a mathematically simpler special case. Suppose the object's dimensions are smaller in size than a, so that it is a good approximation to expand all the exponentials in (6.11) to first order. Upon taking the matrix element of the equation, where $|\mathbf{x}\rangle \equiv |\mathbf{x}_1, \mathbf{x}_2, ...\mathbf{x}_N\rangle$ is the position eigenstate of all the particles, we have:

$$\frac{d}{dt}\langle\mathbf{x}|\hat{\rho}(t)|\mathbf{x}'\rangle \approx -i\langle\mathbf{x}|[\hat{H},\hat{\rho}(t)]|\mathbf{x}'\rangle$$

$$-\frac{\lambda}{2}\sum_{i,j=1}^{N}\left[-\frac{(\mathbf{x}_i - \mathbf{x}_j)^2}{4a^2} - \frac{(\mathbf{x}'_i - \mathbf{x}'_j)^2}{4a^2} + 2\frac{(\mathbf{x}_i - \mathbf{x}'_j)^2}{4a^2}\right]\langle\mathbf{x}|\hat{\rho}(t)|\mathbf{x}'\rangle$$

$$= -i\langle\mathbf{x}|[\hat{H},\hat{\rho}(t)]|\mathbf{x}'\rangle - \frac{\lambda}{4a^2}\sum_{i,j=1}^{N}\left[\mathbf{x}_i \cdot \mathbf{x}_j + \mathbf{x}'_i \cdot \mathbf{x}'_j - 2\mathbf{x}_i \cdot \mathbf{x}'_j\right]\langle\mathbf{x}|\hat{\rho}(t)|\mathbf{x}'\rangle$$

$$= -i\langle\mathbf{x}|[\hat{H},\hat{\rho}(t)]|\mathbf{x}'\rangle - \frac{\lambda N^2}{4a^2}[\mathbf{x}_{cm} - \mathbf{x}'_{cm}]^2\langle\mathbf{x}|\hat{\rho}(t)|\mathbf{x}'\rangle, \tag{13.12}$$

where, in the last step, we have used $\mathbf{x}_{cm} \equiv \frac{1}{N}\sum_{i=1}^{N}\mathbf{x}_i$. Writing $|\mathbf{x}\rangle$ as the direct product of $N-1$ relative coordinate position eigenstates and the center of mass position eigenstate $|\mathbf{x}_{cm}\rangle$, and writing the density matrix as the direct product $\hat{\rho}_{rel}\hat{\rho}_{cm}$ (each of trace 1), and then taking the trace over the relative coordinates yields

$$\frac{d}{dt}\langle\mathbf{x}_{cm}|\hat{\rho}_{cm}(t)|\mathbf{x}'_{cm}\rangle = -i\langle\mathbf{x}_{cm}|[\frac{1}{2NM}\hat{P}^2_{cm},\hat{\rho}_{cm}(t)]|\mathbf{x}'_{cm}\rangle$$

$$-\frac{\lambda N^2}{4a^2}[\mathbf{x}_{cm} - \mathbf{x}'_{cm}]^2\langle\mathbf{x}_{cm}|\hat{\rho}_{cm}(t)|\mathbf{x}'_{cm}\rangle$$

or, removing the basis,

$$\frac{d}{dt}\hat{\rho}_{cm}(t) = \sum_{j=1}^{3}\left\{-i\frac{1}{2NM}[\hat{P}^{j2}_{cm},\hat{\rho}_{cm}(t)] - \frac{\lambda N^2}{4a^2}[\hat{X}^j_{cm},[\hat{X}^j_{cm},\hat{\rho}_{cm}(t)]]\right\}, \tag{13.13}$$

where $\hat{X}^1_{cm}, \hat{X}^2_{cm}, \hat{X}^3_{cm}$ are the center of mass position coordinate operators.

As discussed in Section 6.5, in each of the three directions, this density matrix evolution equation has the same form as the density matrix evolution equation for the one-dimensional motion of a particle of mass m undergoing collapse with $\hat{A} - \hat{Y}$:

$$\frac{d}{dt}\hat{\rho}(t) = -i\frac{1}{2m}[\hat{P}^2,\hat{\rho}(t)] - \frac{\lambda}{2}[\hat{X},[\hat{X},\hat{\rho}(t)]] \tag{13.14}$$

(this is in Section 10.2.1, Eq. (10.38a), discussed in Section 3.5).

We see that (13.13) is the same as (13.14) if we replace m, the particle mass m with NM, the total mass of the object and replace the collapse rate λ (of dimension Length^{-2}Time^{-1}) by $\lambda N^2/2a^2$ (which has the same dimension, since CSL's λ has dimension Time^{-1}).

Now we treat the general case, the density matrix evolution equation for the center of mass without any restriction on the size of the object. (The result reduces to Eq. (13.13) in the limit that all dimensions of the object are $<< a$.) Again, we start with

(13.11) and write its matrix element in terms of relative and center of mass coordinates using $\hat{\mathbf{X}}_i = \hat{\mathbf{R}}_i + \hat{\mathbf{X}}_{cm}$:

$$\frac{d}{dt}\langle\mathbf{x}|\hat{\rho}(t)|\mathbf{x}'\rangle = -i\langle\mathbf{x}|[\frac{\hat{\mathbf{P}}^2_{cm}}{2NM} + \hat{H}_{rel}, \hat{\rho}(t)]|\mathbf{x}'\rangle$$

$$-\frac{\lambda}{2}\sum_{i,j=1}^{N}\langle\mathbf{x}|\left[e^{-\frac{(\hat{\mathbf{R}}_i - \hat{\mathbf{R}}_j)^2}{4a^2}}\hat{\rho}(t) + \hat{\rho}(t)e^{-\frac{(\hat{\mathbf{R}}'_i - \hat{\mathbf{R}}'_j)^2}{4a^2}} - 2e^{-\frac{(\hat{\mathbf{R}}_{iL} + \hat{\mathbf{X}}_{cmL} - \hat{\mathbf{R}}'_{jR} - \hat{\mathbf{X}}'_{cmR})^2}{4a^2}}\hat{\rho}(t)\right]|\mathbf{x}'\rangle.$$

$$(13.15)$$

We want to consider only the center of mass motion. We write $|\mathbf{x}\rangle \equiv |\mathbf{x}_1, ...\mathbf{x}_N\rangle = |\mathbf{r}_1...\mathbf{r}_{N-1}\rangle|\mathbf{x}_{cm}\rangle$. Then, the relative position operator for the Nth particle is not an independent operator, it is a function of the operators for the other particles: $\hat{R}_N = -\sum_{i=1}^{N-1}\hat{R}_i$, and $\hat{R}_N|\mathbf{r}_1...\mathbf{r}_{N-1}\rangle = -\sum_{i=1}^{N-1}R_i|\mathbf{r}_1...\mathbf{r}_{N-1}\rangle$.

We neglect internal excitations. Rather than take the nucleons to be localized at the center of atoms in a lattice, we make the simpler (but equivalent, as far as the calculation is concerned) approximation that the $N - 1$ nucleons are uniformly distributed in the volume V of the object, with a product wave function where each term in the product is the identical relative wave function $\phi(\mathbf{r}) = \frac{1}{\sqrt{V}}$ inside the volume and $\phi(\mathbf{r}) = 0$ outside. The wave function for the Nth particle is quite complicated but, since it is only one particle, we will just omit it. That is, in the sum in Eq.(6.15), we will change the upper limit to $N - 1$.

We take $\hat{\rho}(t) \approx \hat{\rho}_{cm}(t)\hat{\rho}_{rel}$ where $\langle\mathbf{r}_1...\mathbf{r}_{N-1}|\hat{\rho}_{rel}|\mathbf{r}_1...\mathbf{r}_{N-1}\rangle = \prod_{k=1}^{N-1}|\phi(\mathbf{r}_k)|^2$. Neglecting internal evolution means setting $\hat{H}_{rel} = 0$ (or $\hat{H}_{rel}\hat{\rho}_{el} = E\hat{\rho}_{rel}$: in either case, $\hat{H}_{rel}$ plays no role), that is, $\hat{\rho}_{rel}$ describes a time independent object except for its center of mass evolution.

Then, we take the trace of (13.15) over the internal coordinates. That is set $\mathbf{r}'_i = \mathbf{r}_i$ and perform the integrations $\int d\mathbf{r}_1...d\mathbf{r}_{N-1}$. The first two Gaussians in the bracket in (13.15) give the same contribution,

$$-\frac{\lambda}{2}\sum_{i,j=1}^{N-1}\int d\mathbf{r}_1...d\mathbf{r}_{n-1}\langle\mathbf{x}_{cm}|\langle\mathbf{r}_1...\mathbf{r}_{N-1}|e^{-\frac{(\hat{\mathbf{R}}_i - \hat{\mathbf{R}}_j)^2}{4a^2}}\hat{\rho}(t)|\mathbf{r}_1...\mathbf{r}_{N-1}\rangle|\mathbf{x}'_{cm}\rangle$$

$$= -\frac{\lambda}{2}\sum_{i,j=1}^{N-1}\int d\mathbf{r}_1...d\mathbf{r}_{n-1}e^{-\frac{(\mathbf{r}_i - \mathbf{r}_j)^2}{4a^2}}\prod_{k=1}^{N-1}|\phi(\mathbf{r}_k)|^2\langle\mathbf{x}_{cm}|\hat{\rho}_{cm}(t)|\mathbf{x}'_{cm}\rangle$$

$$\approx \frac{N^2\lambda}{2V^2}\int_V\int_V d\mathbf{r}_i d\mathbf{r}_j e^{-\frac{(\mathbf{r}_i - \mathbf{r}_j)^2}{4a^2}}\langle\mathbf{x}_{cm}|\hat{\rho}_{cm}(t)|\mathbf{x}'_{cm}\rangle$$

where we have supposed N is large enough to approximate $(N-1)(N-2) \approx N^2$, when considering the terms in the sum where $i \neq j$, and we have neglected the much smaller number N of terms where $i = j$.

We do the same manipulations for the last Gaussian in (13.15). Thus, the trace of (13.15) over internal coordinates yields

$$\frac{d}{dt}\langle\mathbf{x}_{cm}|\hat{\rho}_{cm}(t)|\mathbf{x}'_{cm}\rangle \approx -i\frac{1}{2NM}\langle\mathbf{x}_{cm}|[\hat{\mathbf{P}}^2_{cm},\hat{\rho}_{cm}(t)]|\mathbf{x}'_{cm}\rangle$$

$$-\lambda N^2\frac{1}{V^2}\int_V d\mathbf{z}\int_V d\mathbf{z}'\left[e^{-\frac{(\mathbf{z}-\mathbf{z}')^2}{4a^2}}-e^{-\frac{(\mathbf{z}-\mathbf{z}'+\mathbf{x}_{cm}-\mathbf{x}'_{cm})^2}{4a^2}}\right]\langle\mathbf{x}_{cm}|\hat{\rho}_{cm}(t)|\mathbf{x}'_{cm}\rangle.$$

$$(13.16)$$

where, for a less cumbersome notation, we have replaced $\mathbf{r}_i,\mathbf{r}_j$ by $\mathbf{z},\mathbf{z}'$ and we now temporarily drop the subscript from $\mathbf{x}_{cm},\mathbf{x}'_{cm},\mathbf{P}_{cm}$.

It is self-consistent (the result agrees with the assumption) to assume that there is low probability for the center of mass to be displaced by more than a small fraction of a. That is, $\langle\mathbf{x}|\hat{\rho}(t)|\mathbf{x}'\rangle$ is large only when $|\mathbf{x}-\mathbf{x}'|<<a$, so the second exponential in the square bracket of (13.16) may be expanded in powers of $\mathbf{x}-\mathbf{x}'$, up to the second order. Eq. (13.16) becomes

$$\frac{d}{dt}\langle\mathbf{x}|\hat{\rho}(t)|\mathbf{x}'\rangle \approx -i\frac{1}{2NM}\langle\mathbf{x}|[\hat{\mathbf{P}}^2,\hat{\rho}(t)]|\mathbf{x}' - \lambda N^2\frac{1}{V^2}\int_V d\mathbf{z}\int_V d\mathbf{z}'e^{-\frac{(\mathbf{z}-\mathbf{z}')^2}{4a^2}}$$

$$\cdot\left[\frac{1}{2a^2}(\mathbf{z}-\mathbf{z}')\cdot(\mathbf{x}-\mathbf{x}') + \frac{1}{4a^2}(\mathbf{x}-\mathbf{x}')^2 - \frac{1}{8a^4}[(\mathbf{z}-\mathbf{z}')\cdot(\mathbf{x}-\mathbf{x}')]^2\right]\langle\mathbf{x}|\hat{\rho}(t)|\mathbf{x}'\rangle$$

$$=\sum_{j=1}^{3}\left\{-i\frac{1}{2NM}\langle\mathbf{x}|[\hat{P}^{j2},\hat{\rho}(t)]|\mathbf{x}'\rangle\right.$$

$$\left.-\frac{\lambda N^2}{4a^2}\frac{1}{V^2}\int_V d\mathbf{z}\int_V d\mathbf{z}'e^{-\frac{(\mathbf{z}-\mathbf{z}')^2}{4a^2}}\left[1-\frac{1}{2a^2}[z^j-z'^j]^2\right][x^j-x'^j]^2\langle\mathbf{x}|\hat{\rho}(t)|\mathbf{x}'\rangle\right\}.$$

$$(13.17)$$

where we have used $\int_V d\mathbf{z}\int_V d\mathbf{z}'e^{-\frac{(\mathbf{z}-\mathbf{z}')^2}{4a^2}}[z^j-z'^j]=0$ and $\int_V d\mathbf{z}\int_V d\mathbf{z}'e^{-\frac{(\mathbf{z}-\mathbf{z}')^2}{4a^2}}[z^j-z'^j][z^k-z'^k]=0$ $(j\neq k)$ (under exchange of unprimed and primed variables, the integrand changes sign).

Eq. (13.17) is the matrix element of (putting back the subscript cm, on the operators)

$$\frac{d}{dt}\hat{\rho}_{cm}(t)=\sum_{j=1}^{3}\left\{-i\frac{1}{2NM}[\hat{P}_{cm}^{j2},\hat{\rho}(t)] - \frac{\lambda N^2 f^j}{4a^2}[\hat{X}_{cm}^j,[\hat{X}_{cm}^{'j},\hat{\rho}(t)]]\right\},\text{ where}$$

$$f^j\equiv\frac{1}{V^2}\int_V d\mathbf{z}\int_V d\mathbf{z}'e^{-\frac{(\mathbf{z}-\mathbf{z}')^2}{4a^2}}\left[1-\frac{1}{2a^2}[z^j-z'^j]^2\right].\qquad(13.18)$$

Again, we see that Eq. (13.18) is the three dimensional version of the one-dimensional evolution equation Eq. (13.14), this time when Eq. (13.14)'s variables are replaced by

$$m\to NM,\quad\lambda\to\frac{\lambda N^2}{2a^2}f^j.\qquad(13.19)$$

We note that, for a small enough object so that $\frac{1}{a^2}[z^j-z'^j]^2\approx 0$, then $f^j\approx 1$, and Eq. (13.18) becomes identical to Eq. (13.13).

Eq. (13.18) appears as Eq. (6.16).

13.4 Decay of Particle Number in a Bose–Einstein Condensate

At a sufficiently low temperature, a large number of Bose–statistics atoms will occupy the same ground state of a potential well. Such a Bose–Einstein condensate is a natural place to look for CSL heating, which excites atoms out of the ground state, depleting it. Here, we calculate the rate of such excitation, to use in the discussion of experiments that measure such depletion, Section 6.6.2.

13.4.1 Lindblad Equation for Identical Atoms

The CSL density matrix evolution Eq.(4.13)$_B$, for N nucleons, expressed in the position representation $|\mathbf{x}\rangle \equiv |\mathbf{x}_1, \mathbf{x}_2, ...\mathbf{x}_N\rangle$,

$$\frac{d}{dt}\langle\mathbf{x}|\hat{\rho}(t)|\mathbf{x}'\rangle = -i\langle\mathbf{x}|[\hat{H}, \hat{\rho}(t)]|\mathbf{x}'\rangle$$

$$-\frac{\lambda}{2}\sum_{i,j=1}^{N}\left[e^{-\frac{(\mathbf{x}_i-\mathbf{x}_j)^2}{4a^2}} + e^{-\frac{(\mathbf{x}'_i-\mathbf{x}'_j)^2}{4a^2}} - 2e^{-\frac{(\mathbf{x}_i-\mathbf{x}'_j)^2}{4a^2}}\right]\langle\mathbf{x}|\hat{\rho}(t)|\mathbf{x}'\rangle,$$

implies that, for a collection of identical atoms, the density matrix describing the centers of mass of these atoms satisfies the same form of this equation. That is, the atoms act like particles with mass MA, where A is the atomic number, the number of nucleons in the nucleus, so $\lambda = \lambda M^2/M^2$ is replaced by $\lambda(MA)^2/M^2 = \lambda A^2$. For this proof, we neglect the electrons and neglect the neutron–proton mass difference.

In the equation above, the total number of nucleons is $N = N_A A$, where N_A is the number of atoms. Label the coordinates of a nucleon $\mathbf{x}_{i\alpha} = \mathbf{r}_{i\alpha} + \mathbf{x}_{cmi}$, where $1 \leq i \leq N_A$ labels an atom, $\mathbf{x}_{cmi}$ is the position coordinate of the center of mass of the atom and $\mathbf{r}_{i\alpha}$ is the position coordinate of a nucleon within the ith atom relative to the center of mass of that atom ($1 \leq \alpha \leq A$).

One may neglect the relative coordinates $\mathbf{r}_{i\alpha}$ in all the exponents. When i, j refer to nucleons in the same nucleus, the center of mass difference vanishes, and the relative coordinate differences within the same nucleus are much smaller than a, so the exponents give, to an excellent approximation, $1 + 1 - 2 = 0$. When i, j refer to nucleons in different nucleii, still, the relative coordinates are of nuclear dimensions, and usually[2] much smaller than the center of mass separation of the nucleii of different atoms:

$$e^{-\frac{(\mathbf{x}_{i\alpha}-\mathbf{x}_{j\alpha})^2}{4a^2}} = e^{-\frac{[[\mathbf{r}_{i\alpha}+\mathbf{x}_{cmi}-\mathbf{r}_{j\alpha}-\mathbf{x}_{cmj}]^2}{4a^2}} \approx e^{-\frac{[[\mathbf{x}_{cmi}-\mathbf{x}_{cmj}]^2}{4a^2}}.$$

We change the labeling of the position state vectors from $|\mathbf{x}\rangle$ to

$$|\mathbf{r}\rangle|\mathbf{x}_{cm}\rangle \equiv |\mathbf{r}_{1,1}, ...\mathbf{r}_{N_A,A}\rangle|\mathbf{x}_{cm1}, ...\mathbf{x}_{cmN_A}\rangle.$$

Thus, we obtain

---

[2]Since the atomic wave functions may overlap (*will* overlap in a BEC) there are regions of configuration space where the center of mass difference *is* comparable to or less than nuclear size, but these regions occupy a negligibly small configuration space volume.

$$\frac{d}{dt}\langle\mathbf{r}|\langle\mathbf{x}_{cm}|\hat{\rho}(t)|\mathbf{x}'_{cm}\rangle|\mathbf{r}'\rangle \approx -i\langle\mathbf{r}|\langle\mathbf{x}_{cm}|[\hat{H}_{cm}+\hat{H}_r,\hat{\rho}(t)]|\mathbf{x}'_{cm}\rangle|\mathbf{r}'\rangle$$

$$-\frac{\lambda}{2}\sum_{i,j}^{N_A}\sum_{\alpha,\beta=1}^{A}\left[e^{-\frac{(\mathbf{x}_{cmi}-\mathbf{x}_{cmj})^2}{4a^2}}+e^{-\frac{(\mathbf{x}'_{cmi}-\mathbf{x}'_{cmj})^2}{4a^2}}-2e^{-\frac{(\mathbf{x}_{cmi}-\mathbf{x}'_{cmj})^2}{4a^2}}\right]$$

$$\cdot\langle\mathbf{r}|\langle\mathbf{x}_{cm}|\hat{\rho}(t)|\mathbf{x}'_{cm}\rangle|\mathbf{r}'\rangle$$

$$(13.20)$$

Setting $\mathbf{r}_{i\alpha}=\mathbf{r}'_{i\alpha}$ and taking the trace over these relative coordinates, yields as claimed,

$$\frac{d}{dt}\langle\mathbf{x}_{cm}|\hat{\rho}_{cm}(t)|\mathbf{x}'_{cm}\rangle = -i\langle\mathbf{x}_{cm}|[\hat{H}_{cm},\hat{\rho}_{cm}(t)]|\mathbf{x}'_{cm}\rangle$$

$$-\frac{\lambda A^2}{2}\sum_{i,j}^{N_A}\left[e^{-\frac{(\mathbf{x}_{cmi}-\mathbf{x}_{cmj})^2}{4a^2}}+e^{-\frac{(\mathbf{x}'_{cmi}-\mathbf{x}'_{cmj})^2}{4a^2}}-2e^{-\frac{(\mathbf{x}_{cmi}-\mathbf{x}'_{cmj})^2}{4a^2}}\right]\langle\mathbf{x}_{cm}|\hat{\rho}_{cm}(t)|\mathbf{x}'_{cm}\rangle.$$

$$(13.21)$$

We can write Eq. (13.21) in coordinate independent form. We note that that if $\hat{D}(\mathbf{x})$ is the atomic number density operator, where $\mathbf{x}$ is the coordinate of the atomic center of mass, then $\hat{D}(\mathbf{x})|\mathbf{x}_{cm}\rangle \equiv \hat{D}(\mathbf{x})|\mathbf{x}_{1cm},\mathbf{x}_{2cm},...\mathbf{x}_{N_Acm}\rangle = \sum_{i=1}^{N_A}\delta(\mathbf{x}-\mathbf{x}_{icm})|\mathbf{x}_{cm}\rangle$. Using this, Eq. (13.21) is equivalent to (dropping the cm subscripts):

$$\langle\mathbf{x}|\frac{d}{dt}\hat{\rho}(t)|\mathbf{x}'\rangle = -i\langle\mathbf{x}|[\hat{H},\hat{\rho}(t)]|\mathbf{x}'\rangle - \frac{\lambda A^2}{2}\langle\mathbf{x}|\int d\mathbf{z}'\int d\mathbf{z}\,e^{-\frac{(\mathbf{z}'-\mathbf{z})^2}{4a^2}}$$

$$\cdot\left[D(\mathbf{z}')D(\mathbf{z})\hat{\rho}(t)+\hat{\rho}(t)D(\mathbf{z}')D(\mathbf{z})-2D(\mathbf{z}')\hat{\rho}(t)D(\mathbf{z})\right]|\mathbf{x}'\rangle$$

or

$$\frac{d}{dt}\hat{\rho}(t) = -i[\hat{H},\hat{\rho}(t)] - \frac{\lambda A^2}{2}\int d\mathbf{x}'\int d\mathbf{x}\,e^{-\frac{(\mathbf{x}-\mathbf{x}')^2}{4a^2}}[D(\mathbf{x}'),[D(\mathbf{x}),\hat{\rho}(t)]]. \qquad (13.22)$$

13.1.2 Creation and Annihilation Operators for One Particle Energy Eigenstates

Section 11.1 showed that the particle number density operator can be written as $\hat{D}(\mathbf{x}) = \hat{\xi}^\dagger(\mathbf{x})\hat{\xi}(\mathbf{x})$ (Eq. (11.3)), where $\xi^\dagger(\mathbf{x})$ creates a particle at location $\mathbf{x}$ when acting on the no-particle (vacuum) state $|0\rangle$, that is, $\xi^\dagger(\mathbf{x})|0\rangle = |\mathbf{x}\rangle$. Here it is useful to show that one can write $\hat{D}(\mathbf{x})$ in terms of operators that create single-atom energy eigenstates $|i\rangle$ ($\hat{H}|i\rangle = E_i|i\rangle$) out of the vacuum.

The Hamiltonian for a single atom (with potential equal to that of the trap as well as other atoms in mean field approximation) has energy eigenfunctions $\phi_i(\mathbf{x})$ (belonging to a complete set). $\phi_0(\mathbf{x})$ is the lowest energy state of the atoms, the state occupied by all the atoms in the BEC. $\phi_i(\mathbf{x})$, for $i > 0$, are the successively higher energy states of an atom in the trap.

One can define a creation operator that creates a one-particle state out of the vacuum:

$$\hat{a}_i^\dagger|0\rangle \equiv |i\rangle = \int d\mathbf{x}|\mathbf{x}\rangle\langle\mathbf{x}|i\rangle = \int d\mathbf{x}|\mathbf{x}\rangle\phi_i(\mathbf{x})$$

$$= \int d\mathbf{x}\phi_i(\mathbf{x})\hat{\xi}^\dagger(\mathbf{x})|0\rangle. \tag{13.23}$$

This operator,

$$\hat{a}_i^\dagger = \int d\mathbf{x}\phi_i(\mathbf{x})\hat{\xi}^\dagger(\mathbf{x}), \tag{13.24}$$

acting on any state adds a particle. One easily verifies, using $[\hat{\xi}^\dagger(\mathbf{x}), \hat{\xi}^\dagger(\mathbf{x}')] = \delta(\mathbf{x}-\mathbf{x}')$, that $[\hat{a}_i, \hat{a}_j^\dagger] = \delta_{ij}$, on account of the orthonormality of the eigenfunctions.

Eq. (13.24) can be inverted by multiplying by $\phi_i^*(\mathbf{x}')$, and summing over i, with use of the completeness of the eigenfunctions:

$$\sum_i \phi_i^*(\mathbf{x}')\hat{a}_i^\dagger = \int d\mathbf{x}\sum_i \phi_i^*(\mathbf{x}')\phi_i(\mathbf{x})\hat{\xi}^\dagger(\mathbf{x})] = \int d\mathbf{x}\delta(\mathbf{x} - \mathbf{x}')\hat{\xi}^\dagger(\mathbf{x}) = \xi^\dagger(\mathbf{x}'). \tag{13.25}$$

To check (13.25), we calculate the commutator:

$$[\xi(\mathbf{x}), \xi^\dagger(\mathbf{x}')] = [\sum_j \phi_j(\mathbf{x})\hat{a}_j, \sum_i \phi_i^*(\mathbf{x}')\hat{a}_i^\dagger] = \sum_{ij}\delta_{ij}\phi_j(\mathbf{x})\phi_i^*(\mathbf{x}') = \delta(\mathbf{x} - \mathbf{x}'). \tag{13.26}$$

Then, we can write the atom number density operator as

$$\hat{D}(\mathbf{x}) \equiv \xi^\dagger(\mathbf{x})\xi(\mathbf{x}) = \sum_{ij}\phi_i^*(\mathbf{x})\phi_j(\mathbf{x})\hat{a}_i^\dagger\hat{a}_j \tag{13.27}$$

The atom number operator is therefore

$$\hat{N} = \int d\mathbf{x}\hat{D}(\mathbf{x}) = \sum_{ij}\delta_{ij}\hat{a}_i^\dagger\hat{a}_j = \sum_i \hat{a}_i^\dagger\hat{a}_i. \tag{13.28}$$

In particular, we identify the operator representing the number of atoms in the ith state as $\hat{N}_i = \hat{a}_i^\dagger\hat{a}_i$.

13.4.3 Decay of Atoms from the BEC

In order to calculate $\overline{N_0}(t)$, the mean number of atoms in the ground state at any time t, we multiply Eq. (13.22) by $\hat{N}_0 \equiv \hat{a}_0^\dagger\hat{a}_0$ and take the trace:

$$\frac{d}{dt}\overline{N_0}(t) \equiv \frac{d}{dt}\mathrm{Tr}\hat{N}_0\hat{\rho}(t) = -i\mathrm{Tr}\hat{\rho}(t)[\hat{N}_0, \sum_j E_j\hat{N}_j]$$

$$-\frac{\lambda A^2}{2}\int d\mathbf{x}'\int d\mathbf{x}e^{-\frac{(\mathbf{x}-\mathbf{x}')^2}{4a^2}}\mathrm{Tr}\hat{\rho}(t)[\hat{D}(\mathbf{x}'), [\hat{D}(\mathbf{x}), \hat{N}_0]]$$

$$= -\frac{\lambda A^2}{2}\int d\mathbf{x}'\int d\mathbf{x}e^{-\frac{(\mathbf{x}-\mathbf{x}')^2}{4a^2}}\mathrm{Tr}\hat{\rho}(t)[\hat{\xi}^\dagger(\mathbf{x}')\hat{\xi}(\mathbf{x}'), [\hat{\xi}^\dagger(\mathbf{x})\hat{\xi}(\mathbf{x}), \hat{a}_0^\dagger\hat{a}_0]],$$

$$\tag{13.29}$$

where we have used the cyclic permutivity of the trace operation to take $\hat{\rho}(t)$ out of the commutators, we have written the Hamiltonian as $\hat{H} = \sum_j E_j \hat{N}_j$, and used $[\hat{N}_i, \hat{N}_j] = 0$.

Next, we have to evaluate the commutators in (13.29). Employing (13.25), we see that

$$[\hat{\xi}(\mathbf{x}), \hat{a}_0^\dagger] = \phi_0(\mathbf{x}), \quad [\hat{\xi}^\dagger(\mathbf{x}), \hat{a}_0] = -\phi_0^*(\mathbf{x}). \tag{13.30}$$

Then, the commutators in (13.29) yield, using the identity $[\hat{A}\hat{B}, \hat{C}] = \hat{A}[\hat{B}, \hat{C}] + [\hat{A}, \hat{C}]\hat{B}$:

$$[\hat{\xi}^\dagger(\mathbf{x})\hat{\xi}(\mathbf{x}), \hat{a}_0^\dagger \hat{a}_0] = \hat{\xi}^\dagger(\mathbf{x})[\hat{\xi}(\mathbf{x}), \hat{a}_0^\dagger \hat{a}_0] + [\hat{\xi}^\dagger(\mathbf{x}), \hat{a}_0^\dagger \hat{a}_0]\hat{\xi}(\mathbf{x})$$
$$= \hat{\xi}^\dagger(\mathbf{x})\phi_0(\mathbf{x})\hat{a}_0 - \hat{a}_0^\dagger \hat{\xi}(\mathbf{x})\phi_0^*(\mathbf{x}),$$

$$[\hat{\xi}^\dagger(\mathbf{x}')\hat{\xi}(\mathbf{x}'), [\hat{\xi}^\dagger(\mathbf{x})\hat{\xi}(\mathbf{x}), \hat{a}_0^\dagger \hat{a}_0]] = \hat{\xi}^\dagger(\mathbf{x}')[\hat{\xi}(\mathbf{x}'), \hat{\xi}^\dagger(\mathbf{x})\phi_0(\mathbf{x})\hat{a}_0 - \hat{a}_0^\dagger \hat{\xi}(\mathbf{x})\phi_0^*(\mathbf{x})] +$$
$$[\hat{\xi}^\dagger(\mathbf{x}'), \hat{\xi}^\dagger(\mathbf{x})\phi_0(\mathbf{x})\hat{a}_0 - \hat{a}_0^\dagger \hat{\xi}(\mathbf{x})\phi_0^*(\mathbf{x})]\hat{\xi}(\mathbf{x}')$$
$$= \hat{\xi}^\dagger(\mathbf{x}')\Big(\delta(\mathbf{x} - \mathbf{x}')\phi_0(\mathbf{x})\hat{a}_0 - \hat{\xi}(\mathbf{x})\phi_0(\mathbf{x}')\phi_0^*(\mathbf{x})\Big) +$$
$$\Big(\delta(\mathbf{x} - \mathbf{x}')\phi_0^*(\mathbf{x})\hat{a}_0^\dagger - \hat{\xi}^\dagger(\mathbf{x})\phi_0^*(\mathbf{x}')\phi_0(\mathbf{x})\Big)\hat{\xi}(\mathbf{x}')$$

Putting this into (13.29), we obtain

$$\frac{d}{dt}\overline{N_0}(t) = -\frac{\lambda A^2}{2}\int d\mathbf{x}'\int d\mathbf{x}\, e^{-\frac{(\mathbf{x}-\mathbf{x}')^2}{4a^2}}\text{Tr}\hat{\rho}(t)$$
$$\cdot\Big[\delta(\mathbf{x} - \mathbf{x}')[\hat{\xi}^\dagger(\mathbf{x}')\hat{a}_0\phi_0(\mathbf{x}) + \hat{a}_0^\dagger \hat{\xi}(\mathbf{x}')\phi_0^*(\mathbf{x})]$$
$$-\hat{\xi}^\dagger(\mathbf{x}')\hat{\xi}(\mathbf{x})\phi_0^*(\mathbf{x})\phi_0(\mathbf{x}') - \hat{\xi}^\dagger(\mathbf{x})\hat{\xi}(\mathbf{x}')\phi_0^*(\mathbf{x}')\phi_0(\mathbf{x})\Big]$$
$$= -\lambda A^2\Big[\text{Tr}\hat{\rho}(t)\hat{a}_0^\dagger \hat{a}_0 - \int d\mathbf{x}'\int d\mathbf{x}\, e^{-\frac{(\mathbf{x}-\mathbf{x}')^2}{4a^2}}\phi_0^*(\mathbf{x})\phi_0(\mathbf{x}')\text{Tr}\hat{\rho}(t)\hat{\xi}^\dagger(\mathbf{x}')\hat{\xi}(\mathbf{x})\Big]. \tag{13.31}$$

In going to the last line, the first term in the square bracket follows from (13.25), and in the second term we have set $\mathbf{x} \leftrightarrow \mathbf{x}'$, in one of the integrals.

Because the BEC wave function is so much larger in scale than a, we can treat the exponential in (13.31) as approximately a delta function, $e^{-\frac{(\mathbf{x}-\mathbf{x}')^2}{4a^2}} \approx (\pi 4a^2)^{3/2}\delta(\mathbf{x} - \mathbf{x}')$:

$$\frac{d}{dt}\overline{N_0}(t) \approx -\lambda A^2\Big[\overline{N_0}(t) - (4\pi)^{3/2}a^3\int d\mathbf{x}|\phi_0(\mathbf{x})|^2\text{Tr}\hat{\rho}(t)\hat{\xi}^\dagger(\mathbf{x})\hat{\xi}(\mathbf{x})\Big]. \tag{13.32}$$

The second bracketed term in (13.32) may be neglected compared to the first term. $\text{Tr}\hat{\rho}(t)\hat{\xi}^\dagger(\mathbf{z})\hat{\xi}(\mathbf{z})$ is the mean number density $\overline{N}(\mathbf{x}, t)$ of all particles at time t. We consider the situation where most of the particles are in the ground state, so $\overline{N}(\mathbf{x}, t) \approx \overline{N_0}(t)/V$, where $\approx V$ is the trap volume where the BEC wave function is large. Thus, the second bracketed term in (13.32) is of order $[a^3/V]\overline{N_0}(t)$, negligible compared to $\overline{N_0}(t)$.

We thus have simply

$$\overline{N_0}(t) = \overline{N_0}(0)e^{-\lambda A^2 t}.$$

(13.33)

which is copied in Eq. (6.23), and used to evaluate the limit on λ based upon experimental data giving the decay of the number of atoms in the BEC.

14

Supplement to Chapter 7

14.1 Observable Reality for an Electron in the Ground State of Hydrogen

Chapter 7 discussed nucleon number-stuff in an arbitrary volume V. To show the variety of applications of the stuff idea, here we discuss particle number-stuff in various volumes V, in situations when there is only one particle (as mentioned in Chapter 7's endnote A).

Consider an electron in the ground state of hydrogen. The wave function is $\psi(\mathbf{x}) = \frac{1}{\sqrt{\pi}a_0^{3/2}}e^{-r/a_0}$, where r is the radial coordinate, and $a_0 \approx .5 \times 10^{-8}$cm is the Bohr radius. Suppose the volume V is a sphere of radius R centered at $r = 0$. The number operator for the particle in V is $\hat{N}_V = \int_V dV |\mathbf{x}\rangle\langle\mathbf{x}|$. Its eigenvectors are $|\mathbf{x}_1\rangle$ $(r_1 \equiv \sqrt{\mathbf{x}_1 \cdot \mathbf{x}_1})$ with eigenvalue 1 if $r_1 < R$ and eigenvalue 0 if $r_1 > R$.

The total amounts of 1-stuff and 0-stuff in V are respectively

$$|_V\langle 1|\psi, t\rangle|^2 = 4\pi \int_0^R r^2 dr |\psi(\mathbf{x})|^2 = 1 - e^{-\frac{2R}{a_0}}\left[1 + \frac{2R}{a_0} + \frac{1}{2}\left[\frac{2R}{a_0}\right]^2\right]. \quad (14.1a)$$

$$|_V\langle 0|\psi, t\rangle|^2 = 4\pi \int_R^\infty r^2 dr |\psi(\mathbf{x})|^2 = e^{-\frac{2R}{a_0}}\left[1 + \frac{2R}{a_0} + \frac{1}{2}\left[\frac{2R}{a_0}\right]^2\right], \quad (14.1b)$$

Section 7.3 gave a criterion for observability. Using the stuff as a probability distribution, one calculates $\bar{n}_V$, the mean value of the particle number in V, and $\bar{\sigma}_V$, the standard deviation of the particle number in V, and takes the fractional standard deviation ratio $\bar{\sigma}_V/\bar{n}_V$. If this ratio is "sufficiently small," one can say that the particle number-stuff has observability status and that the number of particles in V is $\bar{n}_V$. What is meant by "sufficiently small" was defined as that the fractional experimental error in making the best possible measurement of the number of particles in V is larger than $\bar{\sigma}_V/\bar{n}_V$.

So, let's consider what is the smallest value of R permitting the statement that the electron lies within the sphere.

Using

$$\bar{n}_V = 0 \times |_V\langle 0|\psi, t\rangle|^2 + 1 \times |_V\langle 1|\psi, t\rangle|^2 = |_V\langle 1|\psi, t\rangle|^2,$$

$$\sigma_V^2 \equiv \overline{n_V^2} - \bar{n}_V^2 = |_V\langle 1|\psi, t\rangle|^2 - |_V\langle 1|\psi, t\rangle|^4 = |_V\langle 1|\psi, t\rangle|^2|_V\langle 0|\psi, t\rangle|^2,$$

we have

$$\sigma_V/\bar{n}_V = |_V\langle 0|\psi, t\rangle|/|_V\langle 1|\psi, t\rangle|.$$

For example, if $R \approx 1.3a_0$, then $_V\langle 1|\psi,t\rangle \approx _V\langle 0|\psi,t\rangle \approx \frac{1}{\sqrt{2}}$, so $\bar{n}_V \approx 1/2$ and $\sigma_V/\bar{n}_V \approx 1$. Then, our prescription makes no statement about observability of the number of electrons within R, and good thing, too, since a scattering experiment finds the electron within R and outside of R an equal amount of times.

But, if $R \approx 3a_0$, we find $\bar{n}_V \approx 1 - 3 \times 10^{-2}$, $\sigma_V/\bar{n}_V \approx 3 \times 10^{-2}$. Say a 3% accurate experiment is the best that can be done. Then we declare that there is 1 electron within $R \approx 3a_0$ (or any larger radius) and, indeed, that is what is always observed (within experimental error).

14.2   Observable Reality for a Particle Undergoing Two-Slit Interference

Consider a particle in a two-slit interference experiment, which Feynman famously said

> ... has been designed to contain all the mystery of quantum mechanics, to put you up against the paradoxes and mysteries and peculiarities of nature one hundred percent.[1].

Here is a qualitative description of what happens, according to CSL. The state vector describes the particle wave packet impinging on the slit screen, as well as a lot of un-triggered highly localized detectors on a detection screen. The packet is described by the usual Schrödinger dynamics, because the collapse rate is negligible for a single particle. Two wave packets emerge from the slits with equal amplitudes and they spread and overlap.

Upon reaching the detectors, the electron wave function interacts with them. According to usual quantum theory, this ends up as a superposition of many states, each state describing one localized detector triggered by absorbing the particle, and the rest idle. But, according to CSL, as the detector states just begin to evolve toward detection/non-detection status, the CSL term, hitherto negligible, takes over, causing these detector states to rapidly play the collapse version of the Gambler's Ruin game. When one state wins, it describes the event of detection of the particle by one detector somewhere on the screen. The ensemble of such outcomes of course displays the usual two-slit interference pattern.

Now, consider the time when the two wave packets are just emerging from the two slits, not yet overlapping. The wave function description is the usual one, with CSL dynamics playing no role.

The number operator for a volume V is the same as defined in the previous discussion, with eigenvalues either 1 or 0. Let V_L be a volume that lies lie in front of the left slit, large enough to encompass "most" of the emerging left packet, and similarly for V_R. The total $n = 1$-stuff in each volume has magnitude $\approx 1/2$. Since the ratio of standard deviation to mean particle number for either volume is 1, the observability criterion is not satisfied. *We cannot say which slit the electron goes through.*

Now, consider a volume encompassing both slits, $V = V_L + V_R$. Then $\bar{n}_V = 1$, $\sigma_V = 0$,. Thus, according to our observability criterion, it is a fact that *a particle has surely passed through the slits.* The interference pattern on the screen eventually appears because *stuff has gone through both slits.*

---

[1]Feynman (1985)

14.3   Photon Number Operator

A way to interpret CSL-modified quantum theory was discussed in Chapter 7. The idea of nucleon number in V-stuff was introduced. It was used to specify a criterion for observability, for when we can specify the number of nucleons in V.

Here we sketch how to obtain a different kind of stuff that could alternatively be used to define observability,

For, to observe is, of course, to see. And, really, we *see* the world via photons. To obtain a relevant basis, we may imagine a hollow tube of length L that can be held to an eye. Suppose it has a filter that allows light in a small band of wavelengths around $2\pi/k_0$ to pass through. (We would be particularly interested in k_0 lying in the optical spectrum.) Then, we can define an operator $\hat{N}_V(\mathbf{k}_0)$ representing the number of photons contained within the volume of such a tube, traveling along its length with momenta near $\mathbf{k}_0$. Using its eigenvectors, $\hat{N}_V(\mathbf{k}_0)|n, \mathbf{k}_0\rangle_V = n(\mathbf{k}_0)|n, \mathbf{k}_0\rangle_V$, one can obtain its associated stuff $|_V\langle n, \mathbf{k}_0|\psi, t\rangle|^2$.

One may then proceed to define observability in terms of this, and give examples of its use, but we will not go any further in this direction here. Here we concentrate on showing what is involved in constructing $\hat{N}_V(\mathbf{k}_0)$, which is a prime necessity if this approach is to work.

To simplify the discussion, we suppose the tube is oriented along the x-axis, with $-L/2 \le x \le L/2$ (this region of x will hereafter be called "the tube") and we neglect the extension of the tube in the y-z plane. We therefore wish to construct the operator $\hat{N}_L(k_0)$ corresponding to the number of one-dimensional photons contained within the tube in a limited momentum range around k_0.

The annihilation operator for a photon at x is

$$\hat{\xi}(x) = \frac{1}{\sqrt{2\pi}} \int_{-\infty}^{\infty} dk\, e^{ikx} \hat{a}(k) \tag{14.2}$$

where $\hat{a}(k)$ is the annihilation operator for a photon with momentum k. We begin by defining a new annihilation operator for a photon, $\hat{\eta}_{k_0}(x)$. The photon exists in a "small" spatial region around x, characterized by length parameter σ, where small means much smaller than L, but yet much longer than many light wavelengths. The photon is also characterized as occupying a small momentum region around k_0, with width parameter $1/\sigma$, which means that many wavelengths fit into the width σ:

$$\begin{aligned}
\hat{\eta}_{k_0}(x) &\equiv \int_{-\infty}^{\infty} dx'\, e^{ik_0(x-x')} \frac{1}{\sqrt{4\pi\sigma^2}} e^{-(x-x')^2/4\sigma^2} \hat{\xi}(x') \\
&= \int_{-\infty}^{\infty} dx'\, e^{ik_0(x-x')} \frac{1}{\sqrt{4\pi\sigma^2}} e^{-(x-x')^2/4\sigma^2} \frac{1}{\sqrt{2\pi}} \int_{-\infty}^{\infty} dk\, e^{ikx'} \hat{a}(k) \\
&= \frac{1}{\sqrt{2\pi}} \int_{-\infty}^{\infty} dk\, e^{ikx} e^{-(k-k_0)^2\sigma^2} \hat{a}(k).
\end{aligned} \tag{14.3}$$

It follows from (14.3) that

$$[\hat{\eta}_{k_0}(x), \hat{\eta}_{k_0'}^\dagger(x')] = \frac{1}{2\pi} \int_{-\infty}^{\infty} dk dk' e^{i(kx - k'x')} e^{-(k-k_0)^2 \sigma^2} e^{-(k'-k_0')^2 \sigma^2} [\hat{a}(k), \hat{a}^\dagger(k')]$$

$$= \frac{1}{2\pi} \int_{-\infty}^{\infty} dk e^{ik(x-x')} e^{-(k-k_0)^2 \sigma^2} e^{-(k-k_0')^2 \sigma^2}$$

$$= \frac{1}{2\sqrt{2\pi\sigma^2}} e^{-\frac{1}{8\sigma^2}(x-x')^2} e^{-\sigma^2(k_0-k_0')^2} e^{i(x-x')(k_0+k_0')/2}. \tag{14.4}$$

So, the commutator approximately vanishes for $|x-x'|$ larger than a few σ, and $|k_0-k_0'|$ larger than a few $1/\sigma$.

We choose $\sigma \ll L$ and, since we want the momentum range to be much smaller than the momentum k_0, we also choose $\frac{1}{\sigma} \ll k_0$. Since the wavelengths are in the optical spectrum, roughly 4×10^{-5}cm to 7×10^{-5}cm, these choices are readily achieved if L is of centimeter scale if $\sigma \approx 0.01$ cm, which is the application in mind here.

Then, using the first expression in Eq.(14.3) we define the operator to represent the number of photons in the tube with momenta near k_0 as

$$\hat{N}_L(k_0) \equiv \int_{-L/2}^{L/2} dx \hat{\eta}_{k_0}^\dagger(x) \hat{\eta}_{k_0}(x)$$

$$= \int_{-\infty}^{\infty} dx_1 dx_2 \hat{\xi}^\dagger(x_1)\hat{\xi}(x_2) \int_{-L/2}^{L/2} dx e^{-ik_0(x-x_1)} \frac{1}{\sqrt{4\pi\sigma^2}} e^{-(x-x_1)^2/4\sigma^2}$$

$$\cdot e^{ik_0(x-x_2)} \frac{1}{\sqrt{4\pi\sigma^2}} e^{-(x-x_2)^2/4\sigma^2}$$

$$\approx \int_{-L/2}^{L/2} dx_1 dx_2 \hat{\xi}^\dagger(x_1)\hat{\xi}(x_2) e^{ik_0(x_1-x_2)} \int_{-L/2}^{L/2} dx \frac{1}{4\pi\sigma^2} e^{-(x-x_1)^2/4\sigma^2} e^{-(x-x_2)^2/4\sigma^2}$$

$$\approx \int_{-L/2}^{L/2} dx_1 dx_2 \hat{\xi}^\dagger(x_1)\hat{\xi}(x_2) e^{ik_0(x_1-x_2)} \frac{1}{\sqrt{8\pi\sigma^2}} e^{-(x_1-x_2)^2/8\sigma^2}. \tag{14.5}$$

Two approximations have been made in (14.5). The integral over x in the second equation is restricted to being within the tube. Therefore, because of the narrow width of the Gaussians, effectively x_1, x_2 are restricted to being within the tube, so we changed their integral range to $(-L/2, L/2)$ in the third step. Then, the limits on the integral over x have been extended to infinity, since the Gaussians are close to 0 in value anyway for x outside the tube. This allows the integral over x to be performed, resulting in the last line.

The expression (14.5) is the result we have sought. We now want to see that $\hat{N}_L(k_0)$ behaves properly.

Let's first deal with a single photon. The state vector for a single photon has the form

$$|\psi\rangle = \int_{-\infty}^{\infty} dx\phi(x)\hat{\xi}^{\dagger}(x)|0\rangle$$

$$= \int_{-\infty}^{\infty} dx\phi(x)\frac{1}{\sqrt{2\pi}} \int_{-\infty}^{\infty} dk e^{-ikx}\hat{a}^{\dagger}(k)|0\rangle$$

$$= \int_{-\infty}^{\infty} dk\tilde{\phi}(k)\hat{a}^{\dagger}(k)|0\rangle \text{ where } \tilde{\phi}(k) \equiv \frac{1}{\sqrt{2\pi}} \int_{-\infty}^{\infty} dx\phi(x)e^{-ikx} \qquad (14.6)$$

($\phi(x)$ is a normalized wave function).

Let's consider two cases, when there is no photon in the tube and when the photon is in the tube.

When there is no photon in the tube, then $\phi(x) = 0$ for $-L/2 \leq x \leq L/2$. When the photon lies within the tube, then $\phi(x)$ is non–zero only within the tube. From (14.5), (14.6),

$$\hat{N}_L(k_0)|\psi\rangle = \int_{-L/2}^{L/2} dx_1 dx_2 \hat{\xi}^{\dagger}(x_1)\hat{\xi}(x_2)e^{ik_0(x_1-x_2)}$$

$$\cdot \frac{1}{\sqrt{8\pi\sigma^2}}e^{-(x_1-x_2)^2/8\sigma^2} \int_{-\infty}^{\infty} dx\phi(x)\hat{\xi}^{\dagger}(x)|0\rangle$$

$$= \int_{-L/2}^{L/2} dx_1 \hat{\xi}^{\dagger}(x_1)|0\rangle \int_{-\infty}^{\infty} dx e^{ik_0(x_1-x)} \frac{1}{\sqrt{8\pi\sigma^2}}e^{-(x_1-x)^2/8\sigma^2} \cdot \phi(x)$$

$$(14.7)$$

In going from the first line to the second line, we have used

$$\hat{\xi}(x_2)\hat{\xi}^{\dagger}(x)|0\rangle = [\hat{\xi}(x_2), \hat{\xi}^{\dagger}(x)]|0\rangle = \delta(x_2 - x)|0\rangle.$$

If there is no photon in the tube, it is a good approximation to take $e^{-(x_1-x)^2/8\sigma^2}\phi(x) \approx 0$, since the Gaussian ensures x is close to x_1, but x_1 lies within the tube where $\phi(x)$ vanishes. So, $\hat{N}_L(k_0)|\psi, t\rangle \approx 0$, as it should be.

If the photon lies within the tube, we replace $\phi(x)$ in (14.6) by its Fourier transform:

$$\hat{N}_L(k_0)|\psi\rangle = \int_{-L/2}^{L/2} dx_1 \hat{\xi}^{\dagger}(x_1)|0\rangle \int_{-\infty}^{\infty} dx e^{ik_0(x_1-x)}$$

$$\cdot \frac{1}{\sqrt{8\pi\sigma^2}}e^{-(x_1-x)^2/8\sigma^2} \frac{1}{\sqrt{2\pi}} \int_{-\infty}^{\infty} dk\tilde{\phi}(k)e^{ikx}$$

$$= \int_{-L/2}^{L/2} dx_1 \hat{\xi}^{\dagger}(x_1)|0\rangle e^{ik_0 x_1} \frac{1}{\sqrt{2\pi}} \int_{-\infty}^{\infty} dk\tilde{\phi}(k)$$

$$\cdot \int_{-\infty}^{\infty} dx \frac{1}{\sqrt{8\pi\sigma^2}}e^{-(x_1-x)^2/8\sigma^2} e^{i(k-k_0)x}$$

$$= \int_{-L/2}^{L/2} dx_1 \hat{\xi}^{\dagger}(x_1)|0\rangle \frac{1}{\sqrt{2\pi}} \int_{-\infty}^{\infty} dk e^{ikx_1}\tilde{\phi}(k)e^{-(k-k_0)^2 2\sigma^2}. \qquad (14.8)$$

There are two cases to consider: momentum outside the desired range, and momentum within the desired range.

We see from (14.8) that, if $\tilde{\phi}(k)$ is only large for momenta far from k_0 (a few $1/\sigma$'s away), then $\hat{N}_L(k_0)|\psi\rangle \approx 0$.

On the other hand, if $\tilde{\phi}(k)$ is only large for momenta close to k_0 (within $\approx 1/2\sigma$), we may make the approximation $e^{-(k-k_0)^2 2\sigma^2} \approx 1$. In that case, (14.8) becomes

$$\hat{N}_L(k_0)|\psi\rangle = \int_{-L/2}^{L/2} dx_1 \hat{\xi}^\dagger(x_1)|0\rangle\phi(x_1) \approx \int_{-\infty}^{\infty} dx_1 \hat{\xi}^\dagger(x_1)|0\rangle\phi(x_1) = 1 \cdot |\psi\rangle, (14.9)$$

using (14.6).

So, to a good approximation, $\hat{N}_L(k_0)$ acts like the number operator for a photon that lies wholly within the tube, and within a momentum range $k_0 \pm \frac{1}{\sigma}$.

This concludes our analysis of the one-photon case.

For a state vector representing N identical photons within the tube, with momenta in range $k_0 \pm \frac{1}{\sigma}$,

$$|\psi\rangle_N \equiv \frac{1}{\sqrt{N!}} \left[\int_{-\infty}^{\infty} dx\phi(x)\hat{\xi}^\dagger(x) \right]^N |0\rangle, \tag{14.10}$$

we obtain from (14.5):

$$\hat{N}_L(k_0)|\psi\rangle_N = \int_{-L/2}^{L/2} dx_1 dx_2 \hat{\xi}^\dagger(x_1)\hat{\xi}(x_2)e^{ik_0(x_1-x_2)}\frac{1}{\sqrt{8\pi\sigma^2}}e^{-(x_1-x_2)^2/8\sigma^2}$$

$$\cdot \frac{1}{\sqrt{N!}} \left[\int_{-\infty}^{\infty} dx\phi(x)\hat{\xi}^\dagger(x) \right]^N |0\rangle$$

$$= \int_{-L/2}^{L/2} dx_1 dx_2 \hat{\xi}^\dagger(x_1)\phi(x_2)e^{ik_0(x_1-x_2)}\frac{1}{\sqrt{8\pi\sigma^2}}e^{-(x_1-x_2)^2/8\sigma^2}$$

$$\cdot N\frac{1}{\sqrt{N!}} \left[\int_{-\infty}^{\infty} dx\phi(x)\hat{\xi}^\dagger(x) \right]^{N-1} |0\rangle$$

$$\approx \int_{-L/2}^{L/2} dx_1 dx_2 \hat{\xi}^\dagger(x_1)\phi(x_2)e^{ik_0(x_1-x_2)}\delta(x_1 - x_2)$$

$$\cdot N\frac{1}{\sqrt{N!}} \left[\int_{-\infty}^{\infty} dx\phi(x)\hat{\xi}^\dagger(x) \right]^{N-1} |0\rangle$$

$$= N|\psi\rangle_N. \tag{14.11}$$

In going from the first line to the second line, we have let $\hat{\xi}(x_2)$ act on the state vector. In going from the second line to the third line, we have made the approximation $\frac{1}{\sqrt{8\pi\sigma^2}}e^{-(x_1-x_2)^2/8\sigma^2} \approx \delta(x_1 - x_2)$, which is a good approximation since a separation of size σ for x_1, x_2 is quite small on the scale of L. In going from the third line to the last line, we have extended the range of integration from $-L/2, L/2$ to $-\infty, \infty$.

Again we see, in this case of N identical photons, that the number operator we have constructed behaves properly.

As usual, if the stuff distribution $|\langle n_L(k_0)|\psi\rangle|^2$ has a standard deviation that is sufficiently small compared to the mean $\langle\psi, t|\hat{N}_L(k_0)|\psi, t\rangle$, then the mean can be considered to be the observable number of photons in the tube with momentum range $\approx k_0 \pm \frac{1}{\sigma}$.

We note that $[\hat{N}_L(k_0), \hat{N}_L(k_0')] = \int_{-L/2}^{L/2} dx dx' [\hat{\eta}_{k_0}^\dagger(x)\hat{\eta}_{k_0}(x), \hat{\eta}_{k_0'}^\dagger(x)\hat{\eta}_{k_0'}(x)] \approx 0$ for $|k_0 - k_0'|$ larger than a few $1/\sigma$, as stated following Eq. (14.4). Then, for a state vector describing light in a wide range of wavelengths, like sunlight, the stuff $|\langle n_L(k_0)|\psi\rangle|^2$ for different k_0 (separated by a few $1/\sigma$) will be independent and, if the observability condition is satisfied, will give an intensity profile of the light source.

In non–relativistic CSL, photons are not on a par with the other particles because the collapse-generating operator is mass density, $D(\mathbf{x}) = \sum_\alpha m_\alpha \hat{\xi}_\alpha^\dagger(\mathbf{x})\hat{\xi}_\alpha(\mathbf{x})$, and photons, having zero mass, do not contribute to the collapse. However, this can be remedied by replacing the mass density operator by the relativistic energy density operator

$$\sum_\alpha \left([m_\alpha^2 + \nabla^2]^{1/4}\hat{\xi}_\alpha^\dagger(\mathbf{x})\right)\left([m_\alpha^2 + \nabla^2]^{1/4}\hat{\xi}_\alpha(\mathbf{x})\right).$$

The choice

$$\sum_\alpha \frac{1}{2}\left[\hat{\xi}_\alpha^\dagger(\mathbf{x})\left([m_\alpha^2 + \nabla^2]^{1/2}\hat{\xi}_\alpha(\mathbf{x})\right) + \left([m_\alpha^2 + \nabla^2]^{1/2}\hat{\xi}_\alpha^\dagger(\mathbf{x})\right)\hat{\xi}_\alpha(\mathbf{x})\right]$$

is another Hermitian possibility. For non–relativistic momenta, where $m_\alpha c^2$ is much larger than the kinetic energy, these reduce to $D(\mathbf{x})$. For photons, with zero mass, the mass density operator is replaced by

$$[\nabla^{1/2}\hat{\xi}_\alpha^\dagger(\mathbf{x})][\nabla^{1/2}\hat{\xi}_\alpha(\mathbf{x})]$$

and there is collapse dynamics for photons.

As in the case of massive particles, interesting anomalous behavior is predicted.[2]

For example, consider that touchstone of dynamical collapse behavior, an initial superposition of an object in two widely separate places,. If one could somehow generate a pulse of photons in such a superposition, of mean momentum $\mathbf{k}_0$, occupying a volume characterized by length σ, the off-diagonal element of the density matrix is calculated to decay at rate $\lambda N^2 \frac{1}{2}[\frac{k_0}{M}]^2[\frac{a}{\sigma}]^2$.

Other, anomalous, phenomena are predicted, to be compared with the predictions of standard quantum theory. Photons have their energy increased and their direction of motion changed by the collapse dynamics. So, for example, photons are knocked out of a longer beam; one can calculate the distribution of momenta of the ejected photons

A glowing blackbody's radiation spectrum is altered.

Experimental limits on these and other effects place weak upper limits on λ.

---

[2]Pearle (2018)

15

A Stochastic Differential Equation Cookbook

While browsing through the mathematics collection in the small science library of my college in the mid-1970s, by chance I came across a couple of books on stochastic differential equations (SDE's). I had been looking for a mathematical way to describe wave function collapse driven by a randomly fluctuating function of time, and there it was! The random function of time is white noise, the time derivative of Brownian motion, which agreed with my thoughts that the stochastic element behind collapse should be Brownian motion, because of its ubiquity and simplicity.

These were mathematics books, so they were filled with epsilons and deltas, which are not part of my native language. It took me a while to figure out what was going on. In my first paper[1] that utilized an SDE approach to construct collapse equations (not CSL), I figured most readers wouldn't be familiar with SDEs either, so I included an introduction to SDE's.

In this book I do not use SDEs, preferring the equivalent approach based upon two equations: a modified Schrödinger equation that is linear in the state vector, together with a probability rule that is non–linear (quadratic) in the state vector.

One can combine these two and then there is an advantage: the theory is described by a single modified Schrödinger equation. Strictly speaking, there *is* a second equation, a probability rule, but it is the probability of white noise (which appears in the modified Schrödinger equation), and that is well known.

However, there is a disadvantage: the modified Schrödinger equation is non–linear in the state vector. Disadvantage notwithstanding, there are elegant rules for manipulating SDEs. Since the SDE approach to CSL appears in the literature, I thought I would present it in this Supplement.

So as to separate mathematics from physics, here is a short, but reasonably complete, and I hope reader-friendly, course on stochastic differential equations, *independently of any application*. The pedagogical approach taken here does not appear elsewhere to my knowledge. It is suitable for people who wish to learn about SDEs fairly rapidly, and who have no interest in its application to CSL

Then, the following three chapters present the application of SDE theory to CSL.

Chapter 16 shows how, starting from the two-equation approach, to obtain the SDE that is the CSL-modified Schrödinger equation. Then, the Lindblad equation for the density matrix is obtained from that.

---

[1] Pearle (1979)

Chapters 17 and 18 offer a couple of applications. Chapters17 solves the SDE for the free particle problem. tChapters 18 applies the SDE approach to the harmonic oscillator problem.

15.1   SDE Cookbook Ingredients

Sometimes, one encounters an equation of this form:

$$\frac{d}{dt}x(t) = a(x(t)) + b(x(t))\eta(t), \tag{15.1}$$

where, $\eta(t)$ is a wildly fluctuating function of time. Actually, when this is seen, $\eta(t)$ is usually not a single function at all, but one of a number of possible functions $\eta_i(t)$, where the ith function occurs with some probability P_i. For example, this could be an application of Newton's second law, with x a classical particle's momentum, $a(x)$ a force due to a potential, and $b(x)\eta(t)$ a randomly varying force.

Often, one isn't given an actual list of the $\eta_i(t)$ and the P_i, but rather a few statistical facts. Often, these are the mean $\overline{\eta}(t) \equiv \sum_i P_i \eta_i(t) = 0$ (its vanishing can be arranged by adjusting $a(x)$) and autocorrelation function $C(t, t') \equiv \overline{\eta(t)\eta(t')} \equiv \sum_i P_i \eta_i(t)\eta_i(t')$. A most usual situation is that the autocorrelation function depends on the time difference, $\equiv C(t' - t)$, so that the behaviour is time-translation invariant: this is a property called "stationarity." Most usual too is that $C(t - t')$'s maximum is $C(0)$, and it drops to 0 as its argument increases.

How can you solve this? Usually, one is not interested in the individual solutions (and usually one can't get them anyway, one can't integrate (15.1)), but rather one wants statistical information about the whole ensemble of solutions. For example, one would like to know $\overline{x(t)}$, $\overline{x^2(t)}$, and even have a shot at finding the probability $\rho(x, t|x_0, t_0)dx$ that x lies in the interval dx at time t, given its initial value x_0 at time t_0.

One can neatly achieve that if one is willing to approximate $\eta(t)$ by white noise $w(t)$ for which $\overline{w(t)} = 0$ and $\overline{w(t)w(t')} = \sigma^2\delta(t - t')$. (A discussion of white noise appears in Appendix D.) Often one *is* willing, when $C(t)$ behaves as described above.

A cookbook presents recipes. A recipe tells you how to achieve something with minimal explanation of why it works. That's what will be done now. But, of course, that isn't good enough so, after the recipe, the remainder of this section supplies explanations.

We will need the ingredient Brownian motion $B(t)$. It is related to white noise $w(t)$ by $dB(t) \equiv B(t + dt) - B(t) \equiv w(t)dt$.

We will often see the term $[dB(t)]^2$. Using the Brownian motion probability density expression Eq. (C.1), applied to an infinitesimal time interval, the conditional probability density of $B(t + dt)$, given $B(t)$, is

$$P(B(t + dt)|B(t)) = \frac{1}{\sqrt{2\pi\sigma^2 dt}}e^{-\frac{1}{2\sigma^2 dt}[B(t+dt)-B(t)]^2}. \tag{15.2}$$

Multiplying (15.2) by $B(t + dt) - B(t)$ and integrating over $dB(t + dt)$, it follows that

$$\overline{[dB(t)]^2} = \sigma^2 dt. \tag{15.3}$$

However, we argue something stronger can be used,

$$[dB(t)]^2 = \sigma^2 dt, \tag{15.4}$$

no average is involved!

Before getting on with the cookbook, we might as well justify (15.4) here and now (for those uninterested, skip the next few paragraphs.)

Here is an intuitive justification. One can regard Brownian motion as a random walk where each step in time dt is $dB(t) = \pm\sigma\sqrt{dt}$, and the square of this is (15.4).

A reminder about the connection between random walk and Brownian motion. In Appendix B we obtained the expression for the probability of a random walk arriving at x, when the steps are of length $\Delta/2$ and each takes time τ. When the probability is $1/2$ for left or right steps, Eq. (B.3) is:

$$P_t(x) \approx \frac{1}{\sqrt{2\pi(\Delta^2/4\tau)t}} e^{-\frac{1}{2(\Delta^2/4\tau)t}x^2}, \tag{15.5}$$

where the approximation becomes exact in the limit $\Delta \to 0, \tau \to 0$ in such a way that $\Delta^2/\tau \to$ constant. If, in Eq. (15.5) we set $\Delta/2 = \sigma\sqrt{dt}$, $\tau = dt$ and $x = dB(t)$, we obtain the Brownian motion probability distribution (15.2).

Now, we look at the rigorous justification for (15.4). If there are two sums S_1, S_2 involving a stochastic variable, where each sum approximates an integral over dt, and if it can then be shown that $\lim_{dt\to 0} \overline{[S_1 - S_2]^2} = 0$, it is said that they are *equal in mean square*. This is the best definition of equality in stochastic analysis.

Now, apply this to arrive at (15.4). Let $S_1 \equiv \sum_{k=0}^{N-1} [dB(t_k)]^2$ (where $t_k \equiv kdt$ and $Ndt \equiv t$) and $S_2 \equiv \sum_{k=0}^{N-1} \sigma^2 dt$ be the two sums approximating integrals. We can, of course, immediately do the second sum: $S_2 = \sigma^2 t$. Using this, we can evaluate the mean square difference:

$$\overline{[S_1 - S_2]^2} = \overline{\left[\sum_{k=0}^{N-1} [dB(t_k)]^2 - \sigma^2 t \right]^2}$$

$$= \overline{\sum_{j=0}^{N-1} [dB(t_j)]^2 \sum_{k=0}^{N-1} [dB(t_k)]^2 - 2\sigma^2 t \sum_{k=0}^{N-1} [dB(t_k)]^2} + (\sigma^2 t)^2$$

$$= (\sigma^2 dt)^2 [N(N-1) + 3N] - 2(\sigma^2 t)N(\sigma^2 dt) + (\sigma^2 t)^2$$

$$= (\sigma^2 t)^2 + 2(\sigma^2 t)\sigma^2 dt - 2(\sigma^2 t)^2 + (\sigma^2 t)^2 = 2(\sigma^2 t)\sigma^2 dt \xrightarrow[dt\to 0]{} 0.$$

In going from the first line to the second line, we just multiplied out the square. In going to the third line, we have used (15.3) to give

$$\overline{[dB(t_j)]^2 [dB(t_k)]^2} = \overline{[dB(t_j)]^2} \cdot \overline{[dB(t_k)]^2} = (\sigma^2 dt)^2$$

for each of the $N(N-1)$ terms in the double sum where $j \neq k$, since each factor is statistically independent. For the N terms where $j = k$, we have used Eq.(A.4) to obtain $\overline{[dB(t_k)]^4} = 3(\sigma^2 dt)^2$. In going from the third line to the fourth line, we use $Ndt = t$.

We see that all the $(\sigma^2 t)^2$ terms have cancelled, leaving us with a term $\sim dt$ that vanishes in the limit. Thus, (15.4) is true in mean square.

Since all the $[dB(t)]^2$ we encounter will end up in such sums approximating integrals, we can always use the replacement (15.4). We are accustomed to infinitesimal functions of t being proportional to dt. What drives all the following discussion is that the infinitesimal quantity $dB(t)$ is not proportional to dt, its square is!

15.2   Recipe

The recipe consists of four "Steps." They lead one, starting with the stochastic differential equation Eq. (15.1), with $\eta(t)dt$ replaced by $w(t)dt = dB(t)$, to end up with an ordinary partial differential equation for the probability of x at any time t. Explanation for these weird steps will not be given yet. It is just asserted that, as with any recipe, if you follow directions, out will come the desired result. But, afterwards, it will be shown why this works.

1) Start with Eq.(15.1). Multiply it by dt, so we have

$$dx(t) = a(x(t))dt + b(x(t))\eta(t)dt.$$

This is the equation we are interested in approximating. Accordingly, we make the following replacements. Replace $b(x(t))$ by

$$b_S(x(t)) \equiv b\big(\tfrac{1}{2}[x(t + dt) + x(t)]\big).$$

$b_S(x(t))$, and any function of $x(t)$ with the same *peculiar* argument, (named after the Russian mathematician Ruslan Stratonovich) is called a *Stratonovich function*. Replace $\eta(t)dt$ by $dB(t)$ (where $dB(t) \equiv B(t + dt) - B(t)$). We thus end up with a stochastic differential equation that we wish to solve:

$$dx(t) = a(x(t))dt + b_S(x(t))dB(t) \text{ which means that} \qquad (15.6a)$$

$$x(t) - x(0) = \int_0^t dt\, a(x(t)) + \int_0^t dt\, b_S(x(t))dB(t). \qquad (15.6b)$$

Eq.(15.6a) is called a *Stratonovich stochastic differential equation*. Its virtue, as we eventually prove, is that it can be manipulated using the ordinary rules of calculus. Of course, this is true of the original ordinary first order differential equation (15.1) that (15.6a) is approximating. This is why such a standard differential equation is first converted to a Stratonovich SDE. The downside of a Stratonovich stochastic differential equation is that averages of functions of $x(t)$ cannot easily be found directly from it.

The first integral in (15.6b) is an ordinary, Riemann, integral. The second integral is called a *Stratonovich integral*, the limit as $dt \to 0$ of the sum

$$dt \sum_{k=0}^{N-1} b\big(\tfrac{1}{2}[x((k+1)dt) + x(kdt)]\big)\Big[B((k+1)dt) - B(kdt)\Big],$$

where $Ndt = t$.

2) To remedy the Stratonovich equation deficiency, that is, to rewrite Eq. (15.6a) in a form where averages can easily be found, we next convert it to a different type of stochastic differential equation (named after the Japanese mathematician Kiyoshi Itô), an *Itô differential equation*. This is done by changing $b_S(x)$ to $b(x)$, and compensating for this change by adding to (15.6a) the *Itô correction term* $\frac{\sigma^2}{2}b'(x)b(x)dt$ (where $b'(x) \equiv \frac{d}{dx}b(x)$). We now have

$$dx(t) = \left[a(x(t)) + \frac{\sigma^2}{2}b'(x(t))b(x(t))\right]dt + b(x(t))dB(t) \qquad (15.7a)$$

$$\equiv m(x(t))dt + b(x(t))dB(t), \text{ which means that} \qquad (15.7b)$$

$$x(t) - x(0) = \int_0^t dt\, m(x(t)) + \int_0^t dt\, b(x(t))dB(t). \qquad (15.7c)$$

Eq. (15.7b) is an *Itô stochastic differential equation*. The first integral in (15.7c) is an ordinary, Riemann, integral. The second integral is called an *Itô integral*, the limit as $dt \to 0$ of the sum

$$\sum_{k=0}^{N-1} b\Big(x(kdt)\Big)\Big[B((k+1)dt) - B(kdt)\Big].$$

What is important in this sum is that $x(t)$ depends only upon $B(t')$ for $t' \le t$. The Itô equation (15.7b), written as

$$x(t + dt) = x(t) + m(x(t))dt + b(x(t))[B(t + dt) - B(t)]$$

shows that $x(t + dt)$ depends upon $B(t + dt)$, but not upon B at any later time. Thus, the property propagates forward, that x at time t depends upon B at no larger time than t.

A function like $x(t)$, which enjoys this property, is called a *non–anticipating function*. (An anticipating function at time t would have some knowledge of, that is, some dependence on, the value of B in the future of t. Such is a Stratonovich function.)

The virtue of Eq.(15.7b), as will now be seen, is that it is possible to readily find averages directly from it. $b(x(t))$ is a non=-anticipating function, which means it is statistically independent of $dB(t) = B(t + dt) - B(t)$. The vice of the Itô equation is that some ordinary rules of calculus are altered.

So, now that we have our Itô stochastic differential equation, our next Step is to find some averages.

3) Using (15.7b), we now see that we can easily find $\overline{dx(t)}/dt$ and $\overline{[dx(t)]^2}/dt$ when x is known with certainty (and therefore any function of x is known with certainty). For the first of these,

$$\overline{dx(t)} = \overline{m(x(t))}dt + \overline{b(x(t))dB(t)} = \overline{m(x(t))}dt$$

$$\to m(x)dt. \qquad (15.8)$$

Here we have used the statistical independence of $b(x(t))$ and $dB(t)$, since $b(x(t))$ is a non–anticipating function, to obtain $\overline{b(x(t))dB(t)} = \overline{b(x(t))} \times \overline{dB(t)} = b(x(t)) \times 0 = 0$.

For our application, it is assumed that $x(t) = x$ is known, so the stochastic average bar may be removed, $\overline{m(x(t))} = m(x(t))$.

For the second of these,

$$
\begin{aligned}
\overline{[dx(t)]^2} &= \overline{[m(x(t))dt + b(x(t))dB(t)][m(x(t))dt + b(x(t))dB(t)]} \\
&= \overline{m(x(t))^2 dt^2} + \overline{2m(x(t)dtb(x(t))B(t)} + \overline{[b(x(t))dB(t)]^2} \\
&= b^2(x)\overline{[dB(t)]^2} \\
&= b^2(x)\sigma^2 dt.
\end{aligned}
\tag{15.9}
$$

We have discarded the term $\overline{m^2(x)}dt^2$ since it vanishes when divided by dt and the limit $dt \to 0$ is taken. The average of $\overline{m(x(t))b(x(t))dtdB(t)}$ equals 0 since $m(x(t))b(x(t))$ is a non–anticipating function and therefore statistically independent of $dB(t)$. (A different, intuitive, reason for discarding it is if we regard $dB(t) \sim dt^{1/2}$, then this term is $\sim dt^{3/2}$, whereas only terms of order dt contribute.) Finally, we have used $\overline{[dB(t)]^2} = \sigma^2 dt$ (Eq. (15.3)).

4) Our final Step is to put the results of 3) into the *Fokker–Planck equation* (named after the Dutch physicist Adriaan Fokker and the German physicist Max Planck). This is a partial differential equation for the probability density $\rho(x, t|x_0, t_0)$:

$$
\begin{aligned}
\frac{\partial}{\partial t}\rho(x, t|x_0, t_0) &= -\frac{\partial}{\partial x}\frac{\overline{dx(t)}}{dt}\rho(x, t|x_0, t_0) + \frac{1}{2}\frac{\partial^2}{\partial x^2}\frac{\overline{[dx(t)]^2}}{dt}\rho(x, t|x_0, t_0) \\
&= -\frac{\partial}{\partial x}m(x)\rho(x, t|x_0, t_0) + \frac{\sigma^2}{2}\frac{\partial^2}{\partial x^2}b^2(x)\rho(x, t|x_0, t_0).
\end{aligned}
\tag{15.10}
$$

The claim is that the solution $\rho(x, t|x_0, t_0)$ of (15.10)) is the correct probability density for the ensemble of solutions of (15.6a) or the equivalent (15.7b).

End of recipe!

To summarize, the four Steps of the recipe are:

1) Write your equation as a Stratonovich SDE.

2) Convert it to an Itô SDE, adding a correction term.

3) Easily find $\overline{dx(t)}$ and $\overline{[dx(t)]^2}$ using the Itô SDE.

4) Put your expressions for $\overline{dx(t)}/dt$ and $\overline{[dx(t)]^2}/dt$ into the Fokker–Planck equation, which you may then solve (using ordinary calculus) for the probability distribution of x at any time t.

15.3   Examples

It is always good to see examples of how new ideas work!

For each example, we will do the same things. Starting 1) with an equation of the form (15.1) depending upon a random function $\eta(t)$, we display the Stratonovich equation with $\eta(t)$ replaced by white noise $w(t)$, that is, $\eta(t)dt$ replaced by $dB(t)$. In some cases, we can integrate that directly, using the standard rules of calculus, which is what a Stratonovich equation is good for. Regardless of whether that can be done, in all cases, we follow the three remaining Steps. Lastly, we show the solution of the Fokker–Planck equation, and discuss it.

15.3.1   Example 1

Consider the simplest possible equation, $\frac{d}{dt}x = \eta(t)$, where x could be a particle's momentum and $\eta(t)$ a random force. The Stratonovich equation is

$$dx(t) = dB(t). \tag{15.11}$$

The integral of (15.11), under the assumption that $x(0) = B(0) = 0$ is $x(t) = B(t)$. Since the probability density distribution of B is (8.5), we obtain the probability density distribution of x by replacing B there by x:

$$P_t(x) = \frac{1}{\sqrt{2\pi\sigma^2 t}}e^{-\frac{1}{2\sigma^2 t}x^2}. \tag{15.12}$$

Problem completely solved!

Nonetheless, let's follow our recipe. The Itô correction term in (15.7a) vanishes, $\frac{\sigma^2}{2}b'(x)b(x)dt = 0$ (since $b(x) = 1$), so (15.11) is also the Îto equation. We read out directly from (15.11) that $\overline{dx(t)} = \overline{dB(t)} = 0$ and $\overline{[dx(t)]^2} = \overline{[dB(t)]^2} = [dB(t)]^2 = \sigma^2 dt$. Putting these values into the Fokker-Planck equation (15.10) gives

$$\frac{\partial}{\partial t}\rho(x,t|x_0,t_0) = \frac{\sigma^2}{2}\frac{\partial^2}{\partial x^2}\rho(x,t|x_0,t_0). \tag{15.13}$$

This is the well-known Brownian motion diffusion equation Eq.(B.5), which indeed is satisfied by (15.12). For more general initial conditions, $\rho(x,t|x_0,t_0) = P_{t-t_0}(x-x_0)$ is the solution.

15.3.2   Example 2

Consider the equation, $\frac{d}{dt}x = -x\eta(t)$: This might arise from Newton's second law, where x represents a particle's momentum and the force is a viscous force with $\eta(t)$ a time-dependent viscosity. The Stratonovich equation is

$$dx(t) = -x_S dB(t). \tag{15.14}$$

We have claimed that a Stratonovich equation can be manipulated using the ordinary rules of calculus. Just treat x_s as x and integrate (15.14):

$$x(t) = x(0)e^{-B(t)}. \tag{15.15}$$

Using the probability density of $B(t)$ (given in (15.12), with B replacing x), we can calculate moments such as

$$\overline{x(t)} = \int_{-\infty}^{\infty} dB P_t(B)x(t) = \int_{-\infty}^{\infty} dB\frac{1}{\sqrt{2\pi\sigma^2 t}}e^{-\frac{B^2}{2\sigma^2 t}}x(0)e^{-B} = x(0)e^{\frac{\sigma^2}{2}t}.$$

We invert (15.15) to find $-B(t) = \ln\frac{x}{x_0}$. Inserting this into $P_t(-B)$ gives the probability distribution for x,

$$\rho(x,t|x_0,0)dx = d\ln x\frac{1}{\sqrt{2\pi\sigma^2}}e^{-\frac{[\ln x - \ln x_0]^2}{2\sigma^2 t}}.$$

Although we have solved this, let's follow our recipe. The Itô equation is obtained using $b(x) = x$ to construct the Itô correction term $\frac{\sigma^2}{2}b(x)b'(x)dt = \frac{\sigma^2}{2}xdt$, and putting it into (15.14):

$$dx(t) = \frac{\sigma^2}{2}x(t)dt - x(t)dB(t).$$

(15.16)

We read off from (15.16) ,

$$\overline{dx(t)} = \frac{\sigma^2}{2}\overline{x(t)}dt - \overline{x(t)dB(t)} = \frac{\sigma^2}{2}\overline{x(t)}dt - \overline{x(t)} \times \overline{dB(t)} = \frac{\sigma^2}{2}\overline{x(t)}dt.$$

(using the statistical independence of the non-anticipating $x(t)$ and $dB(t)$). Likewise, discarding terms of order $(dt)^2$ and $(dt)^{3/2}$,

$$\overline{[dx(t)]^2} = \overline{[x(t)dB(t)]^2} = \overline{[x(t)]^2}\sigma^2 dt,$$

(again employing the statistical independence).

After removing the average from the right hand side of these expressions, they are put into the Fokker–Planck equation (15.10):

$$\frac{\partial}{\partial t}\rho(x,t|x_0,t_0) = -\frac{\partial}{\partial x}\frac{\sigma^2}{2}x\rho(x,t|x_0,t_0) + \frac{\sigma^2}{2}\frac{\partial^2}{\partial x^2}x^2\rho(x,t|x_0,t_0).$$

(15.17)

(15.17) can be solved by writing $z = \ln x$ and $\rho(x,t|x_0,t_0) = \frac{1}{x}R(z,t|z_0,t_0)$, converting it to the standard diffusion equation for Brownian motion, $\frac{\partial}{\partial t}R(z,t|z_0,t_0) = \frac{\sigma^2}{2}\frac{\partial^2}{\partial z^2}R(z,t|z_0,t_0)$. This solution is what we obtained earlier from consideration of the Stratonovich equation.

15.3.3 Example 3: Langevin Equation

Consider the equation $m\frac{d}{dt}v = -m\gamma v + \sqrt{2\gamma mkT}\eta(t)$. This arises from Newton's second law describing physical Brownian motion. That is, $v(t)$ is the velocity of an object of mass m under a random force due to impacts from molecules at temperature T and a viscous force due to those same molecules, characterized by the viscosity coefficient γ. Replacing $\eta(t)dt$ by $dB(t)$, we get the Stratonovich SDE and, as the Itô correction term vanishes, the Itô SDE is the same:

$$dv(t) = -\gamma vdt + dB(t) \text{ with } \sigma \equiv \sqrt{2\gamma kT/m}.$$

(15.18)

Calculating $\overline{dv(t)} = -\gamma\overline{v}dt$, $\overline{[dv(t)]^2} = \sigma^2 dt$, and inserting these into the Fokker–Planck equation gives

$$\frac{\partial}{\partial t}\rho(v,t|v_0,0) = \gamma\frac{\partial}{\partial v}v\rho(v,t|v_0,0) + \frac{\gamma kT}{m}\frac{\partial^2}{\partial v^2}\rho(v,t|v_0,0).$$

(15.19)

If initially $\rho(v,0|v_0,0) = \delta(v - v_0)$, that is, the object has a definite initial velocity v_0, nonetheless, under the random force, the object can acquire a range of possible final velocities, each with an associated probability of occurring. The solution of (15.19), subject to the initial condition, is

$$\rho(v,t|v_0,0) = \frac{1}{\sqrt{\pi(2kT/m)[1-e^{-2\gamma t}]}} e^{-\frac{[v-v_0 e^{-\gamma t}]^2}{(2kT/m)[1-e^{-2\gamma t}]}}, \tag{15.20}$$

which approaches the Maxwell-Boltzmann velocity distribution (in one dimension), $\sim e^{-\frac{mv^2}{2kT}}$ for large t.

15.4 Plan for the rest of this Supplement

Having given the recipe and a few examples, we now proceed to justify the recipe. We will go through the Steps *backwards*, that is, in reverse order. First, we will give the reasoning behind 4), showing that the Itô equation and the Fokker–Planck equation are equivalent. Step 3) the finding of averages from the Itô equation has already been explained. So, next comes 2), explaining the correction term required to go from a Stratonovich differential equation form to an Itô differential equation form, or vice versa. Finally, we will demonstrate that the Stratonovich calculus rules are the same as ordinary calculus rules, as asserted in Step 1.

Along the way we will acquire a bunch of useful formulas.

15.5 Fokker–Planck Equation

A Gaussian probability distribution in x is completely determined by two numbers, its mean $\bar{x}$ and its variance $\overline{x^2} - \bar{x}^2$. Similarly, the Fokker–Planck equation, and therefore the probability distribution that solves it, is completely determined by two quantities, $d\overline{x(t)}/dt$ and $\overline{[dx(t)]^2}/dt$ as functions of $x(t)$, which depend upon the two Itô equation quantities, $m(x)$ and $b(x)$.

The equivalence of the Itô equation and the Fokker–Planck equation to which it gives rise can be proved by showing that they lead to identical expressions for all moments $\overline{x^n}$. We shall show how that works for $\bar{x}$ and $\overline{x^2}$, and then for $\overline{x^n}$.

To start, multiply the Fokker–Planck Eq. (15.10) by x, and then integrate over all x:

$$\frac{d}{dt}\overline{x(t)} \equiv \frac{\partial}{\partial t}\int_{-\infty}^{\infty} dx\, x\rho(x,t|x_0,t_0)$$

$$= \int_{-\infty}^{\infty} dx\, x\left[-\frac{\partial}{\partial x}m(x)\rho(x,t|x_0,t_0) + \frac{\sigma^2}{2}\frac{\partial^2}{\partial x^2}b^2(x)\rho(x,t|x_0,t_0)\right]$$

$$= \int_{-\infty}^{\infty} dx\, m(x)\rho(x,t|x_0) - \frac{\sigma^2}{2}\frac{\partial}{\partial x}b^2(x)\rho(x,t|x_0,t_0)|_{-\infty}^{\infty} = \overline{m(x(t))} \tag{15.21}$$

using integration by parts, with no contribution from the integration limits since $\rho(\pm\infty, t|x_0) = 0$.

Eq.(15.21) is a consequence of the Itô Eq. (15.7b), $dx(t) = m(x(t))dt + b(x(t))dB(t)$, simply by taking the average of both sides of this equation (as in the first line of Eq. (15.8)).

Next, multiply the Fokker–Planck Eq.(15.10) by x^2, and then integrate over all x:

$$\frac{d}{dt}\overline{x^2(t)} = \frac{\partial}{\partial t}\int_{-\infty}^{\infty} dx x^2 \rho(x,t|x_0,t_0)$$

$$= \int_{-\infty}^{\infty} dx x^2 \left[-\frac{\partial}{\partial x} m(x)\rho(x,t|x_0,t_0) + \frac{\sigma^2}{2}\frac{\partial^2}{\partial x^2} b^2(x)\rho(x,t|x_0,t_0) \right]$$

$$= \int_{-\infty}^{\infty} dx 2x m(x)\rho(x,t|x_0,t_0) + \sigma^2 \int_{-\infty}^{\infty} dx b^2(x)\rho(x,t|x_0,t_0)$$

$$= 2\overline{x(t)m(x(t))} + \sigma^2 \overline{b^2(x(t))}. \tag{15.22}$$

This is a result we have not yet obtained from the Itô equation. We want to obtain an Itô equation for $dx^2(t)$ from the one we have one for $dx(t)$, and then take the average. To do this, we more generally consider how an Itô equation for $df(x(t))$, where $f(x)$ is an arbitrary function of x, can be obtained from the Itô equation for $dx(t)$.

15.6 Itô Chain Rule

We use the Taylor series expansion of $df(x)$, and have to take it to order $(dx)^2$ (but no further) where $dx = m(x)dt + b(x)dB(t)$ since, as pointed out, $(dx)^2$ makes a contribution $\sim (dB(t))^2 = \sigma^2 dt$:

$$df(x(t)) = f'(x(t))dx(t) + \frac{1}{2}f''(x(t))[dx(t)]^2 + \ldots$$

$$= f'(x(t))[m(x(t))dt + b(x(t))dB(t)] + \frac{1}{2}f''(x(t))[m(x(t))dt + b(x(t))dB(t)]^2$$

$$= f'(x(t))dx(t) + \frac{1}{2}f''(x(t))b^2(x(t))\sigma^2 dt. \tag{15.23}$$

In obtaining (15.23), we have dropped terms of order dt^2 and $dtdB \sim dt^{3/2}$, since they vanish when divided by dt and the limit $dt \to 0$ is taken.

The chain rule in ordinary calculus is $df(x) = f'(x)dx$. The chain rule (15.23) for the Itô equation has an extra term.

Taking the average of (15.23), we find

$$\frac{d}{dt}\overline{f(x(t))} = \overline{f'(x(t))m(x(t))} + \frac{\sigma^2}{2}\overline{f''(x(t))b^2(x(t))}. \tag{15.24}$$

Eqs. (15.23), (15.24), and, even more, the method used to obtain them, are useful to know.

Now, let's apply Eq. (15.24) to our particular problem. We set $f(x(t)) = x^2(t)$, and so obtain

$$\frac{d}{dt}\overline{(x^2(t))} = \overline{2xm(x(t))} + \sigma^2\overline{b^2(x(t))}.$$

This is the same as (15.22). So, we have shown that the Fokker–Planck equation and the Itô SDE give the same expressions for $d\bar{x}/dt$ and $d\bar{x}^2/dt$.

To show that this holds for any moment of x, set $f(x(t)) = x^n(t)$ in (15.24):

$$\frac{d}{dt}\overline{x^n(t)} = n\overline{x^{n-1}(t)m(x(t))} + \frac{\sigma^2}{2}n(n-1)\overline{x^{n-2}(t)b^2(x)(t)}. \tag{15.25}$$

We can immediately see that this is the same result obtained by using the Fokker-Planck equation (15.10) to calculate $\frac{d}{dt}\overline{x^n(t)}$, following integrations by parts.

To summarize, we have showed that the Fokker-Planck equation (8.10), constructed as we have specified from the functions $m(x), b(x)$ that appear in the Itô equation, gives all the same expressions for $\overline{x^n(t)}$ that arise directly from the Itô equation. If the moments are the same, their probability density distributions are the same. Thus we have proved the validity of Steps 3 and 4 of our recipe.

15.7 Itô Product Rule

This is a good place to obtain another useful result, an expression for the product rule $d[f(x)g(x)]$ when x satisfies the Îto Eq. (15.7b). We replace $f(x)$ in the chain rule (15.23) by $f(x)g(x)$ (for simplicity, we are writing x instead of $x(t)$):

$$
\begin{aligned}
d[f(x)g(x)] &= \frac{d}{dx}[f(x)g(x)]dx + \frac{1}{2}\frac{d^2}{dx^2}[f(x)g(x)]b^2(x)\sigma^2 dt \\
&= [f'(x)g(x) + f(x)g'(x)]dx + \\
&\quad \frac{1}{2}[f''(x)g(x) + 2f'(x)g'(x) + f(x)g''(x)]b^2(x)\sigma^2 dt \\
&= f(x)\left[g'(x)dx + \frac{\sigma^2}{2}g''(x)b^2(x)dt\right] + g(x)\left[f'(x)dx + \frac{\sigma^2}{2}f''(x)b^2(x)dt\right] \\
&\quad + \sigma^2 f'(x)g'(x)b^2(x)dt \\
&= f(x)dg(x) + g(x)df(x) + df(x)dg(x). \tag{15.26}
\end{aligned}
$$

In the last line of (15.26), we have used (15.23)'s expression for $df(x), dg(x)$, which gives

$$
\begin{aligned}
df(x)dg(x) &= \left[f'(x)[m(x)dt + b(x)dB(t)] + \frac{1}{2}f''(x)b^2(x)\sigma^2 dt\right] \\
&\quad \cdot \left[g'(x)[m(x)dt + b(x)dB(t)] + \frac{1}{2}g''(x)b^2(x)\sigma^2 dt\right] \\
&= \sigma^2 f'(x)g'(x)b^2(x)dt,
\end{aligned}
$$

all other terms but the one $\sim [dB(t)]^2 = \sigma^2$ vanishing as $\sim (dt)^2$ or $\sim (dt)^{3/2}$.

The first two terms of (15.26) are the usual product rule, the last term is called an *Itô correction*. This is a simply stated, elegant result, and worth remembering.

As an example, we can again obtain the result (15.22) using (15.26), setting $f(x) = g(x) = x$:

$$d[x^2] = d[x \cdot x] = 2x[m(x) + b(x)dB(t)] + \sigma^2 b^2(x)dt, \tag{15.27}$$

followed by taking the average.

15.8   Itô and Stratonovich Integrals

Next, we turn to proving the validity of Step 2, the procedure for going from the Stratonovich form to the Itô form. To do that, we have to consider the definition of the Stratonovich and Itô integrals. It is useful to start with the integral we know, the Riemann integral, and see how the other integrals differ.

15.8.1   Riemann Integral

Suppose one has a known function $F(x)$ and a known function $x(t)$ and one wants to do the standard Riemann integral $\int_0^t dt' F(x(t'))$ to find the area under the curve $F(x(t))$.

One can think of the Riemann integral as summing little rectangular areas, a height multiplied by the base of length dt' (left end of the base at point t', right end of the base at point $t' + dt'$). There is freedom in choosing the height of the rectangles. One can take the area to be $dt' F(x(t'))$, where the height of the rectangle is the value of the function at the left end of the rectangle's base. One can take the area to be $dt' F(x(t' + dt'))$, where the height of the rectangle is the value of the function at the right end of the rectangle's base. One can take the area to be $dt' F(\frac{1}{2}[x(t'+dt')+x(t')])$, where the height of the rectangle is the value of the function somewhere in the middle of the rectangle's base. It doesn't matter which you choose because the difference between them is of order $(dt')^2$, and the sum of those terms vanishes as $dt' \to 0$. For example, in the latter case,

$$
\begin{aligned}
\int_0^t dt' F(\tfrac{1}{2}[x(t' + dt') + x(t')]) &= \int_0^t dt F(\tfrac{1}{2}[x(t' + dt') - x(t')] + x(t')) \\
&= \int_0^t dt' F(\tfrac{1}{2} dx(t') + x(t')) \\
&= \int_0^t dt' F(x(t')) + \frac{1}{2} \int_0^t dt' dx(t') \frac{dF(x(t'))}{dx(t')} + \dots \\
&= \int_0^t dt F(x(t')) + \frac{1}{2} dt' [F(x(t)) - F(x(0))] \\
&\to \int_0^t dt' F(x(t')). \qquad\qquad (15.28)
\end{aligned}
$$

In the first line, we have written $\frac{1}{2}x(t') = -\frac{1}{2}x(t') + x(t')$. In going from the second line to the third line, we have used the Taylor's series expansion. In the next line we have taken dt' to be a constant and performed the integral and, in going to the last line, we have let $dt' = 0$.

15.8.2   An Îto Integral Compared to a Stratonovich Integral

The result's independence of how you construct the height of the rectangle is fine for a Riemann integral, but the result is different if you do an Itô integral (rectangle height at left end of interval) or a Stratonovich integral (rectangle height somewhere in the middle of interval). That is, $\int_0^t dB(t') F(x(t'))$ differs from $\int_0^t dB(t') F_S(x(t'))$.

As an example, consider the simple Itô ntegral,

$$I(t) \equiv \int_0^t B(t')dB(t'), \tag{15.29}$$

where we recall, for an Itô integral, that $dB(t') = B(t' + dt) - B(t)$. Suppose we made a guess that the Itô integral result is the same as the standard Riemann integral result. Then, it would follow from (15.29) that $I(t) = \frac{1}{2}B^2(t)$ (where, as usual, we take $B(0) = 0$). If that is true, we find the average, $\overline{I(t)} = \frac{1}{2}\sigma^2 t$. But, this means $\overline{dI(t)} = \frac{1}{2}\sigma^2 dt$. That is incorrect since, directly calculating from (15.29),

$$dI(t) = d\int_0^t B(t')dB(t') = B(t)dB(t)$$

and so

$$\overline{dI(t)} = \overline{B(t)dB(t)} = \overline{B(t)} \cdot \overline{dB(t)} = 0 \times 0 = 0.$$

Therefore, the Îto integral result is not the Riemann integral result.

To do this Itô integral (15.29), as with the Riemann integral, divide t into small intervals $dt' = t/N$, with $t_k = kdt'$, and so write the integral as the limit $N \to \infty$ of

$$I(t) = \sum_{k=0}^{N-1} B(t_k)[B(t_{k+1}) - B(t_k)]$$

$$= \sum_{k=0}^{N-1} \left\{ \frac{1}{2}[B(t_k) + B(t_{k+1})] + \frac{1}{2}[B(t_k) - B(t_{k+1})] \right\}[B(t_{k+1}) - B(t_k)]$$

$$= \frac{1}{2}\sum_{k=0}^{N-1}[B^2(t_{k+1}) - B^2(t_k)] - \frac{1}{2}\sum_{k=0}^{N-1}[dB(t_k)]^2$$

$$= \frac{1}{2}B^2(t) - \frac{1}{2}\sigma^2 t. \tag{15.30}$$

In the second line, $\frac{1}{2}B(t_{k+1})$ has been added and subtracted from each term in the sum. In the third line, each term has been multiplied out, and we have written $B(t_{k+1}) - B(t_k) \equiv dB(t_k)$. In going from the third to the the last line, we have realized that, in the left sum, all terms cancel except $\frac{1}{2}B^2(t)$ And, in the right sum, each term gives an identical contribution $[dB(t_k)]^2 = \sigma^2 dt$.

Taking the average of (15.30), we get $\overline{I(t)} = \frac{1}{2}B^2(t) - \frac{1}{2}\sigma^2 t = 0$, and so $\overline{dI(t)} = 0$, in agreement with our previous result for $\overline{dI(t)}$

Now let's do the comparable Stratonovich integral:

$$\int_0^t B_S(t)dB(t) = \sum_{k=0}^{N-1} \frac{1}{2}[B(t_{k+1}) + B(t_k)][B(t_{k+1}) - B(t_k)]$$

$$= \frac{1}{2}\sum_{k=0}^{N-1}[B^2(t_{k+1}) - B^2(t_k)] = \frac{1}{2}B^2(t). \tag{15.31}$$

Thus, we obtain for the Stratonovich integral (15.31) the usual Riemann integral value, which, as we said earlier, is the usefulness of the Stratonovich integral.

15.8.3 Stratonovich Integral and the Itô Correction Term

The general form of a Stratonovich integral is $S(t) \equiv \int_0^t dB(t')F_S(x(t'))$. The general form of an Itô integral is $I(t) \equiv \int_0^t dB(t')F(x(t'))$. We can relate these two integrals, when $dx(t) = m(x(t)dt + b(x(t))dB(t)$ (that is, when $x(t)$ satisfies the Itô stochastic differential equation) as follows:

$$S(t) = \int_0^t dB(t')F\left(\tfrac{1}{2}[x(t'+dt') + x(t')]\right)$$

$$= \int_0^t dB(t')F\left(\tfrac{1}{2}[x(t'+dt') - x(t')] + x(t')\right) = \int_0^t dB(t')F\left(\tfrac{1}{2}dx(t') + x(t')\right)$$

$$= \int_0^t dB(t')F(x(t')) + \frac{1}{2}\int_0^t dB(t')dx(t')F'(x(t')) + \dots$$

$$= \int_0^t dB(t')F(x(t')) + \frac{1}{2}\int_0^t dB(t')[m(x)dt' + b(x)dB(t')]F'(x(t'))$$

$$S(t) = I(t) + \tfrac{1}{2}\sigma^2 \int_0^t dt'b(x(t'))F'(x(t')) \tag{15.32}$$

In going from the second line to the third line, we have used the Taylor series expansion. In going to the last line, we have dropped the term $\sim dB(t')dt' \sim (dt')^{3/2}$).

The first term on the right hand side in the last line of (15.32) is the Ito integral. The second term, a Riemann integral, is a correction term.

Equating the integrands in (15.32), we obtain the differential form

$$dB(t)F_S(x(t)) = dB(t)F(x(t)) + \tfrac{1}{2}\sigma^2 b(x(t))F'(x(t))dt, \tag{15.33}$$

Thus, this Stratonovich expression can be written in terms of an Itô expression.

And now, we see howStep 2) comes about, that is, to see the origin of Eq. (15.7a) where, to the Stratonovich SDE is added the Itô correction term that allows us to get the Itô SDE. Starting with the Stratonovich stochastic differential equation (15.6a) (reproduced as the first line of Eq. (15.34) below), we apply (15.33) with $F_S = b_S$ and so get the equivalent Itô stochastic differential equation:

$$dx(t) = a(x(t))dt + b_S(x(t))dB(t)$$

$$= a(x(t))dt + [dBb(x(t)) + \frac{1}{2}\sigma^2 b(x(t))b'(x(t))dt].$$

$$\equiv m(x(t))dt + b(x(t))dB(t) \text{ with the definition}$$

$$m(x(t)) \equiv a(x(t)) + \frac{1}{2}\sigma^2 b(x(t))b'(x(t)). \tag{15.34}$$

15.9 Stratonovich Calculus

We now have proven the correctness of Steps 4), 3), 2). What remains to be explained about the recipe is the assertion, in Step 1), that the usual calculus manipulations are valid when one treats the Stratonovich $F_S(x)$ as if it were the Riemann $F(x)$. We have already seen one example of this, Eq. (15.31). Now to see, in general, how the Stratonovich integral gives the Riemann integral result.

15.9.1 Stratonovich Chain Rule

Given an arbitrary function $F(x)$, we calculate $F_S(x(t))dx$, taking similar steps to those used for the integrand in (15.32):

$$dx(t)F_S(x(t)) \equiv dxF\left(\tfrac{1}{2}[x(t+dt)+x(t)]\right)$$

$$= dx(t)F\left(\tfrac{1}{2}[x(t+dt)-x(t)]+x(t)\right) = dx(t)F\left(\tfrac{1}{2}dx(t)+x(t)\right)$$

$$= dx(t)[F(x(t))+\tfrac{1}{2}dx(t)F'(x(t))+...]$$

$$= dx(t)F(x(t))+\tfrac{1}{2}[m(x(t))dt+b(x(t))dB(t)]^2F'(x(t))$$

$$= dx(t)F(x(t))+\tfrac{1}{2}\sigma^2 dtb^2(x(t))F'(x(t)), \tag{15.35}$$

where, in the fourth line, we have chosen to write dx in its Itô form (rather than in its equivalent Stratonovich form) and, in the last step, as usual, terms $\sim (dt)^2, (dt)^{3/2}$ have been dropped.

Now, the Itô chain rule expression (15.23) is

$$df(x(t)) = f'(x(t))dx(t) + \frac{1}{2}\sigma^2 dtb^2(x(t))f''(x(t)). \tag{15.36}$$

In (15.35), set $F_S(x(t)) = f'_S(x(t))$ (i.e., $F(\tfrac{1}{2}[x(t+dt)+x(t)]) = f'(\tfrac{1}{2}[x(t+dt)+x(t)])$, so (15.35) becomes:

$$dx(t)f'_S(x(t)) = dx(t)f'(x(t)) + \tfrac{1}{2}\sigma^2 dtb^2(x(t))f''(x(t)).$$

Comparing the right-hand side of this to the right-hand side of (15.36), we see that they are identical! Therefore, the left-hand sides of these two equations must be identical:

$$df(x(t)) = dx(t)f'_S(x(t)). \tag{15.37}$$

This is the Stratonovich chain rule. It relates the ordinary differential of a function to the derivative of a function whose argument is in Stratonovich form.

The ordinary chain rule is $df(x) = dx f'(x)$, which has the same form as (15.37). Indeed, the integral of (15.37) is

$$f(x(t)) - f(x(0)) = \int_0^{x(t)} f'_S(x)dx \tag{15.38}$$

This shows that the Riemann integral result, the left-hand side of Eq. (15.38), is obtained from the similar looking integral where, however, it is the Stratonovich form $f'_S(x)$ that is being integrated, not the Riemann form. This is what we set out to prove.

Note that Eq. (15.37) is *not* $df_S(x(t)) = dx f'_S(x(t))$, and indeed this is not true. However, $\frac{d}{dx}f_S(x(t)) = f'_S(x(t))$ is true, that is to say, taking the derivative of a function whose argument is the Stratonovich argument is equivalent to taking the ordinary derivative of the function of x, and then replacing x by the Stratonovich argument.

15.9.2   Consequences of the Stratonovich Chain Rule

Here are a few examples, illustrating how the Stratonovich chain rule (15.37) allows one to do the usual calculus manipulations with Stratonovich functions.

1) The Stratonovich product rule follows from replacing $f(x(t))$ in Eq. (15.37) by $f(x(t))g(x(t))$. As explained above, by $f'_S(x)$ is meant that you take the derivative of $f(x)$ using the rules of ordinary calculus and, afterwards, replace its argument $x(t)$ by $\frac{1}{2}[x(t+dt)+x(t)]$, that is, convert the ordinary function to a Stratonovich function. By ordinary calculus, $[f(x)g(x)]'dx = [f'(x)g(x) + f(x)g'(x)]dx$, so $[f(x)g(x)]'_S dx = [f'_S(x)g_S(x) + g'_S(x)f_S(x)]dx$. It then follows that

$$d[f(x(t))g(x(t))] = [f'_S(x(t))g_S(x(t)) + g'_S(x(t))f_S(x(t))]dx(t). \tag{15.39}$$

This has the same form as the usual chain rule $d[f(x)g(x)] = [f'(x)g(x)+g'(x)f(x)]dx$.

2) In this and the following examples, $x(t)$ obeys the Stratonovich SDE $dx(t) = a(x(t)) + b_S(x(t))dB(t)$. We are given a differential equation expressed in terms of Stratonovich functions, and we want to solve it.

Consider the differential equation $dy(t) = x_S^n(t)dx(t)$. The Stratonovich chain rule says $d\frac{x^{n+1}(t)}{n} = x_S^n(t)dx(t)$. Since the right sides of these two equations are identical, the left sides are equal: $dy(t) = d\frac{x^{n+1}(t)}{n}$. Upon integrating we obtain the solution: $y(t) - y(0) = \frac{x^{n+1}(t)}{n} - \frac{x^{n+1}(0)}{n}$. The solution $y(t)$ is the same as if you just replaced the Stratonovich expression on the right side of the differential equation by the ordinary expression, and then did the ordinary Riemann integral.

More generally, consider $dy(t) = g_S(x(t))dx(t)$ where $g(x)$ is an arbitrary function.. The chain rule says $d \int_0^{x(t)} dx'g(x') = g_S(x(t))dx(t)$. Therefore, the differential equation is equivalent to $dy(t) = d \int_0^{x(t)} dx'g(x')$ or, integrating, $y(t) - y(0)) = \int_0^{x(t)} dx'g(x')$. The solution is the same as if you just replaced the Stratonovich expression on the right side of the differential equation by the ordinary expression, and then did the ordinary Riemann integral.

3) Consider the first-order Stratonovich differential equation

$$y'(x(t)) - g_S(x(t))y(x(t)) = 0. \tag{15.40}$$

This is the same as $d[\ln y(x(t))] = g_S(x(t))dx(t)$, which has been solved in example 2) above. The solution is

$$\ln y(t) - \ln y(0) = \int_0^{x(t)} dx'g(x')$$

or

$$y(t) = y(0)e^{\int_0^{x(t)} dx'g(x')},$$

the same result obtained as if you just replaced the Stratonovich expression on the right side of (15.40) by the ordinary expression, and then solved as usual.

4) Consider the general second-order linear differential equation

$$u''(x) + \alpha_S(x)u'(x) + \beta_S(x)u = 0.$$

It is well known that, by a change of variable, this equation can be converted to one where the coefficient of the first derivative term vanishes:

$$y''(x) = r_S(x)y(x)$$

where $r_S(x)$ is a function of $\alpha_S(x), \beta_S(x)$.

To solve this, write it in the form

$$\frac{y''(x)}{y(x)}dx = r_S(x)dx.$$

Now, as in example 2), consider the problem of solving $\frac{dz(x)}{dx} = r_S(x)$, or

$$dz(x) = r_S(x)dx.$$

Using the chain rule, it was shown that $z(x(t)) = \int_0^{x(t)} dx r(x)$ (setting $z(0) = 0$ for simplicity), or $z'(x) = r(x)$.

Since we have two expressions for $r_S(x)dx$, we can equate the left sides, to find that y satisfies $\frac{y''(x)}{y(x)}dx = dz(x)$, or $y''(x) = z'(x)y(x)$ or $y''(x) = r(x)y(x)$. So, the solution is the same as if you just replaced the Stratonovich expression in the differential equation by the ordinary expression, and then solved it as usual.

This finishes our discussion of examples of how Stratonovich equation results are identical to ordinary calculus results.

15.10 Conclusion

This concludes our short, but rather complete, course on one-variable stochastic differential equations. We have justified the recipe for 1) going from the Stratonovich stochastic differential equation for $x(t)$ (manipulated according to the usual rules of calculus) to 2) the equivalent Itô equation, by adding a specific correction term, then using this Itô equation to 3) easily find $\overline{dx}/dt, \overline{dx^2}/dt$ and 4) plugging these into the Fokker–Planck equation whose solution is the probability that x takes on values at time t.

A complete SDE tutorial requires generalization to many variables and many Brownian motions. Using the one-variable results above, it is really straightforward to generalize to many variables $x_i(t)$ (collectively notated as $\mathbf{x}$) and many Brownian functions $B_j(t)$.

15.11 More than One Variable, One Brownian Function

In what follows, for ease of notation, $x(t)$ is replaced by x.

For more than one variable, the recipe becomes:

1) Beginning with the Stratonovich equations,

$$dx_i(t) = a_i(\mathbf{x})dt + b_{Si}(\mathbf{x})dB(t) \tag{15.41}$$

$$\equiv a_i(\mathbf{x})dt + b_i\left(\frac{1}{2}[\mathbf{x}(t+dt) + \mathbf{x}(t)]\right)dB(t)$$

$$= a_i(\mathbf{x})dt + b_i\left(\frac{1}{2}d\mathbf{x}(t) + \mathbf{x}(t)\right)dB(t)$$

$$= a_i(\mathbf{x})dt + \left[b_i(\mathbf{x}) + \frac{1}{2}\sum_j\left(\frac{\partial}{\partial x_j}b_i(\mathbf{x})\right)dx_j + ...\right]dB(t)$$

$$= \left[a_i(\mathbf{x}) + \frac{1}{2}\sum_j b_j\left(\frac{\partial}{\partial x_j}b_i(\mathbf{x})\right)\sigma^2\right]dt + b_i(\mathbf{x})dB(t)$$

$$\equiv m_i(\mathbf{x})dt + b_i(\mathbf{x})dB(t), \tag{15.42}$$

we obtain the Itô correction term, the second term in the square bracket, and end up with 2), the Itô equations (15.42).

3) From the Itô equations, we calculate the mean values

$$\overline{\frac{dx_i}{dt}} = \overline{m_i(\mathbf{x})} \rightarrow m_i(\mathbf{x})$$

$$\overline{\frac{dx_i dx_j}{dt}} = \sigma^2\overline{b_i(\mathbf{x})b_j(\mathbf{x})} \rightarrow \sigma^2 b_i(\mathbf{x})b_j(\mathbf{x}), \tag{15.43}$$

and insert them into the 4) Fokker–Planck equation:

$$\frac{\partial}{\partial t}\rho(\mathbf{x},t|\mathbf{x}_0,t_0) = -\sum_i \frac{\partial}{\partial x_i}\frac{\overline{dx_i}}{dt}\rho(\mathbf{x},t|\mathbf{x}_0,t_0) + \frac{1}{2}\sum_{ij}\frac{\partial}{\partial x_i}\frac{\partial}{\partial x_j}\frac{\overline{dx_i dx_j}}{dt}\rho(\mathbf{x},t|\mathbf{x}_0,t_0)$$

$$= -\sum_i \frac{\partial}{\partial x_i}m_i(\mathbf{x})\rho(\mathbf{x},t|\mathbf{x}_0,t_0) + \frac{\sigma^2}{2}\sum_{ij}\frac{\partial}{\partial x_i}\frac{\partial}{\partial x_j}b_i(\mathbf{x})b_j(\mathbf{x})\rho(\mathbf{x},t|\mathbf{x}_0,t_0).$$
$$\tag{15.44}$$

Recipe completed.

The chain rule for the Itô equations is

$$df(\mathbf{x}) = \sum_i\left(\frac{\partial}{\partial x_i}f(\mathbf{x})\right)dx_i + \frac{1}{2}\sum_{ij}\left(\frac{\partial}{\partial x_i}\frac{\partial}{\partial x_j}f(\mathbf{x})\right)dx_i dx_j + ...$$

$$= \sum_i\left(\frac{\partial}{\partial x_i}f(\mathbf{x})\right)dx_i + \frac{1}{2}\sum_{ij}\left(\frac{\partial}{\partial x_i}\frac{\partial}{\partial x_j}f(\mathbf{x})\right)b_i(\mathbf{x})dB(t)b_j(\mathbf{x})dB(t)$$

$$= \sum_i\left(\frac{\partial}{\partial x_i}f(\mathbf{x})\right)dx_i + \frac{\sigma^2}{2}\sum_{ij}\left(\frac{\partial}{\partial x_i}\frac{\partial}{\partial x_j}f(\mathbf{x})\right)b_i(\mathbf{x})b_j(\mathbf{x})dt. \tag{15.45}$$

where the first line uses the Taylor series, the second line inserts the expressions for dx_i, retaining only the term whose contribution is $\sim dt$ (discarding $dt^{3/2}, dt^2$ terms), the last line uses $[dB(t)]^2 = \sigma^2 dt$. Of course, this reduces to (15.23) in the one variable limit.

The product rule for the Itô equations, using (15.45), is unchanged in form from (15.26):

$$d[f(\mathbf{x})g(\mathbf{x})] = \sum_i \left(\frac{\partial}{\partial x_i} f(\mathbf{x})g(\mathbf{x}) \right) dx_i + \frac{\sigma^2}{2} \sum_{ij} \left(\frac{\partial}{\partial x_i} \frac{\partial}{\partial x_j} f(\mathbf{x})g(\mathbf{x}) \right) b_i(\mathbf{x}) b_j(\mathbf{x}) dt$$

$$= f(\mathbf{x}) dg(\mathbf{x}) + g(\mathbf{x}) df(\mathbf{x}) + \sigma^2 \sum_i \left(\frac{\partial}{\partial x_i} f(\mathbf{x}) \right) b_i(\mathbf{x}) \sum_i \left(\frac{\partial}{\partial x_i} g(\mathbf{x}) \right) b_j(\mathbf{x}) dt$$

$$= f(\mathbf{x}) dg(\mathbf{x}) + g(\mathbf{x}) df(\mathbf{x}) + df(\mathbf{x}) dg(\mathbf{x}), \tag{15.46}$$

where, in going from the first line to the second line, the second derivatives of $f(\mathbf{x}), g(\mathbf{x})$ have been gathered into $df(\mathbf{x}), dg(\mathbf{x})$ with use of (15.45) and, in going to the last line, use has been made of

$$df(\mathbf{x}) dg(\mathbf{x}) = \left[\sum_i \left(\frac{\partial}{\partial x_i} f(\mathbf{x}) \right) dx_i + ... \right] \left[\sum_j \left(\frac{\partial}{\partial x_i} g(\mathbf{x}) \right) dx_j + ... \right]$$

$$= \left[\sum_i \left(\frac{\partial}{\partial x_i} f(\mathbf{x}) \right) [m_i(\mathbf{x}) dt + b_i(\mathbf{x}) dB(t)] \right]$$

$$\cdot \left[\sum_j \left(\frac{\partial}{\partial x_i} g(\mathbf{x}) \right) [m_j(\mathbf{x}) dt + b_j(\mathbf{x}) dB(t)] \right]$$

$$= \left[\sum_i \left(\frac{\partial}{\partial x_i} f(\mathbf{x}) \right) [b_i(\mathbf{x}) dB(t)] \right] \left[\sum_j \left(\frac{\partial}{\partial x_i} g(\mathbf{x}) \right) [b_j(\mathbf{x}) dB(t)] \right]$$

$$= \sigma^2 \sum_i \left(\frac{\partial}{\partial x_i} f(\mathbf{x}) \right) b_i(\mathbf{x}) \sum_i \left(\frac{\partial}{\partial x_i} g(\mathbf{x}) \right) b_j(\mathbf{x}) dt.$$

The chain rule for the Stratonovich equations is obtained from an argument similar to that previously given in Section 15.9.1 above, ending in Eq. (15.37). First, for an arbitrary function $F(\mathbf{x})$, calculate

$$dx_i F_S(\mathbf{x}) = dx_i \left[F(\mathbf{x}) + \frac{1}{2} \sum_j \left(\frac{\partial}{\partial x_j} F(\mathbf{x}) \right) dx_j \right]$$

$$= dx_i F(\mathbf{x}) + \frac{\sigma^2}{2} \sum_j \left(\frac{\partial}{\partial x_j} F(\mathbf{x}) \right) b_i(\mathbf{x}) b_j(\mathbf{x}). \tag{15.47}$$

Then, setting $F_S(\mathbf{x}) = \left(\frac{\partial}{\partial x_i} f(\mathbf{x}) \right)_S$, summing over i in Eq. (15.47), and comparing with the Itô chain rule (15.45), we obtain

$$df(\mathbf{x}) = \sum_i \left(\frac{\partial}{\partial x_i} f(\mathbf{x}) \right)_S dx_i. \tag{15.48}$$

15.12   One Variable, More than One Brownian Function

Now the recipe becomes:

1) Beginning with the Stratonovich equation,

$$dx(t) = a(x)dt + \sum_i b_{Si}(x)dB_i(t) \tag{15.49}$$

$$= a(x)dt + \sum_i \left[b_i(x) + \frac{1}{2}\left(\frac{d}{dx}b_i(x)\right)dx(t)\right]dB_i(t)$$

$$= \left[a(x) + \frac{\sigma^2}{2}\left(\frac{d}{dx}b_i(x)\right)b_i(x)\right]dt + \sum_i b_i(x)dB_i(t)$$

$$\equiv m(x)dt + \sum_i b_i(x)dB_i(t), \tag{15.50}$$

we obtain the Itô correction term (second term in the square bracket) and end up with 2), the Itô equation (15.50).

Something new was involved in this derivation: we took $dB_i(t)dB_j(t) = 0$ for $i \neq j$. (Note, this is the product we take to vanish, not just the mean $\overline{dB_i(t)dB_j(t)} = 0$, which is true since $dB_i(t), dB_j(t)$ are statistically independent.) This occurred in going from the second to the third line, where appeared $dxdB_i(t) = \sum_j b_j(x)dB_j(t)dB_i(t) = \sigma^2 dt b_i(x)$ (also dropping terms of order $(dt)^{3/2}$).

This requires a proof in mean square (as did our assertion in Section 8.1, that $[dB(t)]^2 = \sigma^2 dt$, not just that $\overline{[dB(t)]^2} = \sigma^2 dt$):

$$\overline{\left[\sum_{k=0}^{N-1} dB_1(kdt)dB_2(kdt) - 0\right]^2} = \overline{\sum_{j=0}^{N-1} dB_1(jdt)dB_2(jdt)\sum_{k=0}^{N-1} dB_1(kdt)dB_2(kdt)}$$

$$= \sum_{j=0}^{N-1} \sigma^2 dt \sigma^2 dt = \sigma^4 (dt)^2 N = \sigma^4 t dt \to 0. \tag{15.51}$$

using $Ndt = t$.

So, to summarize, in all manipulations we take $dD_i(t)dD_j(t) = \delta_{ij}\sigma^2 dt$.

3) From the Itô equation, we calculate the mean values

$$\frac{\overline{dx}}{dt} = \overline{m(x)} \to m(x)$$

$$\frac{\overline{dxdx}}{dt} = \sigma^2 \sum_i \overline{b_i^2(\mathbf{x})} \to \sigma^2 \sum_i b_i^2(\mathbf{x}), \tag{15.52}$$

and insert them into the 4) Fokker–Planck equation:

$$\frac{\partial}{\partial t}\rho(x,t|x_0,t_0) = -\frac{\partial}{\partial x}m(x)\rho(x,t|x_0,t_0) + \frac{\sigma^2}{2}\frac{\partial^2}{\partial x^2}\sum_i b_i^2(x)\rho(x,t|x_0,t_0). \tag{15.53}$$

The chain rule for the Itô equation is

$$df(x) = \frac{d}{dx}f(x)dx + \frac{1}{2}\frac{d^2}{dx^2}f(x)(dx)^2$$

$$= \frac{d}{dx}f(x)dx + \frac{\sigma^2}{2}\frac{d^2}{dx^2}f(x)\sum_i b_i^2(x). \tag{15.54}$$

The product rule for the Itô equation, using (15.54),

$$d[f(x)g(x)] = \frac{d}{dx}[f(x)g(x)]dx + \frac{\sigma^2}{2}\frac{d^2}{dx^2}[f(x)g(x)]\sum_i b_i^2(x)$$

$$= g(x)df(x) + f(x)dg(x) + \sigma^2 f'(x)g'(x)\sum_i b_i^2(x)$$

$$= g(x)df(x) + f(x)dg(x) + df(x)dg(x) \tag{15.55}$$

is unchanged in form.

The chain rule for the Stratonovich equations is found by the argument previously given. For an arbitrary function $F(x)$, we find

$$dx F_s(x) = dx\left[F(x) + \frac{1}{2}\frac{d}{dx}F(x)dx\right] = dx F(x) + \frac{\sigma^2}{2}\frac{d}{dx}F(x)\sum_i b_i^2(x).$$

$$\tag{15.56}$$

Replacing $F(x)$ by $\frac{d}{dx}f(x)$, the right hand side of (15.56) becomes identical to the chain rule equation (15.54) and so

$$df(x) = f'_s(x)dx. \tag{15.57}$$

15.13 More than One Variable, More than One Brownian Function

This is the most complicated situation and, with this, our considerations of SDEs ends.

The recipe becomes:

1) Beginning with the Stratonovich equations,

$$dx_i(t) = a_i(\mathbf{x})dt + \sum_j b_{Sij}(\mathbf{x})dB_j(t) \tag{15.58}$$

$$= a_i(\mathbf{x})dt + \sum_j \left[b_{ij}(\mathbf{x})dB_j(t) + \frac{1}{2}\sum_k \left(\frac{\partial}{\partial x_k}b_{ij}(\mathbf{x})\right)dx_k dB_j(t)\right]$$

$$= \left[a_i(\mathbf{x}) + \frac{\sigma^2}{2}\sum_{jk} b_{kj}\frac{\partial}{\partial x_k}b_{ij}(\mathbf{x})\right]dt + \sum_j b_{ij}(\mathbf{x})dB_j(t)$$

$$\equiv m_i(\mathbf{x})dt + \sum_j b_{ij}(\mathbf{x})dB_j(t), \tag{15.59}$$

we obtain the Itô correction term (second term in the square bracket) and end up with 2), the Itô equations (15.59).

3) From the Itô equations, we calculate the mean values

$$\frac{\overline{dx_i}}{dt} = \overline{m_i(\mathbf{x})} \to m_i(\mathbf{x})$$

$$\frac{\overline{dx_i dx_j}}{dt} = \sigma^2 \sum_k \overline{b_{ik}(\mathbf{x})b_{jk}(\mathbf{x})} \to \sigma^2 \sum_k b_{ik}(\mathbf{x})b_{jk}(\mathbf{x}), \qquad (15.60)$$

and insert them into the 4) Fokker–Planck equation:

$$\frac{\partial}{\partial t}\rho(\mathbf{x},t|\mathbf{x}_0,t_0) = -\sum_i \frac{\partial}{\partial x_i} m_i(\mathbf{x})\rho(\mathbf{x},t|\mathbf{x}_0,t_0)$$

$$+\frac{\sigma^2}{2}\sum_{ij}\frac{\partial}{\partial x_i}\frac{\partial}{\partial x_j}\sum_k b_{ik}(\mathbf{x})b_{jk}(\mathbf{x})\rho(\mathbf{x},t|\mathbf{x}_0,t_0). \quad (15.61)$$

Recipe completed.

From here on, no proofs are given as the details have been covered in the previous two sections.

The chain rule for the Itô equations is

$$df(\mathbf{x}) = \sum_i \left(\frac{\partial}{\partial x_i}f(\mathbf{x})\right)dx_i + \frac{\sigma^2}{2}\sum_{ijk}\left(\frac{\partial}{\partial x_i}\frac{\partial}{\partial x_j}f(\mathbf{x})\right)b_{ik}(\mathbf{x})b_{jk}(\mathbf{x}). \quad (15.62)$$

The product rule for the Itô equations is unchanged in form,

$$d[f(\mathbf{x})g(\mathbf{x})] = f(\mathbf{x})dg(\mathbf{x}) + g(\mathbf{x})df(\mathbf{x}) + df(\mathbf{x})dg(\mathbf{x}). \qquad (15.63)$$

Similarly the chain rule for the Stratonovich equations is unchanged in form,

$$df(\mathbf{x}) = \sum_i \left(\frac{\partial}{\partial x_i}f(\mathbf{x})\right)_S dx_i. \qquad (15.64)$$

There are a couple of generalizations that could be made. All functions a_i, b_{ij} could explicitly depend upon t in addition to their dependence upon $\mathbf{x}$ (which depends upon t). And, each different dD_k could have its own variance, $dD_k^2(t) - \sigma_k^2 dt$. We have not chosen to consider these, as they would complicate equations without adding any new conceptual value.

This concludes the SDE cookbook!

16

CSL Expressed as a Schrödinger Stochastic DE

Section 3.4 presented the simplest CSL evolution equation, with operator $\hat{A}$ as collapse-generating operator, in Eq. (3.15). Multiply this by dt, replace $|\psi', t\rangle$ by $|\psi, t\rangle$, replace $w(t)dt$ by $dB_{CSL}(t)$, set $H = 0$ for simplicity, and we have the parallel stochastic Stratonovich Schrödinger equation:

$$d|\psi, t\rangle = [\hat{A}dB_{CSL}(t) - \lambda dt \hat{A}^2]|\psi, t\rangle. \tag{16.1}$$

The Brownian motion in (16.1), $B_{CSL}(t)$, is not the standard Brownian motion appearing in the SDEs of Chapter 15 (we are reserving $B(t)$ for that). $|\psi, t\rangle$ in (16.1) is an unnormalized state vector whose probability of occurring is given by the probability rule.

We begin our treatment by converting 16.1) to the standard Stratonovich form, as a Schrödinger equation for a normalized state vector. Then, starting with this Stratonovich equation, we end up with the equivalent Itô Schrödinger equation for the normalized state vector. From there it is straightforward to obtain the density matrix evolution equation.

Finally, as done in Chapters 3 and 4, we generalize from this formulation with one collapse-generating operator to many collapse-generating operators, with their associated many Brownian motions and choose these operators as smeared mass-density operators so we end up with the Itô equation describing non–relativistic CSL.

16.1 Stratonovich Schrödinger Equation

Eq. (16.1) is expressed in terms of a state vector, and all the SDE equations in Chapter 15 involve functions. We may convert Eq. (16.1) to an equation involving functions by taking its scalar product with the eigenstates of $\hat{A}$:

$$d\langle a_i|\psi, t\rangle = [a_i dB_{CSL}(t) - \lambda dt a_i^2]\langle a_i|\psi, t\rangle. \tag{16.2}$$

With identification of $x_i \equiv \langle a_i|\psi, t\rangle$, this is the case of more than one variable, one Brownian function discussed in Section 8.11. When it comes to converting this to an Itô SDE, using the techniques in Chapter 15, we will use these state vector components. However, in this section we deal only with Stratonovich equations and, since the manipulations are all those of standard calculus, we can do those with the state vector.

Since this is a Stratonovich equation, wherever we see $x_i(t)$ it actually means $x_{Si}(t) = \frac{1}{2}[x_i(t+dt)+x_i(t)]$ so, wherever we see $|\psi,t\rangle$, it actually means $|\psi,t\rangle_S = \frac{1}{2}[|\psi,t+dt\rangle+|\psi,t\rangle]$. However, since all state vectors in this section are Stratonovich, we drop the S subscript here.

To begin with, to have the kind of Stratonovich equation discussed in Chapter 15, Eq. (16.2) needs to be expressed in terms of $dB(t)$, not in terms of $dB_{CSL}(t)$. We use the probability rule to relate the two.

The probability rule states that the probability that $B_{CSL}(t)$, and the state vector $|\psi,t\rangle$ that evolves under its influence, is proportional to $\langle\psi,t|\psi,t\rangle$, the proportionality factor ensuring that the probability sums to 1 for all possible $B_{CSL}(t)$.

Let us focus on the infinitesimal time-evolution, governed by $dB_{CSL}(t)$. The state vector $|\psi,t+dt\rangle$ evolves over the infinitesimal time interval dt from a normalized state vector $|\psi,t\rangle_N$ to an un-normalized state vector $|\psi,t+dt\rangle$ according to Eq. (3.1) (specialized to these initial and final times):

$$|\psi,t+dt\rangle = e^{-\frac{1}{4\lambda dt}[dB_{CSL}(t)-2\lambda dt\hat{A}]^2}|\psi,t\rangle_N$$

The probability for $dB_{CSL}(t)$ is therefore

$$P(dB_{CSL}(dt)) \sim \langle\psi,t+dt|\psi,t+dt\rangle$$
$$= {}_N\langle\psi,t|e^{-\frac{1}{2\lambda dt}[dB_{CSL}(t)-2\lambda dt\hat{A}]^2}|\psi,t\rangle_N$$
$$= e^{-\frac{1}{2\lambda dt}[dB_{CSL}(t)]^2}{}_N\langle\psi,t|[1+2dB_{CSL}(t)\hat{A}+...]|\psi,t\rangle_N$$
$$= e^{-\frac{1}{2\lambda dt}[dB_{CSL}(t)]^2}[1+2dB_{CSL}(t)_N\langle\psi,t|\hat{A}|\psi,t\rangle_N+...]$$
$$= e^{-\frac{1}{2\lambda dt}\{[dB_{CSL}(t)-2\lambda dt_N\langle\psi,t|\hat{A}|\psi,t\rangle_N]^2+...\}}. \tag{16.3}$$

In the third, fourth, and fifth lines, terms of order $(dt)^2$ and higher have been omitted.

We see, if in (16.3) we set

$$dB(t) = dB_{CSL}(t) - 2\lambda dt_N\langle\psi,t|\hat{A}|\psi,t\rangle_N, \tag{16.4}$$

then the probability density for $dB(t)$ is that of the usual Brownian motion.

Next, we need to convert the Stratonovich Schrödinger equation (16.1) for the un-normalized state vector to a Stratonovich Schrödinger equation for the normalized state vector ((16.4) is already expressed in terms of the normalized state vector $|\psi,t\rangle_N$). That is straightforward to do, using the usual calculus manipulations, since (16.1) is a Stratonovich equation:

$$d|\psi,t\rangle_N = d\frac{|\psi,t\rangle}{\sqrt{\langle\psi,t|\psi,t\rangle}}$$
$$= \frac{1}{\sqrt{\langle\psi,t|\psi,t\rangle}}d|\psi,t\rangle - |\psi,t\rangle\frac{1}{2\langle\psi,t|\psi,t\rangle^{3/2}}d\langle\psi,t|\psi,t\rangle$$
$$= \frac{1}{\sqrt{\langle\psi,t|\psi,t\rangle}}[\hat{A}dB_{CSL}(t)-\lambda dt\hat{A}^2]|\psi,t\rangle$$
$$\qquad - \frac{1}{2\langle\psi,t|\psi,t\rangle^{3/2}}2\langle\psi,t|[\hat{A}dB_{CSL}(t)-\lambda dt\hat{A}^2]|\psi,t\rangle|\psi,t\rangle$$
$$= [\hat{A}-\langle\hat{A}\rangle_N]|\psi,t\rangle_N dB_{CSL}(t) - [\hat{A}^2-\langle\hat{A}^2\rangle_N]|\psi,t\rangle_N\lambda dt, \tag{16.5}$$

where, in going from the second to the third line we have used

$$d[\langle\psi,t|\psi,t\rangle] = \langle\psi,t|d|\psi,t\rangle + [d\langle\psi,t|]|\psi,t\rangle$$

and the last line uses the shorthand notation ${}_N\langle\psi,t|\hat{O}|\psi,t\rangle_N \equiv \langle\hat{O}\rangle_N$. So, this Stratonovich equation for the change of the normalized state vector is expressed totally in terms of the normalized state vector.

We now can put Eq. (16.4) into Eq. (16.5) to get *the Stratonovich Schrödinger equation for the normalized state vector* (we remove the subscript N and also put the Hamiltonian back in):

$$d|\psi,t\rangle = -i\hat{H}|\psi,t\rangle + [\hat{A} - \langle\hat{A}\rangle]|\psi,t\rangle dB(t) + \left[-\hat{A}^2 + \langle\hat{A}^2\rangle + 2(\hat{A} - \langle\hat{A}\rangle)\langle\hat{A}\rangle\right]|\psi,t\rangle\lambda dt.$$

$$= -i\hat{H}|\psi,t\rangle + [\hat{A} - \langle\hat{A}\rangle]|\psi,t\rangle dB(t) + \left[-[\hat{A} - \langle\hat{A}\rangle]^2 + \langle\hat{A}^2\rangle - \langle\hat{A}\rangle^2\right]|\psi,t\rangle\lambda dt.$$

$$= -i\hat{H}|\psi,t\rangle + [\hat{A} - \langle\hat{A}\rangle]|\psi,t\rangle dB(t) - \left[[\hat{A}^2 - \langle\hat{A}^2\rangle] - 2\langle\hat{A}\rangle[\hat{A} - \langle\hat{A}\rangle]\right]|\psi,t\rangle\lambda dt,$$

$$(16.6)$$

(We will keep in mind that this is a Stratonovich equation, that is, $|\psi,t\rangle \equiv |\psi,t\rangle_S$).

We have given here three different groupings of the last term, as one or another facilitates future manipulations. For example, the last equation in (16.6) makes it easier to see that the state vector remains normalized to 1 throughout its evolution, if it is normalized to 1 initially. Since $d\langle\psi,t\ |\psi,t\rangle = d(\langle\psi,t|)|\psi,t\rangle + \langle\psi,t|d(|\psi,t\rangle)$, we calculate

$$\langle\psi,t|d|\psi,t\rangle = -i\langle\psi,t|\hat{H}|\psi,t\rangle + \langle\psi,t|[\hat{A} - \langle\hat{A}\rangle]|\psi,t\rangle dB(t)$$
$$- \langle\psi,t|\left[\hat{A}^2 - \langle\hat{A}^2\rangle\right]|\psi,t\rangle\lambda dt + 2\langle\hat{A}\rangle\langle\psi,t|\left[\hat{A} - \langle\hat{A}\rangle\right]|\psi,t\rangle\lambda dt$$
$$= -i\langle\psi,t|\hat{H}|\psi,t\rangle + 0 + 0 + 0,$$

which is the same as the expression obtained from the usual Schrödinger equation. Adding this to its complex conjugate, we see that

$$d\langle\psi,t\ |\psi,t\rangle = -i\langle\psi,t|\hat{H}|\psi,t\rangle + i\langle\psi,t|\hat{H}|\psi,t\rangle = 0.$$

16.2   Itô Schrödinger Equation

Next, we need to calculate the Itô correction term to add to the Stratonovich Eq. (16.6), thereby converting it to the equivalent Itô equation. The Itô correction term is given in Eq. (15.43). There one starts with the Stratonovich equation $dx_i = a_i + b_i dB(t)$ and so obtains the correction term

$$\frac{1}{2}\sum_j b_j\left(\frac{\partial}{\partial x_j}b_i(\mathbf{x})\right)\lambda\right] dt.$$

(Note, in Chapter 15, $[dB(t)]^2 = \sigma^2 dt$, but here $[dB(t)]^2 = \lambda dt$, so in Eq. (15.42) we have replaced σ^2 by λ.)

Taking the scalar product of (16.6) with $\langle a_i |$, we identify $x_i \equiv \langle a_i | \psi, t \rangle = \langle \psi, t | a_i \rangle$ (this will be real since we take $\hat{H} = 0$ for this process, and at the end put $\hat{H}$ back again) and identify b_i with the coefficient of $dB(t)$ in (16.6):

$$b_i(\mathbf{x}) = \langle a_i | [\hat{A} - \langle \hat{A} \rangle] | \psi, t \rangle = \Big[a_i - \sum_k a_k x_k^2 \Big] x_i. \tag{16.7}$$

Then, the Itô correction term is:

$$\frac{\lambda dt}{2} \sum_j b_j \Big(\frac{\partial}{\partial x_j} b_i(\mathbf{x}) \Big) = \frac{\lambda dt}{2} \sum_j \Big[a_j - \sum_k a_k x_k^2 \Big] x_j \frac{\partial}{\partial x_j} \Big[a_i - \sum_k a_k x_k^2 \Big] x_i.$$

$$= \frac{\lambda dt}{2} \sum_j \Big[a_j - \sum_k a_k x_k^2 \Big] x_j \Big[-2a_j x_j x_i + \Big[a_i - \sum_k a_k x_k^2 \Big] \delta_{ij} \Big]$$

$$= \frac{\lambda dt}{2} \Big\{ -2 \sum_j a_j^2 x_j^2 + a_i^2 - 2a_i \sum_k a_k x_k^2 + 3 \Big[\sum_k a_k x_k^2 \Big]^2 \Big\}$$

$$= \frac{\lambda dt}{2} \Big\{ -2 \sum_j a_j^2 x_j^2 + 2 \Big[\sum_k a_k x_k^2 \Big]^2 + \Big[a_i - \sum_k a_k x_k^2 \Big]^2 \Big\} x_i$$

$$= \frac{\lambda dt}{2} \langle a_i | \Big\{ -2[\langle \hat{A}^2 \rangle - \langle \hat{A} \rangle^2] + [\hat{A} - \langle \hat{A} \rangle]^2 \Big\} | \psi, t \rangle \tag{16.8}$$

Adding the correction term (16.8) without the scalar product factor $\langle a_i |$ to the second line of Eq. (16.6), we obtain *the Itô Schrödinger equation for the normalized state vector*:

$$d|\psi, t\rangle = -i\hat{H}|\psi, t\rangle dt + [\hat{A} - \langle \hat{A} \rangle]|\psi, t\rangle dB(t) - \frac{1}{2}[\hat{A} - \langle \hat{A} \rangle]^2 |\psi, t\rangle \lambda dt. \tag{16.9}$$

In using CSL as a SDE, this Itô form is most useful. We can calculate many things directly from Eq. (16.9), without expressing it in a basis.

For example, we can check that the state vector norm is indeed constant, using the Itô product rule (15.64) in the form:

$$d\langle \psi, t | \psi, t \rangle = \langle \psi, t | d[|\psi, t\rangle] + d[|\psi, t\rangle] \langle \psi, t | + d[\langle \psi, t |] d[|\psi, t\rangle]$$

$$= 2\langle [\hat{A} - \langle \hat{A} \rangle] \rangle dB(t) - \lambda dt \langle [\hat{A} - \langle \hat{A} \rangle]^2 \rangle + \lambda dt \langle [\hat{A} - \langle \hat{A} \rangle]^2 \rangle = 0. \tag{16.10}$$

16.3 Density Matrix Evolution Equation

Using the Itô product rule, we can find the evolution equation for the density matrix $\hat{\rho}(t) \equiv \overline{|\psi, t\rangle \langle \psi, t|}$. First we calculate the evolution equation for $|\psi, t\rangle \langle \psi, t|$, and afterwards we will take the ensemble average:

$$d\big[|\psi,t\rangle\langle\psi,t|\big] = d\big[|\psi,t\rangle\big]\langle\psi,t| + |\psi,t\rangle d\big[\langle\psi,t|\big] + d\big[|\psi,t\rangle\big]d\big[\langle\psi,t|\big]$$

$$= -i[H,|\psi,t\rangle\langle\psi,t|]dt + \Big\{[\hat{A}-\langle\hat{A}\rangle]|\psi,t\rangle\langle\psi,t| + |\psi,t\rangle\langle\psi,t|[\hat{A}-\langle\hat{A}\rangle]\Big\}dB(t)$$

$$+\Big\{-\frac{1}{2}[\hat{A}-\langle\hat{A}\rangle]^2|\psi,t\rangle\langle\psi,t| - \frac{1}{2}|\psi,t\rangle\langle\psi,t|[\hat{A}-\langle\hat{A}\rangle]^2$$

$$+[\hat{A}-\langle\hat{A}\rangle]|\psi,t\rangle\langle\psi,t|[\hat{A}-\langle\hat{A}\rangle]\Big\}\lambda dt$$

$$= -i[H,|\psi,t\rangle\langle\psi,t|]dt + \Big\{[\hat{A}-\langle\hat{A}\rangle]|\psi,t\rangle\langle\psi,t| + |\psi,t\rangle\langle\psi,t|[\hat{A}-\langle\hat{A}\rangle]\Big\}dB(t)$$

$$-\frac{\lambda}{2}\Big\{\hat{A}^2|\psi,t\rangle\langle\psi,t| + |\psi,t\rangle\langle\psi,t|\hat{A}^2 - 2\hat{A}|\psi,t\rangle\langle\psi,t|\hat{A}\Big\}dt. \qquad (16.11)$$

Upon taking the ensemble average, that is, $\overline{|\psi,t\rangle\langle\psi,t|} = \hat{\rho}(t)$ and $\overline{dB(t)} = 0$, we get the Lindblad equation (3.12):

$$d\hat{\rho}(t) = -i[H,\hat{\rho}(t)]dt - \frac{\lambda}{2}dt[\hat{A},[\hat{A},\hat{\rho}(t)]]. \qquad (16.12)$$

As an aside, one can write, in Itô form, a more general (non-CSL) state vector evolution, where $\hat{A}$ is *not* Hermitian. Defining $R \equiv \frac{1}{2}\langle[\hat{A}+\hat{A}^\dagger]\rangle$, this is:

$$d|\psi,t\rangle = -i\hat{H}|\psi,t\rangle dt + [\hat{A}-R]|\psi,t\rangle dB(t) - \frac{1}{2}[\hat{A}^\dagger\hat{A} - 2\hat{A}R + R^2|\psi,t\rangle\lambda dt.$$

(which reduces to (16.9) when A is Hermitian). One can readily check that this state vector remains normalized to 1, following the steps in (16.10). Then, following the steps in (16.11), (16.12), one arrives at the most general single operator Lindblad equation (as discussed in Appendix E):

$$d\hat{\rho}(t) = -i[H,\hat{\rho}(t)]dt - \frac{\lambda}{2}dt[\hat{A}^\dagger\hat{A}\rho(t) + \rho(t)\hat{A}^\dagger\hat{A} - 2\hat{A}\rho(t)\hat{A}^\dagger].$$

16.4 Non–relativistic CSL

As in Section 3.6, the generalization to many mutually commuting operators $\hat{A}_i$ requires many independent Brownian motion functions $dB_i(t)$. The Stratonovich Schrödinger Eq. (16.1) is replaced by

$$d|\psi,t\rangle = \Big[\sum_i \hat{A}_i dB_{iCSL}(t) - \lambda dt \sum_i \hat{A}_i^2\Big]|\psi,t\rangle. \qquad (16.13)$$

The argument leading to the Itô Eq. (16.9) is unchanged, since every term is independent of every other term, so the Itô Schrödinger equation for the normalized state vector is:

$$d|\psi,t\rangle = -i\hat{H}|\psi,t\rangle dt + \sum_i[\hat{A}_i - \langle\hat{A}_i\rangle]|\psi,t\rangle dB_i(t) - \frac{1}{2}\sum_i[\hat{A}_i - \langle\hat{A}_i\rangle]^2|\psi,t\rangle\lambda dt. \quad (16.14)$$

Taking the scalar product of (16.14) with the joint eigenstate of the operators satisfying $\hat{A}_i|a_1...a_i...\rangle = a_i|a_1...a_i...\rangle$, we see that (16.14) becomes a case of "more than one variable, more than one Brownian motion" in Section 15.13. Here, $x_{a_1...a_i...}(t) \equiv \langle a_1...a_i...|\psi, t\rangle$, a different variable x for every set of possible eigenvalues for all operators.

To obtain the density matrix, because every term is independent of every other term. the analogy to Eq.(16.11) is

$$d\big[|\psi, t\rangle\langle\psi, t|\big] = -i[H, |\psi, t\rangle\langle\psi, t|]dt$$
$$+ \sum_i \left\{ [\hat{A}_i - \langle\hat{A}\rangle_i]|\psi, t\rangle\langle\psi, t| + |\psi, t\rangle\langle\psi, t|[\hat{A}_i - \langle\hat{A}\rangle]_i \right\}dB_i(t)$$
$$- \frac{\lambda}{2} \sum_i \left\{ \hat{A}_i^2|\psi, t\rangle\langle\psi, t| + |\psi, t\rangle\langle\psi, t|\hat{A}_i^2 - 2\hat{A}_i|\psi, t\rangle\langle\psi, t|\hat{A}_i \right\}dt.$$

$$(16.15)$$

The Lindblad equation is the generalization of Eq. (16.12):

$$d\hat{\rho}(t) = -i[H, \hat{\rho}(t)]dt - \frac{\lambda}{2}dt \sum_i [\hat{A}_i, [\hat{A}_i, \hat{\rho}(t)]]. \qquad (16.16)$$

Again, a more general (non-CSL) state vector evolution, where $\hat{A}_i$ is not Hermitian, is

$$d|\psi, t\rangle = -i\hat{H}|\psi, t\rangle dt + \sum_i [\hat{A}_i - R_i]|\psi, t\rangle dB(t) - \frac{1}{2}\sum_i [\hat{A}_i^\dagger \hat{A}_i - 2\hat{A}_i R_i + R_i^2]|\psi, t\rangle \lambda dt,$$

where $R_i \equiv \frac{1}{2}\langle[\hat{A}_i + \hat{A}_i^\dagger]\rangle$. One can check that the norm of this state vector remains at 1. Using it leads to *the most general Lindblad equation* (Appendix E):

$$d\hat{\rho}(t) = -i[H, \hat{\rho}(t)]dt - \frac{\lambda}{2}dt \sum_i [\hat{A}_i^\dagger \hat{A}_i\rho(t) + \rho(t)\hat{A}_i^\dagger \hat{A}_i - 2\hat{A}_i\rho(t)\hat{A}_i^\dagger].$$

The next step toward non-relativistic CSL, as in Section 3.7, is to replace the discrete-labeled operators by continuum-labeled operators, $\hat{A}_i \rightarrow \sqrt{d\mathbf{x}}A(\mathbf{x})$ and the discrete-labeled Brownian motions by continuum-labeled Brownian motions, $dB_i(t) \rightarrow \sqrt{d\mathbf{x}}dB(\mathbf{x}, t)$. Then, the Stratonovich Schrödinger equation for the un-normalized state vector (16.13) becomes:

$$d|\psi, t\rangle = \int d\mathbf{x}[\hat{A}(\mathbf{x})dB_{CSL}(\mathbf{x}, t) - \hat{A}^2(\mathbf{x})\lambda dt]|\psi, t\rangle, \qquad (16.17)$$

the Itô Schrödinger equation for the normalized state vector (16.14) becomes

$$d|\psi, t\rangle = -i\hat{H}|\psi, t\rangle dt + \int d\mathbf{x}\left[[\hat{A}(\mathbf{x}) - \langle\hat{A}(\mathbf{x})\rangle]dB(\mathbf{x}, t) - \frac{1}{2}[\hat{A}(\mathbf{x}) - \langle\hat{A}(\mathbf{x})\rangle]^2\lambda dt\right]|\psi, t\rangle.$$

$$(16.18)$$

For all Itô manipulations,

$$dB(\mathbf{x}, t)dB(\mathbf{x}', t) = \delta(\mathbf{x} - \mathbf{x}')\lambda dt \qquad (16.19)$$

and $dB(\mathbf{x}, t)dB(\mathbf{x}', t') = 0$ if $t \neq t'$. Thus, the proof that the state vector norm is constant follows as before from the Itô product rule,

$$d\langle\psi, t|\psi, t\rangle = \langle\psi, t|d[|\psi, t\rangle] + d[\langle\psi, t\rangle]\langle\psi, t| + d[\langle\psi, t|]d[|\psi, t\rangle]$$

$$= -\lambda dt \int d\mathbf{x}\langle[\hat{A}(\mathbf{x}) - \langle\hat{A}(\mathbf{x})\rangle]^2\rangle$$

$$+ \int d\mathbf{x}\langle[\hat{A}(\mathbf{x}) - \langle\hat{A}(\mathbf{x})\rangle] \int d\mathbf{x}'[\hat{A}(\mathbf{x}') - \langle\hat{A}(\mathbf{x}')\rangle]\rangle\lambda dt\delta(\mathbf{x} - \mathbf{x}') = 0.$$

$$(16.20)$$

The density matrix evolution equation is obtained similarly by the product rule applied to $d[|\psi, t\rangle\langle\psi, t|]$ followed by taking the ensemble average, as in Eqs. (16.11), (16.12), ending with Eq. (3.31):

$$\frac{d}{dt}\hat{\rho}(t) = -i[H, \hat{\rho}(t)]dt - \frac{\lambda}{2}\int d\mathbf{x}[\hat{A}(\mathbf{x}), [\hat{A}(\mathbf{x}), \hat{\rho}(t)]]. \qquad (16.21)$$

When $\hat{A}(\mathbf{x})$ is specialized to the "smeared" mass density operator Eq. (4.3), we have in Eq. (16.18) the Itô equation for the state vector, and in Eq. (16.21) the density matrix evolution equations, in non–relativistic CSL.

This concludes our discussion of the SDE formulation of non–relativistic CSL, which is often used in the literature.

17

Applying the CSL Stratonovich Equation to the Free Particle Undergoing Collapse in Position

Here we solve the CSL Stratonovich equation (16.6), where $\hat{H} = \frac{1}{2m}\hat{P}^2$ and $\hat{A} = \hat{X}$:

$$d|\psi, t\rangle = -i\frac{1}{2m}\hat{P}^2|\psi, t\rangle dt + [\hat{X} - \langle\hat{X}\rangle)]|\psi, t\rangle db(t)$$

$$-\lambda\left[[\hat{X}^2 - \langle\hat{X}^2\rangle] - 2\langle\hat{X}\rangle[\hat{X} - \langle\hat{X}\rangle]\right]|\psi, t\rangle dt$$

or, in the position representation,

$$\frac{d}{dt}\psi(x, t) = \frac{i}{2m}\frac{\partial^2}{\partial x^2}\psi(x, t) + [x - \langle\hat{X}\rangle]\psi(x, t)\frac{db(t)}{dt}$$

$$-\lambda\left[x^2 - 2x\langle\hat{X}\rangle + 2\langle\hat{X}\rangle^2 - \langle\hat{X}^2\rangle\right]\psi(x, t)$$

(17.1)

(we have made the symbol replacement $B(t) \to b(t)$ in rewriting (16.6) to avoid a notational duplication apparent in Eq.(17.2)). This is the same as the problem treated in Chapter 10 and discussed in Section 3.5. The results will be shown to be the same.

As before, we are looking for a solution in the form,

$$\psi(x, t) = e^{-A(t)x^2 + B(t)x + C(t)}$$

(17.2)

(Eq.(10.3) or Eq.(3.17)), with initial conditions $A(0) = 1/4\sigma^2, B(0) = 0$. The new feature of (17.1) is its nonlinearity in $|\psi, t\rangle$, expressed in $\langle\hat{X}\rangle, \langle\hat{X}^2\rangle$. We calculate, from (17.2),

$$\langle\hat{X}\rangle \equiv \frac{\int dx\, x|\psi(x, t)|^2}{\int dx|\psi(x, t)|^2} = \frac{B_R(t)}{2A_R(t)}.$$

(17.3)

where the subscripts R, I refer to real and imaginary parts. (We will not need to calculate $\langle\hat{X}^2\rangle$, since it only plays a role in the calculation of the normalization term $C(t)$ which we will not bother to find.)

As before, we substitute (17.2), (17.3) into (17.1), obtaining as coefficients of x^2, x, respectively the equations

$$-\frac{d}{dt}A(t) = \frac{2i}{m}A^2(t) - \lambda \tag{17.4a}$$

$$\frac{d}{dt}B(t) = -\frac{2i}{m}A(t)B(t) + \frac{db(t)}{dt} + \lambda\frac{B_R(t)}{A_R} \tag{17.4b}$$

This results in Eq.(3.9) for $A(t)$, which evolves to an asymptotically constant width. As before, we assume that the packet initial width is this equilibrium width so, setting $\frac{d}{dt}A(t) = 0$ in (17.4a), we obtain (3.10):

$$A(t) = \frac{1-i}{2s^2} \tag{17.5}$$

where, as before, $s^2 \equiv 1/\sqrt{m\lambda}$ and $\tau \equiv \sqrt{m/\lambda}$.

Eq. (17.4b) differs from Eq. (10.4b) in the additional last term. As in Eqs. (10.11), the solution of this equation is (using $\lambda\frac{B_R(t)}{A_R} = 2\frac{B_R(t)}{\tau}$):

$$B(t) = \int_0^t dt' \left[\frac{db(t')}{dt'} + \frac{2B_R(t')}{\tau}\right]e^{-(1+i)\frac{(t-t')}{\tau}} \equiv B_R(t) + iB_I(t), \text{ where}$$

$$B_R(t) \equiv \int_0^t dt' \left[\frac{db(t')}{dt'} + \frac{2B_R(t')}{\tau}\right]e^{-\frac{(t-t')}{\tau}}\cos\frac{(t-t')}{\tau},$$

$$B_I(t) \equiv -\int_0^t dt' \left[\frac{db(t')}{dt'} + \frac{2B_R(t')}{\tau}\right]e^{-\frac{(t-t')}{\tau}}\sin\frac{(t-t')}{\tau}. \tag{17.6}$$

The time derivative of Eqs. (10.6) is

$$\frac{d}{dt}B_R(t) = \frac{db(t)}{dt} + \frac{2B_R(t)}{\tau} - \frac{B_R(t)}{\tau} + \frac{B_I(t)}{\tau}$$

$$= \frac{db(t)}{dt} + \frac{1}{\tau}[B_R(t) + B_I(t)]. \tag{17.7a}$$

$$\frac{d}{dt}B_I(t) = -\frac{1}{\tau}[B_R(t) + B_I(t)]. \tag{17.7b}$$

Adding Eq. (17.7a) and Eq. (17.7b), it follows that

$$\frac{d}{dt}[B_R(t) + B_I(t)] = \frac{db(t)}{dt}, \text{ or } [B_R(t) + B_I(t)] = b(t). \tag{17.8}$$

Putting (17.8) into (17.7b) and integrating gives

$$B_I(t) = -\frac{1}{\tau}\int_0^t dt'b(t'), \tag{17.9}$$

which is the same as Eq. (10.34). Putting this into (17.8) gives

$$B_R(t) = b(t) + \frac{1}{\tau}\int_0^t dt'b(t'). \tag{17.10}$$

which is the same as Eq. (10.28).

Thus, the same results that were obtained in Chapter 10 using the two CSL equations have been obtained here using the single CSL Stratonovich equation.

18

Applying the CSL Stratonovich Equation to the Harmonic Oscillator Undergoing Collapse in Position

Here we solve the CSL Stratonovich equation (16.6), where $\hat{H} = \frac{1}{2m}\hat{P}^2 + \frac{m\omega^2}{2}\hat{X}^2$ and $\hat{A} = \hat{X}$:

$$d|\psi,t\rangle = -i\left[\frac{1}{2m}\hat{P}^2 + \frac{m\omega^2}{2}\hat{X}^2\right]|\psi,t\rangle dt + [\hat{X} - \langle\hat{X}\rangle]|\psi,t\rangle db(t)$$
$$-\lambda\left[[\hat{X}^2 - \langle\hat{X}^2\rangle] - 2\langle\hat{X}\rangle[\hat{X} - \langle\hat{X}\rangle]\right]|\psi,t\rangle dt$$

or, in the position representation,

$$\frac{d}{dt}\psi(x,t) = -i\left[-\frac{1}{2m}\frac{\partial^2}{\partial x^2} + \frac{m\omega^2}{2}x^2\right]\psi(x,t) + [x - \langle\hat{X}\rangle]\psi(x,t)\frac{db(t)}{dt}$$
$$-\lambda\left[x^2 - 2x\langle\hat{X}\rangle + 2\langle\hat{X}\rangle^2 - \langle\hat{X}^2\rangle\right]\psi(x,t). \tag{18.1}$$

As before, we are looking for a solution in the form,

$$\psi(x,t) = e^{-A(t)x^2 + B(t)x + C(t)}, \tag{18.2}$$

with initial condition that this is the vacuum state, so $A(0) = m\omega^2/2, B(0) = 0$. We repeat (17.3),

$$\langle\hat{X}\rangle \equiv \frac{\int dx\, x|\psi(x,t)|^2}{\int dx|\psi(x,t)|^2} = \frac{B_R(t)}{2A_R(t)}. \tag{18.3}$$

As before, we substitute (18.2), (18.3) into (18.1), obtaining as coefficients of x^2, x, respectively, the equations

$$-\frac{d}{dt}A(t) = \frac{2i}{m}A^2(t) - i\frac{m\omega^2}{2} - \lambda \tag{18.4a}$$
$$\frac{d}{dt}B(t) = -\frac{2i}{m}A(t)B(t) + \frac{db(t)}{dt} + \lambda\frac{B_R}{A_R} \tag{18.4b}$$

(we will not bother finding the normalization constant $C(t)$).

As before, Eq. (18.4a) is a Ricatti equation that may be solved and one finds, after some algebra, that here too $A(t)$ asymptotically approaches a constant. As before, we then take this asymptotic value to be the value of A. Setting $\frac{d}{dt}A(t) = 0$ in (18.4a):

$$A^2 = \left[\frac{m\omega}{2}\right]^2 - \frac{i}{2s^4} \text{ so, since } A^2 = A_R^2 - A_I^2 + 2iA_RA_I, \text{ then,}$$

$$A_RA_I = -\frac{1}{4s^4}, A_R^2 - A_I^2 = \left[\frac{m\omega}{2}\right]^2, \text{ so, solving for } A_R^2, \tag{18.5a}$$

$$A_R^2 = \frac{1}{2}\left(\frac{m\omega}{2}\right)^2\left[1 + \sqrt{1 + \frac{4}{(\omega\tau)^4}}\right]. \tag{18.5b}$$

(As before, $s^2 \equiv \frac{1}{\sqrt{m\lambda}}, \tau \equiv \sqrt{\frac{m}{\lambda}}$). We note that (18.5b) has correct limits: if there is no collapse, so $\tau \to \infty$, we get $A_R = \frac{m\omega}{2}$, while in the free particle limit, $\omega \to 0$, we get $A_R = \frac{1}{2s^2}$.

Eq. (18.4b) is a linear inhomogeneous equation for a complex variable, equivalent to two real equations,

$$\frac{d}{dt}B_R(t) = \frac{2}{m}[A_IB_R(t) + A_RB_I(t)] + \frac{\lambda}{A_R}B_R(t) + \dot{b}(t)$$

$$= -\frac{2}{m}A_IB_R(t) + \frac{2}{m}A_RB_I(t) + \dot{b}(t), \tag{18.6a}$$

$$\frac{d}{dt}B_I(t) = -\frac{2}{m}A_RB_R(t) + \frac{2}{m}A_IB_I(t) \tag{18.6b}$$

(where in the second step of (18.6a) we have used the expression for A_RA_I in (18.5a)) and therefore readily solved. Looking first at the homogeneous solution (setting $\dot{b}(t) = 0$ in Eq. (18.6a), we make the Ansatz $B_R(t) = a\cos\Omega t + b\sin\Omega t$, $B_R(t) = a'\cos\Omega t + b'\sin\Omega t$. We then get four homogeneous equations for a, b, a', b'. Eliminating all these variables gives as condition for solution $[\Omega^2 - (\frac{2}{m})^2(A_R^2 - A_I^2)]^2 = 0$, so

$$\Omega^2 = \left[\frac{2}{m}A_R\right]^2 - \left[\frac{2}{m}A_I\right]^2 = \omega^2 \tag{18.7}$$

where we have used the second equation in (18.5a).

We therefore write the solution of the inhomogeneous equation as

$$B(t) = b(t) + \alpha\int_0^t dt'b(t')\cos\omega(t - t') + \beta\int_0^t dt'b(t')\sin\omega(t - t'), \tag{18.8}$$

and determine the complex constants α, β by putting (18.8) into Eqs. (18.6a), (18.6b):

$$\dot{b}(t) + \alpha_R \left[b(t) - \omega \int_0^t dt' b(t') \sin \omega(t - t') \right] + \beta_R \omega \int_0^t dt' b(t') \cos \omega(t - t')$$

$$= \dot{b}(t) - \frac{2}{m} A_I \left[b(t) + \alpha_R \int_0^t dt' b(t') \cos \omega(t - t') + \beta_R \int_0^t dt' b(t') \sin \omega(t - t') \right]$$

$$+ \frac{2}{m} A_R \left[+ \alpha_I \int_0^t dt' b(t') \cos \omega(t - t') + \beta_I \int_0^t dt' b(t') \sin \omega(t - t') \right], \qquad (18.9a)$$

$$\alpha_I \left[b(t) - \omega \int_0^t dt' b(t') \sin \omega(t - t') \right] + \beta_I \omega \int_0^t dt' b(t') \cos \omega(t - t')$$

$$= -\frac{2}{m} A_R \left[b(t) + \alpha_R \int_0^t dt' b(t') \cos \omega(t - t') + \beta_R \int_0^t dt' b(t') \sin \omega(t - t') \right]$$

$$+ \frac{2}{m} A_I \left[+ \alpha_I \int_0^t dt' b(t') \cos \omega(t - t') + \beta_I \int_0^t dt' b(t') \sin \omega(t - t') \right]. \qquad (18.9b)$$

Equating the factors multiplying $b(t)$ and the cosine and sine integrals we have respectively

$$\alpha_R = -\frac{2}{m} A_I = \frac{\lambda}{2A_R}, \alpha_I = -\frac{2}{m} A_R \text{ so}$$

$$\alpha_R \alpha_I = -\frac{1}{\tau^2}, \alpha_I^2 - \alpha_R^2 = \omega^2, \qquad (18.10a)$$

$$\beta_R \omega = -\frac{2}{m} A_I \alpha_R + \frac{2}{m} A_R \alpha_I = \alpha_R^2 - \alpha_I^2 = -\omega^2 \text{ so } \beta_R = -\omega,$$

$$\beta_I \omega = -\frac{2}{m} A_R \alpha_R + \frac{2}{m} A_I \alpha_I = 0 \text{ so } \beta_I = 0, \qquad (18.10b)$$

$$-\alpha_R \omega = -\frac{2}{m} A_I \beta_R + \frac{2}{m} A_R \beta_I \text{ or } \frac{2}{m} A_I \omega = \frac{2}{m} A_I \omega, \text{ identically true,}$$

$$-\alpha_I \omega = -\frac{2}{m} A_R \beta_R + \frac{2}{m} A_I \beta_I \text{ or } \frac{2}{m} A_R \omega = \frac{2}{m} A_R \omega, \text{ identically true.} (18.10c)$$

Thus we have completely solved the Stratonovich SDE (18.1). The solution (18.2) has A_R, A_I given by (18.5a), (18.5b), while $B(t)$, from Eq. (18.8) and Eqs. (18.10), is:

$$B(t) = b(t) + \left[\frac{\lambda}{2A_R} - i\frac{2}{m} A_R \right] \int_0^t dt' b(t') \cos \omega(t - t') - \omega \int_0^t dt' b(t') \sin \omega(t - t'). \quad (18.11)$$

In closing, we note that the mean position (18.3) and mean momentum,

$$\langle \hat{X} \rangle = \frac{B_R(t)}{2A_R}, \langle \hat{P} \rangle = B_I - \frac{A_I}{A_R} B_R, \qquad (18.12)$$

consist of the sum of Brownian motion added to an oscillating Brownian-modulated motion. We also note that (18.11) reduces to the free particle results (17.9), (17.10) in the limit $\omega \to 0$, where $A_R \to \frac{1}{2s^2}$:

$$B(t) \to b(t) + \left[\lambda s^2 - i \frac{1}{ms^2} \right] \int_0^t dt' b(t') = b(t) + \frac{1 - i}{\tau} \int_0^t dt' b(t'). \qquad (18.13)$$

Also correct is the limiting case of no collapse where $b(t) \to 0$ so $B(t) \to 0$.

Appendix A
Gaussians

Integrals of Gaussians are really all the integrals that appear in this chapter. A Gaussian is the function

$$e^{-ax^2+bx}.$$ (A.1)

The integral of (A.1) is worth memorizing, because all Gaussian integrals can be obtained from it:

$$\int_{-\infty}^{\infty} dx\, e^{-ax^2+bx} = \sqrt{\frac{\pi}{a}} e^{b^2/4a},$$ (A.2)

with a and b complex, and the real part of $a > 0$.

By changing variables in Eq. (A.2) to $a \equiv 1/2\sigma^2$, $b \equiv x_1/\sigma^2$, one obtains the integral of the normalized Gaussian:

$$\int_{-\infty}^{\infty} dx\, \frac{1}{\sqrt{2\pi\sigma^2}} e^{-\frac{1}{2\sigma^2}(x-x_1)^2} = 1.$$ (A.3)

This form is also worth memorizing because the integrand is a frequently encountered probability density. The even moments of this density are:

$$\int_{-\infty}^{\infty} dx\, \frac{1}{\sqrt{2\pi\sigma^2}} e^{-\frac{1}{2\sigma^2}(x-x_1)^2} (x-x_1)^{2m} = \frac{(2m)!}{2^m m!} \sigma^{2m}$$

$$= (2m-1)(2m-3)...1\sigma^{2m} \equiv (2m-1)!!\sigma^{2m}.$$ (A.4)

This is proved using $\int_{-\infty}^{\infty} dx\, e^{-ax^2} x^{2n} = (-1)^n \left[\frac{d}{da}\right]^n \int_{-\infty}^{\infty} dx\, e^{-ax^2} = (-1)^n \left[\frac{d}{da}\right]^n \sqrt{\frac{\pi}{a}}$.

Thus, two parameters completely characterize the normalized Gaussian (A.3): its mean $\bar{x} = x_1$ and its variance $\overline{(x-x_1)^2} = \sigma^2$ (another terminology is standard deviation $\sqrt{\overline{(x-x_1)^2}} = \sigma$).

The integral of the product of two normalized Gaussians is

$$\int_{-\infty}^{\infty} dx\, \frac{1}{\sqrt{2\pi\sigma_2^2}} e^{-\frac{1}{2\sigma_2^2}(x-x_2)^2} \frac{1}{\sqrt{2\pi\sigma_1^2}} e^{-\frac{1}{2\sigma_1^2}(x-x_1)^2} = \frac{1}{\sqrt{2\pi[\sigma_2^2+\sigma_1^2]}} e^{-\frac{1}{2[\sigma_2^2+\sigma_1^2]}(x_1-x_2)^2},$$ (A.5)

that is, the means subtract and the variances add.

Sometimes one needs the approximation:

$$\int_{L}^{\infty} dx\, \frac{1}{\sqrt{2\pi\sigma^2}} e^{-\frac{1}{2\sigma^2}x^2} \approx \frac{1}{\sqrt{2\pi\sigma^2}} \frac{\sigma^2}{L} e^{-\frac{1}{2\sigma^2}L^2}$$ (A.6)

for large L. Actually, the right side of (A.6) is an upper limit, and the lower limit has (σ^2/L) replaced by $(\sigma^2/L)[1-(\sigma/L)^2]$.

The N variable Gaussian that generalizes (A.1) may be written in matrix notation as

$$e^{-xAx+bx} \tag{A.7}$$

where $xAx \equiv \sum_{i,j=1}^{N} x_i A_{ij} x_j$, $bx \equiv \sum_{i=1}^{N} b_i x_i$, A is a complex symmetric $N \times N$ matrix whose eigenvalues have positive real parts, and b is an arbitrary complex vector. Its integral,

$$\int_{-\infty}^{\infty} Dx e^{-xAx+bx} = \sqrt{\pi^N \det A^{-1}} e^{\frac{1}{4}bA^{-1}b}, \tag{A.8}$$

($Dx \equiv dx_1...dx_N$) may be obtained by repeated use of Eq. (A.2). The integral of the normalized N variable Gaussian,

$$\int_{-\infty}^{\infty} Dx \frac{1}{\sqrt{(2\pi)^N \det A^{-1}}} e^{-\frac{1}{2}[x-x_1]A[x-x_1]} = 1 \tag{A.9}$$

follows from (A.8). By symmetry, $\overline{[x-x_1]}_i = 0$, and one can show that the covariance matrix $\overline{[x-x_1]_i[x-x_1]_j} = A_{ij}^{-1}$.

The normalized Gaussian can be written as a Fourier transform:

$$\frac{1}{\sqrt{(2\pi)^N \det A^{-1}}} e^{-\frac{1}{2}[x-x_1]A[x-x_1]} = \frac{1}{(2\pi)^N} \int_{-\infty}^{\infty} Dk e^{ik[x-x_1]} e^{-\frac{1}{2}kA^{-1}k} \tag{A.10}$$

($Dk \equiv dk_1...dk_N$). This is useful when it is easier to work with the Fourier transform's exponent linear in x than with the Gaussian's quadratic exponent.

Appendix B
Random Walk

An ordinary variable, which is a function of time, $f(t)$, takes on a single value at time t. In a probabilistic situation, a variable is not represented by one function, but rather by a set of functions, $\{f_\alpha(t)\}$, called sample functions or realizations. Each sample function is assigned a certain probability of occurrence. Thus, the variable takes on one of various possible values at time t, each with the probability assigned to its realization.

Such a variable is called a random, or stochastic (the Greek word stokhastikos may be translated as "conjectural" or "contemplative") variable. It also is called a random or stochastic *process*, because it *proceeds* in time. In this and the next two appendices, we consider three stochastic processes, which naturally lead from one to the next.

The first process is random (or drunkard's) walk, the basis of the Gambler's Ruin game, where a coin (not necessarily a fair coin!) is tossed and, depending on the result, a step of fixed length is taken to the right or to the left. We will consider a limiting case of random walk, where the number of steps is eventually taken as infinite, while the step size and time interval between steps are taken to be infinitesimal, in such a way that a finite distance is covered in a finite time.

The second process starts with the limiting case of the first. It is variously called diffusion (after a Latin word meaning "to spread out") or mathematical Brownian motion or Wiener process (named after the mathematician Norbert Wiener (1894–1964) who first described it rigorously). It is a random variable that represents a one-dimensional mathematically idealized description of the physical process of Brownian motion, the subject of one of Einstein's four 1905 *annus mirabilis* papers.

The third process, called white noise, was first named in an engineering paper[1] discussing the sensory assault in airplanes: "But inside the plane it is different; there we hear all frequencies added together at once, producing a noise which is to sound what white light is to light."

In the analogy to light, white noise $w(t)$ may be thought of as a random process with a flat frequency spectrum, and random phase. We will approach it as the velocity of Brownian motion.

B.1 Limit of Random Walk

Consider a "particle" moving continuously along the x-axis with constant speed. It starts at $x = 0$ at time $t = 0$. Every τ sec it makes a random decision. For the next τ sec, either it moves to the right with speed $\Delta/(2\tau)$, covering a distance $\Delta/2$ with

---

[1] Carson, Miles and Stevens (1943)

probability $(\frac{1}{2} + \epsilon)$ or similarly to the left with probability $(\frac{1}{2} - \epsilon)$. If the x–axis as marked off in units of $\Delta/2$, a decision to reverse motion or proceed is made whenever the particle reaches one of these markers. For example, if $\epsilon = 0$, in 2τ sec it will be at $x = \pm\Delta$ with probability $1/4$, and at $x = 0$ with probability $1/2$.

The case of interest to us will be the limit of an infinite (even) number of such decisions with $\tau, \Delta, \epsilon \to 0$, the limits to be taken in such a way that, in a finite time t, the particle will have traveled a finite distance.

At time $t = 2M\tau$, where M is an integer, the particle will have made $2M$ decisions. As a result, the particle can be at $x = n\Delta$ where n is an integer (positive or negative, $-M \le n \le M$). This occurs if the particle has made $M - n$ decisions to move left and $M + n$ decisions to move right. This can occur in $(2M)!/(M - n)!(M + n)!$ different ways (different sequences of choices) so the probability of ending up at $x = n\Delta$ is a term in the binomial probability distribution expression for the expansion of $[(\frac{1}{2} + \epsilon) + (\frac{1}{2} - \epsilon)]^{2M}$:

$$P(n) = \frac{(2M)!}{(M - n)!(M + n)!}\left[\frac{1}{2} - \epsilon\right]^{M-n}\left[\frac{1}{2} + \epsilon\right]^{M+n} \tag{B.1}$$

We are interested in the case of large M and n but small n/M, so Stirling's formula $N! = \sqrt{2\pi N}(N/e)^N[1 + O(1/N)]$ can be applied to all the factorials in (B.1):

$$P(n) \approx \frac{1}{\sqrt{2\pi(M/2)}}\left(\frac{2M}{e}\right)^{2M}\left(\frac{e}{M-n}\right)^{M-n}\left(\frac{e}{M+n}\right)^{M+n}$$
$$\cdot 2^{-2M}(1 - 2\epsilon)^{M-n}(1 + 2\epsilon)^{M+n}$$

(with the approximation $\sqrt{M \pm n} \approx \sqrt{M}$). We notice that the powers of 2 and e cancel. We write the first parenthesis $M^{2M} = M^{M-n}M^{M+n}$ and insert that into the second and third parentheses, which then becomes $(1 - \frac{n}{m})^{-(M-n)}(1 + \frac{n}{m})^{-(M+n)}$. Then, we use $x^a = e^{a\ln x}$, to obtain

$$P(n) = \frac{1}{\sqrt{2\pi(M/2)}}e^{-(M-n)[\ln(1-\frac{n}{M})-\ln(1-2\epsilon)]}e^{-(M+n)[\ln(1+\frac{n}{M})-\ln(1+2\epsilon)]}.$$

Next, we expand the logarithms, using $\ln(1 - x) = x + \frac{1}{2}x^2 + ...$, dropping higher order terms as these will vanish in the limit taken later, resulting in

$$P(n) \approx \frac{1}{\sqrt{2\pi(M/2)}}e^{-(M-n)[-\frac{n}{M}-\frac{1}{2}(\frac{n}{M})^2+2\epsilon+2\epsilon^2]}e^{-(M+n)[\frac{n}{M}-\frac{1}{2}(\frac{n}{M})^2-2\epsilon+2\epsilon^2]}.$$

Gathering terms, we end up with

$$P(n) \approx \frac{1}{\sqrt{2\pi(M/2)}}e^{M[(\frac{n}{M})^2-4\epsilon^2]}e^{-n(2\frac{n}{M}-4\epsilon)}$$

$$= \frac{1}{\sqrt{2\pi(M/2)}}e^{-\frac{1}{M}[n-2\epsilon M]^2}. \tag{B.2}$$

Now, go back to the original variables in terms of which the problem was couched, making replacements $n = x/\Delta$, $2M = t/\tau$. We also write $P(n) \times 1 = P(n)dn$ as equal to

$P_t(x)dx$, where $P_t(x)$ is the probability per unit length (probability density) of x being the occupied position at time t. Since $n\Delta = x$, so $dx/dn = \Delta$ and $P(n) = P_t(x)\Delta$, (B.2) yields the probability density distribution of the particle's position x at time t:

$$P_t(x) \approx \frac{1}{\sqrt{2\pi(\Delta^2/4\tau)t}}e^{-\frac{1}{2(\Delta^2/4\tau)t}[x-(\epsilon\Delta/\tau)t]^2}. \tag{B.3}$$

Finally, we take the limit $\tau, \Delta, \epsilon \to 0$ in such a way that the two constants $\Delta^2/4\tau \equiv 2D$ and $\epsilon\Delta/\tau \to u$ are finite so

$$P_t(x) = \frac{1}{\sqrt{4\pi Dt}}e^{-\frac{1}{4Dt}[x-ut]^2}. \tag{B.4}$$

Since $M = t/2\tau$, this means that $M \to \infty$ as $\tau \to 0$, where Stirling's approximation, and therefore (B.4), becomes exact. This also means that the range of x becomes, in this limit, $(-M\Delta, M\Delta) = t\sqrt{2D/\tau}(-1,1) \to (-\infty, \infty)$.

Eq. (B.4) is the result we have sought. The individual trajectories are continuous but, in this limit, are everywhere non-differentiable (see Fig. 2.1 in Section 2.1 for a depiction of some trajectories). The initial probability density is $P_0(x) = \delta(x)$. Then, as time increases, the complete ensemble of possible random walks spreads out from the mean $\bar{x} = ut$, which is constant speed motion. The "drift velocity" u arises from the slightly greater probability of moving in one direction rather than the other, characterized by the small probability difference ϵ. Although in the limit $\tau \to 0$, ϵ vanishes as $\epsilon = u\tau^{1/2}/2\sqrt{2D} \to 0$, its influence remains for $u \neq 0$.

This spreading out of the collection of random walks is characterized by the variance of this distribution, $\overline{x^2} = 2Dt$, or its square root, the standard deviation $\sigma = \sqrt{2Dt}$. That is a feature worth noting: the underlying finite random walk motion in either direction is constant speed motion which covers distance $\sim t$ but, because of its back-and-forth nature, the net motion is $\sim \sqrt{t}$, which spreads out much more slowly.

D is called the "diffusion constant," and $P_t(x)$ satisfies the diffusion equation

$$\frac{\partial}{\partial t}P_t(x) = -u\frac{\partial}{\partial x}P_t(x) + D\frac{\partial^2}{\partial x^2}P_t(x), \tag{B.5}$$

(which can readily be checked by putting (B.4) into (B.5)). There are two parameters in this equation reflecting the fact that its solution is a Gaussian, which is completely characterized by its mean $\sim ut$ and variance $\sim D$.

Since the process is invariant under time translation or space translation, for a walk beginning at time t_0 at location x_0, Eq. (B.4) generalizes to

$$P_t(x) = \frac{1}{\sqrt{4\pi D(t-t_0)}}e^{-\frac{1}{4D(t-t_0)}[x-x_0-u(t-t_0)]^2} \tag{B.6}$$

Appendix C
Brownian Motion/Wiener Process

The Wiener process utilizes the Gaussian probability density distribution (B.4) with $u = 0$ (that is, $\epsilon = 0$: the underlying random walk is not biased toward either direction). Most generally, the Wiener process over the time interval (t, t_0) is characterized by two conditions.

The first is a Gaussian conditional probability density distribution of $B(t)$, given $B(t_0)$, obtained by change of notation in (B.6), with x, x_0 replaced by $B(t), B(0)$, setting $D = \sigma^2$ and $u = 0$:

$$P(B(t)|B(t_0)) = \frac{1}{\sqrt{2\pi\sigma^2(t - t_0)}} e^{-\frac{1}{2\sigma^2(t - t_0)}[B(t) - B(t_0)]^2}. \tag{C.1}$$

The second condition is that the process over one time interval is statistically independent of the process over another time interval, if the two time intervals do not overlap.

In particular, this means that the Wiener process is *Markovian*. Named after the Russian mathematician Andrei Markov (1856–1922),[1] a Markov process is one in which the probability of the future value of $B(t_0)$ depends only upon $B(t_0)$, not upon earlier values ($B(t)$ for $t < t_0$). Because of this, the joint probability density that, given $B(t_0)$, the process takes on the successive values $B(t_1)$, $B(t_2)$ ... $B(t_N)$ at respective times $t_1 < t_2 < ... < t_N$ is just the product of the successive conditional probability densities:

$$P(B(t_N), B(t_{N-1}), ...B(t_1)|B(t_0)) =$$

$$\prod_{n=1}^{N} \frac{1}{\sqrt{2\pi\sigma^2(t_n - t_{n-1})}} e^{-\frac{1}{2\sigma^2(t_n - t_{n-1})}[B(t_n) - B(t_{n-1})]^2}, \tag{C.2}$$

Using the Gaussian integral formula (A.5), integration of (C.2) over the variables $B(t_1)$, $B(t_2)$ ... $B(t_{N-1})$ results in Eq. (C.1), with t_N replacing t, that is, just the Brownian motion probability density over the complete interval.

It follows from (C.1) that the mean and variance are

$$\overline{B(t) - B(t_0)} = 0, \tag{C.3a}$$
$$\overline{[B(t) - B(t_0)]^2} = \sigma^2(t - t_0). \tag{C.3b}$$

---

[1] Actually, Markov himself discussed random variables which take on values at discrete instants of time (Feller, 1950), not the continuous-time process associated with his name.

To obtain the *correlation function*, because of the statistical independence property cited above, one does not need to do any more integrations: just use (C.3a), (C.3b), as follows. When there is no overlap between the time intervals $t_3 - t_2, t_1 - t_0$, with $t_3 \geq t_2 \geq t_1 \geq t_0$,

$$\overline{[B(t_3) - B(t_2)][B(t_1) - B(t_0)]} = \overline{[B(t_3) - B(t_2)]} \times \overline{[B(t_1) - B(t_0)]} = 0, \quad (C.4)$$

With overlap, write the difference in B as the sum of non–overlapping and overlapping parts, for example, with $t \geq t' \geq t_0$,

$$\overline{[B(t) - B(t_0)][B(t') - B(t_0)]} = \overline{[B(t) - B(t') + B(t') - B(t_0)][B(t') - B(t_0)]},$$
$$= \overline{[B(t) - B(t')][B(t') - B(t_0)]} + \overline{[B(t') - B(t_0)]^2}$$
$$= \sigma^2 [t' - t_0] \qquad (C.5)$$

since the non–overlap part gives 0 (Eq. (C.4)), while the overlap part is the variance (Eq. (C.3b)).

Appendix D
White Noise

That white noise is annoying needs little argument. No one has been found who really enjoys it.[1]

In view of the ubiquity of white noise in CSL, I get a chuckle from this quote!

Consider a Brownian-moving particle, a continuous but non–differentiable one-dimensional trajectory $x(t)$. The usual velocity of a particle is the time derivative of its position, but Brownian motion is non–differentiable. So, one may define a different kind of velocity, a velocity at time t based upon a fixed interval Δt (during which, in the random walk model, there are in fact many steps):

$$w_{\Delta t}(t) \equiv \frac{B(t) - B(t - \Delta t)}{\Delta t}. \tag{D.1}$$

We want an expression for the probability of $w_{\Delta t}(t)$. We substitute Eq. (D.1) into Eq. (C.1) (with t_0 replaced by $t - \Delta t$):[A]

$$P(w_{\Delta t}(t)) = \sqrt{\frac{\Delta t}{2\pi\sigma^2}} e^{-\frac{1}{2\sigma^2}\Delta t w_{\Delta t}^2(t)}, \tag{D.2}$$

satisfying $\int_{-\infty}^{\infty} dw_{\Delta t}(t) P(w_{\Delta t}(t)) = 1$. The mean and variance follow from (D.2): $\overline{w}_{\Delta t}(t) = 0$, $\overline{w}_{\Delta t}^2(t) = \sigma^2/\Delta t$.

The variance is very large, since $\Delta t << \sigma^2$. Therefore, $w_{\Delta t}(t)$ takes on values in the range $\pm \sigma/\sqrt{\Delta t}$ with roughly equal probability. One can think of a sample function of $w_{\Delta t}(t)$ as oscillating wildly up and down, taking on any value in this range (see Fig. 3.1). Now, we intend to let Δt become infinitesimal, $\Delta t \to dt$, which means the variance $\to \infty$. What does that mean?

To understand it, consider the correlation function of the Brownian particle's velocity at two different times. This can be found using (C.4) and the method of (C.5):

$$\overline{w_{\Delta t}(t + \tau)w_{\Delta t}(t)} = \frac{1}{(\Delta t)^2}\overline{[B(t + \tau) - B(t + \tau - \Delta t)][B(t) - B(t - \Delta t)]}$$

$$= 0 \text{ if } \tau > \Delta t \text{ or } \tau < -\Delta t,$$

$$= \overline{[B(t) - B(t + \tau - \Delta t)]^2} = \frac{\sigma^2}{(\Delta t)^2}[\Delta t - \tau] \text{ for } \Delta t > \tau > 0,$$

$$= \overline{[B(t - |\tau|) - B(t - \Delta t)]^2} = \frac{\sigma^2}{(\Delta t)^2}[\Delta t - |\tau|] \text{ for } \Delta t > -\tau > 0. \tag{D.3}$$

---

[1] Carson, Miles and Stevens (1943).

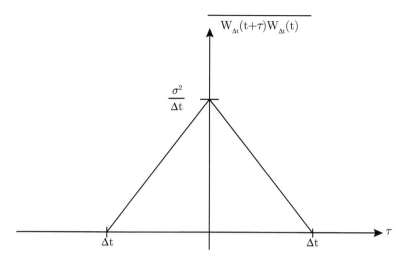

Fig. D.1 The ensemble average of the product $w_{\Delta t}(t+\tau)w_{\Delta t}(t)$. The area under the triangle is $\frac{1}{2} \times 2\Delta t \times \sigma^2/\Delta t = \sigma^2$.

This function is graphed in Fig. D.1. In the limit $\Delta t \to 0$, the width shrinks to 0 while the height grows to ∞, but the area under the triangle remains constant at σ^2. Therefore, in the limit, this is a representation of a delta function: the correlation function approaches $\sigma^2\delta(\tau)$. Writing $\lim_{\Delta t \to 0} w_{\Delta t}(t) \to w(t)$, we have

$$\overline{w(t')w(t)} = \sigma^2\delta(t-t').\tag{D.4}$$

Another approach to white noise is simply to define it as a Gaussian stochastic "function" $w(t)$ satisfying (D.4) and $\overline{w}(t) = 0$.

To see what is the probability density distribution for white noise, consider the sequence of statistically independent functions, $w_{\Delta t}(t_k) = \frac{1}{\Delta t}[B(t_k) - B(t_k - \Delta t)]$ where $t_{k+1} = t_k + \Delta t$. Their joint probability density follows from (C.2) and the Jacobian $\prod_k dB(t_k) = \prod_k dw_{\Delta t}(t_k)\Delta t$:

$$P(w_{\Delta t}(t_{N-1})...w_{\Delta t}(t_0)) = \frac{1}{\sqrt{[2\pi\sigma^2/\Delta t]^N}}e^{-\frac{\Delta t}{2\sigma^2}[w^2_{\Delta t}(t_{N-1})+...+w^2_{\Delta t}(t_0)]}.\tag{D.5}$$

Eq. (D.5) invites one to use the notation

$$P(w(t)) = Ce^{-\frac{1}{2\sigma^2}\int_0^T dt w^2(t)}\tag{D.6}$$

This often provides simplifications in manipulating formal expressions, but one must remember that (D.6) really means (D.5), with the probability density integral satisfying $\int_{-\infty}^{\infty} dw_{\Delta t}(t_{N-1})... \int_{-\infty}^{\infty} dw_{\Delta t}(t_0)P(w(t)) = 1$.

D.1 Frequency Spectrum of the White Noise Function

The results of this section is not be used in this book. It is here in order to explain the sobriquet "white noise."

As with any function of time, the Fourier transform of any sample function $w(t)$ is its frequency spectrum $a(\omega)$, and vice versa:

$$w(t) = \frac{1}{\sqrt{2\pi}} \int_{-\infty}^{\infty} d\omega e^{i\omega t} a(\omega). \tag{D.7}$$

Putting Parseval's theorem ($\int_{-\infty}^{\infty} dt w^2(t) = \int_{-\infty}^{\infty} d\omega |a(\omega)|^2$, which follows from (D.7)) into the probability density expression (D.6) for $w(t)$ (applied for the infinite time interval), we get the probability density associated to the spectrum:

$$P(a(\omega)) = C'e^{-\frac{1}{2\sigma^2}\int_{-\infty}^{\infty} d\omega |a(\omega)|^2} = C'e^{-\frac{1}{\sigma^2}\int_{0}^{\infty} d\omega |a(\omega)|^2} = C'e^{-\frac{1}{\sigma^2}\int_{0}^{\infty} d\omega [a_R^2(\omega)+a_I^2(\omega)]} \tag{D.8}$$

(the second step occurring because reality of $w(t)$ implies $a^*(\omega) = a(-\omega)$, so $|a(\omega)|^2 = |a(-\omega)|^2$).

Eq. (D.8) is quite analogous in form to Eq. (D.6).[B] Just as (D.6) says that the probabilities are the same for $w(t)$ at any time t to take on any numerical value, so the probability of each $a_R(\omega), a_I(\omega)$ is identical for each frequency, according to (D.8):

$$P(a(\omega)) da_R(\omega) da_I(\omega) = \frac{1}{\sqrt{\pi\sigma^2/d\omega}} e^{-\frac{1}{\sigma^2} d\omega |a(\omega)|^2} da_R(\omega) da_I(\omega.) \tag{D.9}$$

As white light may be thought of as containing equal amounts of all colors, so white noise has equal amounts of all frequencies.

Appendix E
White Noise Field

For more than one white noise function $w_i(t)$, all statistically independent, the probability density is just the product of the independent probability densities (D.6),

$$P(w_i(t)) = Ce^{-\frac{1}{2\sigma^2}\int_0^T dt \sum_i w_i^2(t)}, \text{ that is,}$$

$$P(w_i(t))Dw = \prod_{t=0,}^{T} \prod_i \frac{dw_i(t)}{\sqrt{2\pi\sigma^2/dt}} e^{-\frac{1}{2\sigma^2}dtw_i^2(t)} \tag{E.1}$$

The generalization of (D.4) to many indices is

$$\overline{w_i(t)w_j(t')} = \sigma^2\delta_{ij}\delta(t-t'). \tag{E.2}$$

Thus, if $i \neq j$ then $\overline{w_i(t)w_j(t')} = \overline{w_i(t)} \cdot \overline{w_j(t')} = 0$.

Now, we divide space into small volumes $d\mathbf{x}$, labeling them by the index i. We make the replacement $w_i(t) \to \sqrt{d\mathbf{x}}w(\mathbf{x},t)$. Eq. (E.1) becomes

$$P(w(\mathbf{x},t)) = Ce^{-\frac{1}{2\sigma^2}\int_0^T dt \int d\mathbf{x}w^2(\mathbf{x},t)}, \text{that is,}$$

$$P(w(\mathbf{x},t))Dw = \prod_{t=0}^{T} \prod_{\mathbf{x}} \frac{dw(\mathbf{x},t)}{\sqrt{2\pi\sigma^2/dtd\mathbf{x}}} e^{-\frac{1}{2\sigma^2}dtd\mathbf{x}w^2(\mathbf{x},t)}. \tag{E.3}$$

It follows from (E.3) that $\overline{w(\mathbf{x},t)} = 0$ and Eq. (E.2) becomes

$$\overline{w(\mathbf{x},t)w(\mathbf{x}',t')} = \frac{\sigma^2}{\Delta x}\delta_{ij}\delta(t-t') \to \sigma^2\delta(\mathbf{x}-\mathbf{x}')\delta(t-t'). \tag{E.4}$$

The white noise field w($\mathbf{x}$,t) described above has 0 mean. A white noise field with non–zero mean has probability density distribution

$$P(w(\mathbf{x},t)) = Ce^{-\frac{1}{2\sigma^2}\int_0^T dt \int d\mathbf{x}[w(\mathbf{x},t)-A(\mathbf{x},t)]^2}. \tag{E.5}$$

CSL uses such distributions.

Appendix F
Density Matrix

In quantum theory, a physical state of Nature at time t is represented by a state vector, say $|\psi, t\rangle$. However, it also could be represented by the projection operator on the state vector,

$$\hat{\rho}(t) \equiv |\psi, t\rangle\langle\psi, t|, \tag{F.1}$$

since, obviously, the state vector can be recovered from the projection operator. This projection operator is called a pure density matrix. This is a very special case of a density matrix.

The density matrix is most useful as a tool for discussing an ensemble (collection) of state vectors. Suppose you wish to consider such a collection, $|\psi_\alpha, t\rangle$, each with known probability p_α. (These (unit normalized) state vectors do *not* have to be orthogonal.) It is then useful to construct the density matrix

$$\hat{\rho}(t) \equiv \Sigma_\alpha p_\alpha |\psi_\alpha, t\rangle\langle\psi_\alpha, t| \equiv \overline{|\psi, t\rangle\langle\psi, t|}, \tag{F.2}$$

where the last form in (F.2) denotes the ensemble average of the state vectors. (F.2) is called a "mixed" density matrix: often it is called a "mixture." The state vectors have been "mixed up," in the sense that it is impossible to recover the original items from the mixture since there are many different ways (different sets of state vectors and different probabilities) the same density matrix could have been constructed.

F.1 The Trace Operation

One of the things most useful about the density matrix is that it is readily used to calculate, for the collection of state vectors that comprise it, the expectation value of any operator.

The expectation value of an operator $\hat{A}$ for a physical system described by the state vector $|\psi, t\rangle$ is $\langle\hat{A}\rangle \equiv \langle\psi, t|\hat{A}|\psi, t\rangle$.

This can be written as the trace of the product of the pure density matrix (G.1) and $\hat{A}$.

Recall that the trace of a matrix is the sum of its diagonal elements, so

$$\text{Tr}\hat{A}\hat{\rho}(t) = \Sigma_i \langle i|\hat{A}\hat{\rho}|i\rangle = \Sigma_i \langle i|\hat{A}|\psi, t\rangle\langle\psi, t|i\rangle = \Sigma_i \langle\psi, t|i\rangle\langle i|\hat{A}|\psi, t\rangle$$
$$= \langle\psi, t|\hat{A}|\psi, t\rangle, \tag{F.3}$$

where the resolution of the identity operator, $\hat{1} = \Sigma_i |i\rangle\langle i|$, has been used in the last step.

For a mixture, it is easy to see, following the same steps, that one gets $\mathrm{Tr}A\rho(t) = \Sigma_j p_j \langle \psi_j, t | \hat{A} | \psi_j, t \rangle$, the appropriately weighted sum of expectation values for the ensemble of state vectors.

A useful property of the trace of a product of operators is that the operators can be cyclically permuted:

$$\mathrm{Tr}\hat{A}\hat{B} = \Sigma_i \langle i | \hat{A}\hat{B} | i \rangle = \Sigma_i \langle i | \hat{A}\hat{1}\hat{B} | i \rangle = \Sigma_{i,j} \langle i | \hat{A} | j \rangle \langle j | \hat{B} | i \rangle = \Sigma_{i,j} \langle j | \hat{B} | i \rangle \langle i | \hat{A} | j \rangle$$
$$= \Sigma_j \langle j | \hat{B}\hat{A} | j \rangle = \mathrm{Tr}\hat{B}\hat{A}. \tag{F.4}$$

F.2 Properties of the Density Matrix

The density matrix possesses three properties, obtainable from (F.2). They are:

1) The trace of $\hat{\rho}(t)$ is 1:

$$\mathrm{Tr}\hat{\rho}(t) = \Sigma_\alpha p_\alpha \mathrm{Tr}|\psi_\alpha, t\rangle\langle\psi_\alpha, t| = \Sigma_{i,\alpha} p_\alpha \langle i|\psi_\alpha, t\rangle\langle\psi_\alpha, t|i\rangle = \Sigma_\alpha p_\alpha \langle\psi_\alpha, t|\psi_\alpha, t\rangle = \Sigma_\alpha p_\alpha = 1$$

2) $\hat{\rho}(t)$ is Hermitian, $\hat{\rho}^\dagger(t) = \hat{\rho}(t)$;

$$\hat{\rho}^\dagger(t) = \Sigma_\alpha p_\alpha^* [|\psi_\alpha, t\rangle\langle\psi_\alpha, t|]^\dagger = \Sigma_\alpha p_\alpha |\psi_\alpha, t\rangle\langle\psi_\alpha, t| = \hat{\rho}(t).$$

3) $\hat{\rho}(t)$ is positive semi-definite, that is, for any vector $|\phi\rangle$, $\langle\phi|\hat{\rho}(t)|\phi\rangle \geq 0$;

$$\langle\phi|\hat{\rho}(t)|\phi\rangle = \Sigma_\alpha p_\alpha \langle\phi|\psi_\alpha, t\rangle\langle\psi_\alpha, t|\phi\rangle = \Sigma_\alpha p_\alpha |\langle\phi|\psi_\alpha, t\rangle|^2 \geq 0.$$

This only vanishes if $|\phi\rangle$ is orthogonal to all of the states $|\psi_\alpha, t\rangle$ (which therefore do not span the vector space).

Any operator that possesses these three properties is called a density operator or density matrix, regardless of whether or not it is obtained from an expression of the form (F.2).

F.3 Partial Trace

Suppose the physical system under consideration is described by a direct product of two Hilbert spaces $H_a \times H_b$. For example, some particles may be in H_a and the rest in H_b. Or, H_a may describe the center of mass of a collection of particles, and H_b describe the relative coordinates of the particles. Suppose a complete basis for H_a is $|a_i\rangle$ and for H_b it is $|b_j\rangle$. If the state vector at time t is $|\psi, t\rangle$, so the density matrix describing it is a pure density matrix, the probability of the system being in the joint state $|a_i\rangle|b_j\rangle$ is $|\langle a_i|\langle b_j|\psi, t\rangle|^2$.

Often, one is interested in the probability $P(a_i)$ of the system being in the state $|a_i\rangle$ *regardless of the state of the other system*. For a pure density matrix that may be written as

$$P(a_i) \equiv \sum_j |\langle a_i|\langle b_j|\psi, t\rangle|^2 = \sum_j \langle a_i|\langle b_j|\psi, t\rangle\langle\psi, t|b_j\rangle|a_i\rangle$$
$$= \langle a_i| \Big[\sum_j \langle b_j|\psi, t\rangle\langle\psi, t||b_j\rangle\Big] |a_i\rangle \equiv \langle a_i| \Big[\mathrm{Tr}_b|\psi, t\rangle\langle\psi, t|\Big] |a_i\rangle. \tag{F.5}$$

For a general (non–pure) density matrix $\rho(t)$, the partial trace is defined similarly, $\mathrm{Tr}_b \rho(t) \equiv \Sigma_j \langle b_j|\rho(t)|b_j\rangle$. And similarly, since the density matrix is just the sum of possible state vectors at time t multiplied by their associated probabilities, the probability of the system being in the state $|a_i\rangle$ regardless of the state of the other system is the diagonal matrix element $\langle a_i|\text{Tr}_b\rho(t)|a_i\rangle$.

F.4   The Lindblad Equation

In standard quantum theory, each state vector $|\psi_\alpha, t\rangle$ in an ensemble of state vectors obeys the Schrödinger equation $\frac{d}{dt}|\psi_\alpha, t\rangle = -i\hat{H}|\psi_\alpha, t\rangle$, so the associated density matrix obeys the evolution equation

$$\frac{d}{dt}\hat{\rho}(t) = \Sigma_\alpha p_\alpha\left[\left(\frac{d}{dt}|\psi_\alpha, t\rangle\right)\langle\psi_\alpha, t| + |\psi_\alpha, t\rangle\left(\frac{d}{dt}\langle\psi_\alpha, t|\right)\right]$$

$$= \Sigma_\alpha p_\alpha\left[-i\hat{H}|\psi_\alpha, t\rangle\langle\psi_\alpha, t| + |\psi_\alpha, t\rangle\langle\psi_\alpha, t|i\hat{H}t\right]$$

$$= -i[\hat{H}, \hat{\rho}], \tag{F.6}$$

where $[\hat{A}, \hat{B}]$ is the commutator of the operators $\hat{A}, \hat{B}$.

A broader class of state vector evolutions, such as might occur for a physical system that evolves under an ensemble of Hamiltonians whose occurrence is statistically known, or which occurs in an alteration of quantum theory such as CSL, can have its density matrix evolution described by a broader class of equations than (F.6).

Remarkably, one can derive a general form for such an evolution equation subject to five constraints, the three already mentioned (unit trace, Hermitian, positive semi-definite), and two more.

One of these additional constraints is that future values of $\hat{\rho}(t)$ only depend upon it, not upon the value of $\hat{\rho}(t)$ at an earlier time. This is called the Markovian constraint, discussed in Appendix C as a property of Brownian motion. In this case, it means that $\frac{d}{dt}\hat{\rho}(t)$ is equal to an expression that only depends upon $\hat{\rho}(t)$.

The second additional constraint is that this dependence upon $\hat{\rho}(t)$ is linear.

It can be shown that these five constraints limit the evolution of the density matrix in an N-dimensional Hilbert space to the form

$$\frac{d}{dt}\hat{\rho}(t) = -i[\hat{H}, \hat{\rho}] - \sum_{\alpha=1}^{N^2-1}[\hat{L}_\alpha^\dagger\hat{L}_\alpha\hat{\rho}(t) + \hat{\rho}(t)\hat{L}_\alpha^\dagger\hat{L}_\alpha - 2L_\alpha\hat{\rho}(t)\hat{L}_\alpha^\dagger] \tag{F.7}$$

where the operators L_α are completely arbitrary, they do not even have to be Hermitian. (One can have more such operators than $N^2 - 1$, but one can show that is equivalent to $N^2 - 1$ different operators).

Eq. (F.7) is called the Lindblad equation, or the Lindblad-Gorini-Kossakowski-Sudarshan equation after the authors who proved this.[1]

One can immediately see from (F.7) that its Hermitian conjugate is also (F.7), and that the trace of both sides vanishes. The proof that $\langle\phi|\hat{\rho}(t)|\phi\rangle \geq 0$ is too complicated to be treated here.

As an illustration of how Eq. (F.7) works, consider the case of just one Lindblad operator, the Hermitian operator $\sqrt{\frac{\lambda}{2}}\hat{A}$, and $\hat{H} = 0$:

[1]Lindblad (1976), Gorini, Kossakowski and Sudarshan (1976), Pearle (2012).C

$$\frac{d}{dt}\hat{\rho}(t) = -\frac{\lambda}{2}[\hat{A}^2\hat{\rho}(t) + \hat{\rho}(t)\hat{A}^2 - 2\hat{A}\hat{\rho}(t)\hat{A}] = -\frac{\lambda}{2}[\hat{A},[\hat{A},\hat{\rho}(t)]]. \qquad \text{(F.8)}$$

Taking the matrix element $\langle a_i|...|a_j\rangle$ of this equation we obtain:

$$\frac{d}{dt}\langle a_i|\hat{\rho}(t)|a_j\rangle = -\frac{\lambda}{2}[a_i^2\langle a_i|\hat{\rho}(t)|a_j\rangle + \langle a_i|\hat{\rho}(t)|a_j\rangle a_j^2 - 2a_i\langle a_i|\hat{\rho}(t)|a_j\rangle a_j] \quad \text{(F.9a)}$$

so

$$\langle a_i|\hat{\rho}(t)|a_j\rangle = \langle a_i|\hat{\rho}(0)|a_j\rangle e^{-\frac{\lambda}{2}t(a_i-a_j)^2}. \qquad \text{(F.9b)}$$

In this evolution of the ensemble of state vectors, the diagonal matrix element, the probability that the state $|a_i\rangle$ is occupied, is constant: $\langle a_i|\hat{\rho}(t)|a_i\rangle = \langle a_i|\hat{\rho}(0)|a_i\rangle$. (Of course, this ensures that the total probability of all states being occupied, $\mathrm{Tr}\hat{\rho}(t)$, maintains the initial value 1.) The off-diagonal elements decay to 0 (assuming no two eigenvalues have the same value).

This density matrix evolution equation was obtained in Eq.(2.18): it describes collapse dynamics in CSL, where $\hat{A}$ is the collapse-generating operator.

Appendix G
Theoretical Constraint Calculations

Here we give the calculations behind Fig. 6.1, completing the discussion in Section 6.1.

We consider that the states $|L\rangle, |R\rangle$ describe not only the sphere to the left and right, but also the light source illuminating the sphere, as well as the light bouncing off, heading among other places toward the eye.[1]

For simplicity, we suppose that there are precisely N photons reflecting each second off the sphere and coming toward the eye. That is, there is an operator $\hat{N}_L$ representing the photons/s coming toward the eye from the left sphere that satisfies $\hat{N}_L|L\rangle = N|L\rangle$, $\hat{N}_L|R\rangle = 0$ and an operator $\hat{N}_R$ similarly representing the light coming toward the eye from the right sphere that satisfies $\hat{N}_R|R\rangle = N|R\rangle$, $\hat{N}_R|L\rangle = 0$.

Then, for a state vector $|\psi, t\rangle = \sqrt{1 - \epsilon^2}|L\rangle + \epsilon|R\rangle$, the observable number of photons coming toward the eye from the left sphere is $\langle\psi, t|\hat{N}_L|\psi, t\rangle \approx N$, and the observable number of photons coming from the R region is $\langle\psi, t|\hat{N}_R|\psi, t\rangle = \epsilon^2 N$. (Of course, a similar statement holds for the state vector $|\psi, t\rangle = \sqrt{1 - \epsilon^2}|R\rangle + \epsilon|L\rangle$.)

To get a rough handle on N, we suppose the illumination is such that the tiny sphere looks like Polaris, one of the brightest stars in the sky (magnitude ≈ 2) which, it has been estimated,[2] delivers $N \approx 10^5$ photons/s to the eye. Therefore, the collapsed sphere delivers $\epsilon^2 10^5$ photons/s to the eye.

Now, while the *retina* responds to a single photon, it has been estimated that the eye needs to perceive at least nine photons in 0.1s (human perception time) to register a *conscious* perception.

If the collapsed sphere delivers less than 90 photons/s, it delivers less than 9 photons in 0.1s. If there are more than 90 photons/s, two spheres could be seen, which the theory can't allow. This disallowed situation obtains if $\epsilon^2 10^5 \geq 90$ or $\epsilon \geq 0.03$.

In Section 6.1, we showed that

$$\bar{\epsilon} \approx \frac{1}{2}e^{-\Gamma t}, \tag{G.1}$$

where expressions for the the collapse rate Γ were obtained in Section 4.4.2 for a superposition of two images for various object sizes and image separations. Putting in this value for ϵ at time $t = 0.1$s, we get the condition that $\Gamma \lesssim 30$ is not permitted.

---

[1] We emphasize, that we are not putting the eye and the rest of a human visual apparatus into the state vector. We wish to adopt, for this example, the most stringent criterion, that the world should be described as what we would see even when no one is there to see it: see endnote A of Chapter 6. Of course, the relevant human structures could be considered if one wished. If that was done, the collapse rate would be much faster than calculated here once perception began.

[2] Yu (2016).

Snce Γ depends upon λ, a, this inequality provides a condition these parameters must satisfy for the theory to predict two spheres are visible..

To enable numerical estimates, we consider the just-visible sphere of diameter 4×10^{-5}cm (the wavelength of blue light) to be made of carbon, of density $D \approx 2$ gm/cc. Then, its volume is $V = \frac{4\pi}{3}(2 \times 10^{-5})^3 \approx 3 \cdot 10^{-14}$ cc. Since the mass of a nucleon is $M \approx \frac{5}{3} \cdot 10^{-24}$ gm, then the number of nucleons in the sphere is $N = DV/M \approx 4 \cdot 10^{10}$ and so $N/V \approx 10^{24}$ nucleons/cc.

As stated in Section 6.1, we consider three different configurations:

1) a is much less than the sphere diameter.

2) The sphere diameter is smaller than a which is smaller than the sphere separation ℓ.

3) The sphere diameter is smaller than the sphere separation ℓ which is smaller than a.

We calculate:

1. With $a \leq 4 \times 10^{-6}$cm, less than 0.1 the sphere diameter, Eq. (4.12) is applicable and gives the rate

$$\Gamma = \lambda N (4\pi)^{3/2} \frac{Na^3}{V}. \tag{G.2}$$

so the constraint is

$$\lambda \times 4 \cdot 10^{10} \times (4\pi)^{3/2} \times 10^{24} \text{cc}^{-1} a^3 \lesssim 30 \text{ or } \lambda a^3 \lesssim 2 \times 10^{-35}. \tag{G.3}$$

The boundary is shown in Fig. 6.1 as the left-most solid line of negative slope.

For the two other situations, with sphere radius $R \lesssim a$, Eq. (4.11) is applicable:

$$\Gamma = \lambda N^2 \left[1 - e^{-\frac{\ell^2}{4a^2}} \right]. \tag{G.4}$$

2. With $\ell >> a$, so the exponential in (G.4) may be neglected), $\Gamma = \lambda N^2$ so the constraint is

$$\lambda [4 \times 10^{10}]^2 \lesssim 30 \text{ or } \lambda \lesssim 2 \times 10^{-20}. \tag{G.5}$$

The boundary of this constraint is the horizontal line in Fig. 6.1.

3. Suppose that the spheres are in contact so $4 \times 10^{-5} = \ell << a$. We can expand the exponential in (G.4), obtaining $\Gamma = \lambda N^2 \frac{\ell^2}{4a^2}$. so the constraint is

$$\lambda [4 \times 10^{10}]^2 \frac{[4 \times 10^{-5}]^2}{4a^2} \lesssim 30 \text{ or } \frac{\lambda}{a^2} \lesssim 5 \times 10^{-11}. \tag{G.6}$$

The boundary is shown in Fig.6.1 as the rightmost solid line, of positive slope.

Notes

[A] Since, from (D.1), $dB(t) = \Delta t dw_{\Delta t}(t)$, we have
$$P(B(t)|B(t-\Delta t))dB(t) = P(B(t)|B(t-\Delta t))\Delta t dw_{\Delta t}(t)$$
$$\equiv P(w_{\Delta t}(t))dw_{\Delta t}(t).$$

[B] The means are $\overline{a_R(\omega)} = 0, \overline{a_I(\omega)} = 0$, the covariance is most easily found using the white noise covariance:

$$\overline{a^*(\omega)a(\omega')} = \frac{1}{2\pi} \int_{-\infty}^{\infty} dt dt' e^{i\omega t} e^{-i\omega' t'} \overline{w(t)w(t')} = \frac{1}{2\pi} \int_{-\infty}^{\infty} dt dt' e^{i\omega t} e^{-i\omega' t'} \sigma^2 \delta(t-t') = \sigma^2 \delta(\omega-\omega').$$

In the same way it can be shown that $\overline{a_R(\omega)a_R(\omega')} = \overline{a_I(\omega)a_I(\omega')} = \frac{\sigma^2}{2}\delta(\omega-\omega')$, $\overline{a_R(\omega)a_I(\omega')} = 0$.

[C]The derivations of the Lindblad-Gorini-Kossakowski-Sudarshan equation by the authors involve rather sophisticated mathematics. I wrote a pedagogical paper (Pearle (2012)) which derives it using only something well known from quantum mechanics, the expression of a Hermitian matrix in terms of its orthonormal eigenvectors and real eigenvalues.

References

Aalseth, C. E. et. al. (2002), "IGEX 76Ge neutrinoless double-beta decay experiment: Prospects for next generation experiments," *Phys. Rev. D* **65**, 092007.

Adler, S. L. (2005), "Stochastic collapse and decoherence of a non-dissipative forced harmonic oscillator," *J. Phys. A* **38**, 2729.

Adler, S. L. (2007), "Lower and upper bounds on CSL parameters from latent image formation and IGM heating," *J. Phys. A* **40**, 2935.

Adler, S. L. and Ramazanoglu, F. M. (2007), "Photon emission rate from atomic systems in the CSL model," *J. Phys. A* **40**, 13395.

Adler, S. L. (2016), "Gravitation and the noise needed in objective reduction models," in *Quantum Nonlocality and Reality, 50 years of Bell's Theorem*, M. Bell and S. Gao (eds), Cambridge: Cambridge University Press, p.390.

Adler, S. L. (2017), "Minimum interior temperature for solid objects implied by collapse models," arXiv:1712.01071.

Adler, S. L. (2018), "Heating through phonon excitation implied by collapse models," ArXiv: 1801.00509.

Adler S. L. and Vinante, A. (2018), "Bulk heating effects as tests for collapse models," *Phys. Rev. A* **97**, 052119.

Aicardi, F., Borsellino, A., Ghirardi, G. C., and Grassi, R. (1991), "Dynamical models for state-vector reduction: Do they ensure that measurements have outcomes?," *Found. Phys. Lett.* **4**, 109.

Albert, D. Z. and Vaidman, L. (1989), "On a proposed postulate of state-reduction," *Phys. Lett. A* **139**, 1.

Arnquist, I. J. et al (Majorana Collaboration) (2022), "Search for spontaneous radiation from wavefunction collapse at the Majorana Demonstrator," *Phys. Rev. Lett* **129**, 080401.

Bahrami, M., Smirne, A. and Bassi, A. (2014), "Gravity and the collapse of the wave function: a probe into Diósi-Penrose model," *Phys.Rev. A* **90**, 062105.

Bahrami, M. (2018) "Testing collapse models by a Thermometer," *Phys. Rev. A* **97**, 052118.

Ballentine, L. E. (1970), "The statistical interpretation of quantum mechanics," *Revs. Mod. Phys.* **42**, 358.

Ballentine, L. E. (1990), "Limitations of the projection postulate," *Found. Phys.* **20**, 1329.

Bedingham, D. J. (2010), 'Relativistic state reduction dynamics', *Found. Phys.* **41**, 686.

Bedingham, D. J., Dürr, D., Ghirardi, G. C., Goldstein, S., Tumulka, R. and Zanghi, J. (2014), "Matter density and relativistic models of wave function collapse," *Stat. Phys.* **154**, 623.

Bedingham, D. J. and Pearle, P. (2019), "On the CSL scalar field relativistic collapse model," *Phys. Rev. Res.* **1**, 033040.

Bassi, A. (2005), "Collapse models: analysis of the free particle dynamics," *J. Phys. A: Math. Gen.* **38**, 3173.

Bell, J. S. (1987), "Are there quantum jumps?," in *Schrödinger: Centenary Celebration of a Polymath*, C. W. Kilmister (ed), Cambridge: Cambridge University Press, p. 99. Reprinted in J. S. Bell, *Speakable and Unspeakable in Quantum Mechanics*, Cambridge. Cambridge University Press, p. 201.

Bell, J. S. (1989), "Towards an exact quantum mechanics," in *Themes in Contemporary Physics II, Essays in Honor of Julian Schwinger's 80th Birthday*, S. Deser and R. J. Finkelstein (eds), Singapore: World Scientific, p. 1.

Bell, J. S. (1990), "Against Measurement," in *Sixty-Two Years of Uncertainty*, Arthur I. Miller (ed), New York: Plenum Press, p. 17.

Benatti, F., Ghirardi, G.C. and Grassi, R. (1995), "Describing the macroscopic world: Closing the circle within the dynamical reduction program," *Found. Phys.* **25**, 5.

Bengochea, G. R., Leon, G., Pearle, P., Sudarsky, D. (2020), "Discussions about the landscape of possibilities for treatment of cosmic inflation involving continuous spontaneous localization models," *Euro Phys J C* **80**, 1021.

Bera, S., Motwani, B., Singh, T. P., and Ulbricht, H. (2015) "A proposal for the experimental detection of CSL induced random walk," *Sci. Rep.* **5**, 7664.

Bilardello, M., Donadi, S., Vinante, A., and Bassi, A. (2016), "Bounds on collapse models from cold-atom experiments," *Phys. A: Stat. Mech. and App.* **462**, 764.

Bohm, D, and Bub, J. (1966) "A proposed solution of the measurement problem in quantum mechanics by a hidden variable theory," *Revs. Mod. Phys.* **38**, 453.

Canate, P., Pearle, P., and Sudarsky, D. (2013) "Continuous spontaneous localization wave function collapse model as a mechanism for the emergence of cosmological asymmetries in inflation," *Phys. Rev. D* **87**, 104024.

Carlesso, M., Bassi, A., Falferi, P., and Vinante, A. (2016) "Experimental bounds on collapse models from gravitational wave detectors," *Phys. Rev. D* **94**, 124036.

Carlesso, M., Paternostro, M., Ulbricht, H., Vinante A., and Bassi, A. (2018), "Non-interferometric test of the continuous spontaneous localization model based on the torsional motion of a cylinder," *New Journ. Phys.* **20**, 083022.

Carlesso, M., Donadi, S., Ferialdi, L., Paternostro, M., Ulbricht, H., and Bassi, A. (2022), "Present status and future challenges of non-interferometric tests of collapse models," *Nature Phys.* **18**, 243.

Carson, L. D., Miles W. R., and Stevens, S. S. (1943), "Vision, hearing and aeronautical design," *Sci. Mon.* **56**, 446-451.

Collett, B., Pearle, P., Avignone, F. T., and Nussinov, S. (1995), "Constraint on collapse models by limit on spontaneous X-ray emission in Ge," *Found. Phys.* **25**, 1399.

Collett, B., and Pearle, P. (2003), "Wavefunction collapse and random walk," *Found. Phys.* **33**, 1495.

Diosi, L. (1989), "Models for universal reduction of macroscopic quantum fluctuations," *Phys. Rev. A* **40** 1165. Diosi suggested a limit of the GRW (Ghirardi et. al. (1986)) SL theory discussed in Chapter 5 not as a hitting process, but rather as a continuous process described by a stochastic differential equation for many particles in three dimensions. He called it Quantum Mechanics with Universal Position Localization (QMUPL), where the collapse-generating operators are particle positions. However, like SL from which it was derived, the collapse evolution destroys wave function symmetry. This is cured by the QMUDL model also introduced in this paper, and discussed in Chapter 6.

Dirac, P. A. M. (1958), *The Principles of Quantum Mechanics*, Oxford: Clarendon Press, p. 36.

Dirac, P. A. M. (1963), "The evolution of the physicist's picture of Nature," *Sci. Am.* **208** (May), 45.

Donadi, S., Piscicchia, K., Del Grande, R., Curceanu, C., Laubenstein, M., and Bassi, A. (2021), "Novel CSL bounds from the noise-induced radiation from atoms," *Eur. Phys. J. C* **81**, 8.

Dove, C. and Squires, E. (1995), "Symmetric versions of explicit collapse models,"*Found. Phys.* **25**, 1267.

Elbenberger, S., Gerlich, S., Arndt, M., Mayor, M., and Tuxen, J. (2013) "Matterwave interference of particles selected from a molecular library with masses exceeding 10 000 amu," *Phys. Chem. Chem. Phys.* **15**, 14696.

Feldmann, W. and Tumulka, R. (2012), "Parameter diagrams of the GRW and CSL theories of wave function collapse," *J. Phys. A Math. Theor.* **45**, 065304. What I called the parameter range not satisfying the "theoretical constraint" was, in this paper, renamed the "philosophically unsatisfactory" parameter range.

Feller, W. (1950), *An Introduction to Probability Theory and its Applications*, New York: Wiley, Chap. 14.

Feynman, R. P. (1965), "The development of the space-time view of quantum electrodynamics," Feynman's Nobel Prize Lecture. This is available on–line as item 6 at https://www.nobelprize.org/search/?query=feynman

Feynman, R. P. (1985), *The Character of Physical Law*, Cambridge: M.I.T. Press, p 171.

Fine, A. (1986), *The Shaky Game: Einstein, Realism and the Quantum Theory*, Chicago: University of Chicago Press, p. 78, 82.

Fu, Q. (1997), "Spontaneous radiation of free electrons in a nonrelativistic collapse model," *Phys. Rev. A* **56**, 1806. This was the senior project at Hamilton College of Qijia Fu. In the summer of 1996, following his graduation as Valedictorian of his class, looking forward in the fall to attending Harvard to study string theory and to competing as one among the handful of finalists for the Apker Award (given by the American Physical Society for the best undergraduate research in the nation) he was hiking with his brother in the mountains of Utah when he was struck and killed by a freak jump of dry lightning, an overwhelming tragedy. He was a most remarkable individual, humane and brilliant.

Fuentes, I. and Penrose, R. (2018), "Quantum state reduction via gravity, and possible tests using Bose–Einstein condensates," in *Collapse of the Wave Function*, Gao, S. (ed), Cambridge: Cambridge University Press, p. 187.

Garcia, E. et. al. (1995), "Results of a dark matter search with a germanium detector in the Canfranc tunnel," *Phys Rev. D* **51**, 1458.

Gasbarri, G., Toros, M., Donadi, S., and Bassi, A. (2017), "Gravity induced wave function collapse," *Phys. Rev. D* **96**, 104013.

Ghirardi, G. C., Rimini, A., and Weber, T. (1986) "Unified dynamics for macroscopic and microscopic systems," *Phys. Rev. D* **34**, 470 ; (1987), "Disentanglement of quantum wave functions: Answer to Comment on 'Unified dynamics for macroscopic and microscopic systems,'" *Phys. Rev. D* **36**, 3287. Actually, this work was written in terms of density matrices. John Bell (1987) expressed it in terms of state vectors.

Ghirardi, G. C., and Pearle, P. (1990), "Dynamical reduction theories: Changing quantum theory so the statevector represents reality," in *Proceedings of the Philosophy of Science Association, Volume 2*, A. Fine, M. Forbes, and L. Wessels (eds), East Lansing: PSA Association, p. 19. My term "hit," which I introduced to spare one from saying "spontaneous localizations," has been termed a "flash" by Tumulka (2006a).

Ghirardi, G. C., Pearle, P., and Rimini, A. (1990a), "Markov processes in Hilbert space and continuous spontaneous localization of systems of identical particles," *Phys. Rev. A* **42**, 78.

Ghirardi, G. C., Grassi, R., and Rimini, A. (1990b), "Continuous-spontaneous-reduction model involving gravity," *Phys. Rev. A* **42**, 1057. See also Bahrami et al., (2014).

Ghirardi, G. C., Grassi, R., and Pearle, P. (1990c) "Relativistic dynamical reduction models: General framework and examples," *Found. Phys.* **20**, 1271.

Ghirardi, G. C. and Weber, T. (1997), "An interpretation which is appropriate for dynamical reduction theories," in *Potentiality, Entanglement and Passion-at-a-Distance, Quantum Mechanical Studies for Abner Shimony, Vol. 2*, R. S. Cohen, M. Horne and J. Stachel (eds), Dordrecht: Kluwer, p. 89.

Goldwater, D., Paternostro, M., and Barker, P. F. (2016) "Testing wavefunction collapse models using parametric heating of a trapped nanosphere," *Phys. Rev. A* **94**, 010104.

Gorini, V., Kossakowski, A., and Sudarshan, E. C. G. (1976), "Completely positive dynamical semigroups of N-level systems," *J. Math. Phys.* **17**, 821.

Green Day (1997), "Time of Your Life," from the album *Nimrod*. I thank Ashton Parrish-Pearle for bringing this song to my attention.

Gundhi, A., Gaona-Reyes, J. L., Carlesso, M., Bassi, A. (2021), "Impact of dynamical collapse models on inflationary cosmology," *Phys. Rev. Lett.* **127**, 091302.

Heisenberg, W. (1971), *Physics and Beyond; Encounters and Conversations*, Harper & Row, New York, Chap. 17.

Irastorza, I. G. et. al. (2002) "Present status of IGEX dark matter search at Canfranc Underground Laboratory," *Nucl. Phys. B* **110**, 55 and arXiv:hep-ex/0111073.

Jones, G. Pearle, P., and Ring, J. (2004), "Consequence for wavefunction collapse model of the Sudbury Neutrino Observatory experiment," *Found. Phys.* **34**, 1467.

Josset, T., Perez, A., and Sudarsky, D. (2017), "Dark energy from violation of energy conservation," *Phys. Rev. Lett.* **118**, 021102.

Károlyházy, F. (1966), "Gravitation and quantum mechanics of macroscopic bodies," *Il Nuovo Cimento A* **42**, 390.

Károlyházy, F., Frenkel A., and Lukács, B. (1986), "On the possible role of gravity in the reduction of the wave function," in *Quantum Concepts in Space and Time*, R. Penrose and C. J. Isham (eds), Oxford, Clarendon, p. 109.

Kim, I (2022), "New limits on spontaneous wave function collapse models with the XENONnT data," arXiv:2209.15082 [quant-ph].

Laloé, F., Mullin, W. J., and Pearle, P. (2014), "Heating of trapped ultracold atoms by collapse dynamics," *Phys. Rev. A* **90**, 052119.

Li, J., Zippilli, S., Zhang, J., and Vitali, D. (2016), "Discriminating the effects of collapse models from environmental diffusion with levitated nanospheres," *Phys. Rev. A* **93**, 050102.

Li, Y., Steane, A. M., Bedingham, D., and Briggs, G. A. D. (2017), "Detecting continuous spontaneous localization with charged bodies in a Paul trap," *Phys. Rev. A* **95**, 032112.

Lindblad, G. (1976), "On the generators of quantum dynamical semigroups," *Commun. Math. Phys.* **48**, 119.

Marshall, W., Simon, C., Penrose, R., and Bouwmeester, D. (2002) "Towards quantum superpositions of a mirror," *Phys Rev. Lett.* **91**, 130401.

Martin, J. and Vennin V. (2020), "Cosmic microwave background constraints cast a shadow on continuous spontaneous localization models," *Phys. Rev. Lett.* **124**, 080402.

Martin, J. and Vennin, V. (2021) "On the choice of the collapse operator in cosmological continuous spontaneous localisation (CSL) theories," *Eur. Phys. J. C* **81**, 516.

McQueen, K. J. (2015), "Four tails problems for dynamical collapse theories," *Studies in the History and Philosophy of Modern Physics* **49**, 10, and also available in http://arxiv.org/abs/ 1501.05778

Miley, H.S., Avignone, F. T., Brodzinski, R.L., Collar, J.I., and Reeves, J. H. (1990), "Suggestive evidence for the two-neutrino double-decay of ^{76}Ge," *Phys. Rev. Lett.* **65**, 3092.

Modak, S. K., Ortiz, L., Pena, I., and Sudarsky, D. (2015) "Nonparadoxical loss of information in black hole evaporation in a quantum collapse.model," *Phys. Rev. D* **91**, 124009.

Morales, A., et al. (2002), "Improved constraints on WIMPs from the international germanium experiment IGEX," *Phys. Lett. B* **532**, 8.

Myrvold, W. C. (2017), "Relativistic Markovian dynamical collapse theories must employ nonstandard degrees of freedom," *Phys. Rev. A* **96**, 062116.

Myrvold, W. C. (2018), "Ontology for collapse theories," in *Collapse of the Wave Function*, S. Gao ed., Cambridge: Cambridge University Press, p.97.

Nicrosini, O. and Rimini, A. (2003), "Relativistic spontaneous localization: A proposal," *Found. Phys.* 33, 1061.

Okon, E. and Sudarsky, D., (2014), "Benefits of objective collapse models for cosmology and quantum gravity," *Found Phys* 44,114.

Pearle, P. (1967), "Alternative to the orthodox interpretation of quantum theory," *Am. Journ. Phys.* **35**, 742.

Pearle, P (1976), "Reduction of the state vector by a nonlinear Schrödinger equation," *Phys. Rev. A* **113**, 857.

Pearle, P. (1979), "Toward explaining why events occur," *Int. J. Theor. Phys.* **18**, 489.

Pearle, P. (1982), "Might God toss coins?," *Found. Phys.* **12**, 249.

Pearle, P. (1984a), "Models for reduction," presented at a conference at Lincoln College, Oxford, subsequently published in *Quantum Concepts in Space and Time*, P. Penrose and C. J. Isham (eds), Oxford: Clarendon Press, p. 84..

Pearle, P. (1984b), "Experimental tests of dynamical state-vector reduction', *Phys. Rev. D* **29**, 235.

Pearle, P. (1986), "Suppose the state vector is real: the description and consequences of dynamical reduction," in *New Techniques and Ideas in Quantum Measurement Theory*, D. M. Greenberger (ed), New York: New York Academy of Sciences, p. 539. This paper introduced the term "preferred basis," referring to the end states of dynamical collapse.

Pearle, P. (1989), "Combining stochastic dynamical state-vector reduction with spontaneous localization," *Phys. Rev. A* **39**, 2277.

Pearle, P. (1990), "Toward a relativistic theory of statevector reduction," in *Sixty-Two Years of Uncertainty*, A. Miller (ed), New York: Plenum, p. 193.

Pearle, P. (1994) "True Collapse and False Collapse." in *Quantum Classical Correspondence: Proceedings of the 4th Drexel Symposium on Quantum Nonintegrability*, D. H. Feng and B. L. Hu (eds), Cambridge: International Press, p. 51.

Pearle, P., and Squires, E. (1994) "Bound state excitation, nucleon decay experiments, and models of wave function collapse," *Phys. Rev. Lett.* **73**, 1.

Pearle, P. and Squires, E. (1996), "Gravity, energy conservation, and parameter values in collapse models," *Found. Phys.* **26**, 291.

Pearle, P. (1999a), "Relativistic collapse model with tachyonic features," *Phys. Rev. A* **59**, 80.

Pearle, P., Ring, J., Collar, J.I., and Avignone, F. T. (1999b), "The CSL collapse model and spontaneous radiation: An update," *Found. Phys.* **29**, 465.

Pearle,P. (2005), "Quasirelativistic quasilocal finite wave-function collapse model," *Phys. Rev. A* **71**, 032101.

Pearle, P. (2007), "How stands collapse I," J. Phys. A: Math. Theor. **40**, 3189.

Pearle, P. (2009), "How stands collapse II," in *Quantum Reality, Relativistic Causality, and Closing the Epistemic Circle: Essays in Honour of Abner Shimony*, W. C. Myrvold and J. Christian (eds), New York: Springer, p. 257.

Pearle, P., Bart, K., Bilderback, D., Collett, B., Newman, D. and Samuels, S. (2010), "What Brown saw, and you can too," *Am. J. Phys.* **78**, 1278.

Pearle, P. (2012), "Simple derivation of the Lindblad equation," *Eur. J. Phys* **33**, 805.

Pearle, P. (2013), "Collapse miscellany," in *Quantum Theory: A Two Time Success Story. Yakir Aharonov Festschrift*, D. Struppa, and J. Tollakson eds., Milan: Springer, p.131.

Pearle, P. (2015), "Relativistic dynamical collapse model," *Phys. Rev. D* **91**, 105012.

Pearle, P. (2018), "Dynamical collapse for photons" in *Collapse of the Wave Function*, S. Gao (ed), Cambridge: Cambridge University Press, p. 23.

Penrose, R. (1981), "Time-asymmetry and quantum gravity," in *Quantum Gravity 2 A Second Oxford Symposium* C. J. Isham, R. Penrose and D. W. Sciama (eds), Oxford: Clarendon Press, p. 244.

Penrose, R. (1986), "Gravity and state vector reduction," in *Quantum Concepts in Space and Time*, R. Penrose and C. J. Isham (eds), Oxford: Clarendon., p. 129.

Penrose, R. (1994a), "Non-locality and objectivity in quantum state reduction," in *Quantum Coherence and Reality*, J. Anandan and J. L. Safko eds., Singapore: World Scientific, p. 238.

Penrose, R. (1994b), *Shadows of the Mind; An Approach to the Missing Science of Consciousness*, Oxford: Oxford University Press.

Penrose, R. (1996), "On gravity's role in quantum state reduction," *Gen. Rel. Grav.* **28**, 581.

Piscicchia, K. , Bassi, A., Curceanu, C., Del Grande, R., Donadi, S., Hiesmayr, B. C., and Pichler, A. (2017), "CSL collapse model mapped with the spontaneous radiation," *Entropy* **19**, 319.

Schrinski, B., Stickler, B. A., and Hornberger, K. (2017), "Collapse-induced orientation of rigid rotors," *J. Opt. Soc. Am. B* **34**, C1.

Schrödinger, E. (1935), "Die gegenwärtige Situation in der Quantenmechanik," (The Present Status of Quantum Mechanics), *Die Naturwissenschaften* **23**, 807-812, 823-828, 844-849. An English translation by J. D. Trimmer can be obtained on-line.

Shakespeare, W. (1623) *Mr. William Shakespeare's Comedies, Histories, & Tragedies, First Folio*, London: E. Blount, W. and I. Jaggar, *Hamlet*, Act 5. Scene 2.

Shimony, A. (1991), "Desiderata for a modified quantum dynamics," A. Fine, M. Forbes and L. Wessels (eds), East Lansing: Philosophy of Science Association, p. 17.

SNO Collaboration (2004) "Measurement of the total active 8B solar neutrino flux at the Sudbury Neutrino Observatory with enhanced neutral current sensitivity," *Phys. Rev. Lett.* **92**, 181301.

Toros, M., and Bassi, A. (2018), "Bounds on collapse models from matter-wave interferometry: Calculational details," *J. Phys. A: Math. Theor.* **51**, 115302.

Tumulka, R. (2006a), "Collapse and relativity," in *Quantum Mechanics: Are there Quantum Jumps? and On the Present Status of Quantum Mechanics*, A. Bassi, D. Duerr, T. Weber and N. Zanghi (eds), American Institute of Physics Proceedings **844**, 340.

Tumulka, R. (2006b), "On spontaneous wave function collapse and quantum field theory," *Proc. Royal Soc.* **462**, 1897.

Tumulka, R. (2018), "Paradoxes and primitive ontology in collapse theories of quantum mechanics," in *Collapse of the Wave Function*, S. Gao and M. Bell (eds), Cambridge: Cambridge University Press, 134.

Vinante, A., Bahrami, M., Bassi, A., Usenko, O., Wijts, G., and Oosterkamp, T. H. (2016), "Upper bounds on spontaneous wave-function collapse models using millikelvin-cooled nanocantilevers," *Phys. Rev. Lett* **116**, 0090402.

Vinante, A., Carlesso, M., Bassi, A., Chiasera, A., Varas, S., Falferi, P., Margosin, B.,

Mezzona, R., and Ulbricht, H. (2020), "Narrowing the parameter space of collapse models with ultracold layered force sensors," *Phys. Rev. Lett.* **125**, 100404.

Wiener, N. (1954), *The Human Use Of Human Beings: Cybernetics And Society*, Boston: Houghton Mifflin.

Yu, H. (2016), "How many photons get into your eyes?," https://medium.com /cortically-magnified/ estimating-the-number-of-photons-that-hit-the-eye -c0208e7e0b64

Zeilinger, A. (1986), "Testing quantum superposition with cold neutrons," in *Quantum Concepts in Space and Time*, R. Penrose and C. J. Isham (eds), Oxford: Clarendon, p.16.

Zheng, D., Leng, Y., Kong, X., Li, R., Wang, Z., Luo, X., Zhao, J., Duan, C., Huang, P., Du, J., Carlesso, M., and Bassi, A. (2020), "Room temperature test of the continuous spontaneous localization model using a levitated micro-oscillator," *Phys. Rev. Res.* **2**, 013057.

Index